The GNU Octave 3.8 Reference Manual - Part 2/2

A catalogue record for this book is available from the Hong Kong Public Libraries.

Published by Samurai Media Limited.

Email: info@samuraimedia.org

ISBN 978-988-13277-4-1

Table of Contents

18 Linear Algebra

This chapter documents the linear algebra functions provided in Octave. Reference material for many of these functions may be found in Golub and Van Loan, *Matrix Computations, 2nd Ed.*, Johns Hopkins, 1989, and in the LAPACK *Users' Guide*, SIAM, 1992. The LAPACK *Users' Guide* is available at: *http://www.netlib.org/lapack/lug/*

A common text for engineering courses is G. Strang, *Linear Algebra and Its Applications, 4th Edition*. It has become a widespread reference for linear algebra. An alternative is P. Lax *Linear Algebra and Its Applications*, and also is a good choice. It claims to be suitable for high school students with substantial mathematical interests as well as first-year undergraduates.

18.1 Techniques Used for Linear Algebra

Octave includes a polymorphic solver that selects an appropriate matrix factorization depending on the properties of the matrix itself. Generally, the cost of determining the matrix type is small relative to the cost of factorizing the matrix itself. In any case the matrix type is cached once it is calculated so that it is not re-determined each time it is used in a linear equation.

The selection tree for how the linear equation is solved or a matrix inverse is formed is given by:

1. If the matrix is upper or lower triangular sparse use a forward or backward substitution using the LAPACK xTRTRS function, and goto 4.
2. If the matrix is square, Hermitian with a real positive diagonal, attempt Cholesky factorization using the LAPACK xPOTRF function.
3. If the Cholesky factorization failed or the matrix is not Hermitian with a real positive diagonal, and the matrix is square, factorize using the LAPACK xGETRF function.
4. If the matrix is not square, or any of the previous solvers flags a singular or near singular matrix, find a least squares solution using the LAPACK xGELSD function.

The user can force the type of the matrix with the `matrix_type` function. This overcomes the cost of discovering the type of the matrix. However, it should be noted that identifying the type of the matrix incorrectly will lead to unpredictable results, and so `matrix_type` should be used with care.

It should be noted that the test for whether a matrix is a candidate for Cholesky factorization, performed above, and by the `matrix_type` function, does not make certain that the matrix is Hermitian. However, the attempt to factorize the matrix will quickly detect a non-Hermitian matrix.

18.2 Basic Matrix Functions

AA = balance (*A*) [Built-in Function]
AA = balance (*A*, *opt*) [Built-in Function]
[*DD*, *AA*] = balance (*A*, *opt*) [Built-in Function]
[*D*, *P*, *AA*] = balance (*A*, *opt*) [Built-in Function]

`[CC, DD, AA, BB] = balance (A, B, opt)` [Built-in Function]

Compute `AA = DD \ A * DD` in which AA is a matrix whose row and column norms are roughly equal in magnitude, and `DD = P * D`, in which P is a permutation matrix and D is a diagonal matrix of powers of two. This allows the equilibration to be computed without round-off. Results of eigenvalue calculation are typically improved by balancing first.

If two output values are requested, `balance` returns the diagonal D and the permutation P separately as vectors. In this case, `DD = eye(n)(:,P) * diag (D)`, where n is the matrix size.

If four output values are requested, compute `AA = CC*A*DD` and `BB = CC*B*DD`, in which AA and BB have non-zero elements of approximately the same magnitude and CC and DD are permuted diagonal matrices as in DD for the algebraic eigenvalue problem.

The eigenvalue balancing option *opt* may be one of:

`"noperm"`, `"S"`
: Scale only; do not permute.

`"noscal"`, `"P"`
: Permute only; do not scale.

Algebraic eigenvalue balancing uses standard LAPACK routines.

Generalized eigenvalue problem balancing uses Ward's algorithm (SIAM Journal on Scientific and Statistical Computing, 1981).

`cond (A)` [Function File]
`cond (A, p)` [Function File]

Compute the *p*-norm condition number of a matrix.

`cond (A)` is defined as $\| A \|_p * \| A^{-1} \|_p$.

By default, `p = 2` is used which implies a (relatively slow) singular value decomposition. Other possible selections are `p = 1, Inf, "fro"` which are generally faster. See `norm` for a full discussion of possible *p* values.

The condition number of a matrix quantifies the sensitivity of the matrix inversion operation when small changes are made to matrix elements. Ideally the condition number will be close to 1. When the number is large this indicates small changes (such as underflow or round-off error) will produce large changes in the resulting output. In such cases the solution results from numerical computing are not likely to be accurate.

See also: [condest], page 510, [rcond], page 441, [norm], page 439, [svd], page 450.

`det (A)` [Built-in Function]
`[d, rcond] = det (A)` [Built-in Function]

Compute the determinant of A.

Return an estimate of the reciprocal condition number if requested.

Routines from LAPACK are used for full matrices and code from UMFPACK is used for sparse matrices.

The determinant should not be used to check a matrix for singularity. For that, use any of the condition number functions: `cond`, `condest`, `rcond`.

See also: [cond], page 436, [condest], page 510, [rcond], page 441.

`lambda = eig (A)` [Built-in Function]
`lambda = eig (A, B)` [Built-in Function]
`[V, lambda] = eig (A)` [Built-in Function]
`[V, lambda] = eig (A, B)` [Built-in Function]

Compute the eigenvalues (and optionally the eigenvectors) of a matrix or a pair of matrices

The algorithm used depends on whether there are one or two input matrices, if they are real or complex and if they are symmetric (Hermitian if complex) or non-symmetric.

The eigenvalues returned by `eig` are not ordered.

See also: [eigs], page 513, [svd], page 450.

`g = givens (x, y)` [Built-in Function]
`[c, s] = givens (x, y)` [Built-in Function]

Return a 2×2 orthogonal matrix

$$G = \begin{bmatrix} c & s \\ -s' & c \end{bmatrix}$$

such that

$$G \begin{bmatrix} x \\ y \end{bmatrix} = \begin{bmatrix} * \\ 0 \end{bmatrix}$$

with x and y scalars.

For example:

```
givens (1, 1)
   ⇒   0.70711   0.70711
      -0.70711   0.70711
```

`[g, y] = planerot (x)` [Function File]

Given a two-element column vector, returns the 2×2 orthogonal matrix G such that `y = g * x` and `y(2) = 0`.

See also: [givens], page 437.

`x = inv (A)` [Built-in Function]
`[x, rcond] = inv (A)` [Built-in Function]

Compute the inverse of the square matrix A. Return an estimate of the reciprocal condition number if requested, otherwise warn of an ill-conditioned matrix if the reciprocal condition number is small.

In general it is best to avoid calculating the inverse of a matrix directly. For example, it is both faster and more accurate to solve systems of equations ($A*x = b$) with `y = A \ b`, rather than `y = inv (A) * b`.

If called with a sparse matrix, then in general x will be a full matrix requiring significantly more storage. Avoid forming the inverse of a sparse matrix if possible.

See also: [ldivide], page 139, [rdivide], page 140.

`x = linsolve (A, b)` [Function File]
`x = linsolve (A, b, opts)` [Function File]
`[x, R] = linsolve (...)` [Function File]

Solve the linear system `A*x = b`.

With no options, this function is equivalent to the left division operator (`x = A \ b`) or the matrix-left-divide function (`x = mldivide (A, b)`).

Octave ordinarily examines the properties of the matrix *A* and chooses a solver that best matches the matrix. By passing a structure *opts* to `linsolve` you can inform Octave directly about the matrix *A*. In this case Octave will skip the matrix examination and proceed directly to solving the linear system.

Warning: If the matrix *A* does not have the properties listed in the *opts* structure then the result will not be accurate AND no warning will be given. When in doubt, let Octave examine the matrix and choose the appropriate solver as this step takes little time and the result is cached so that it is only done once per linear system.

Possible *opts* fields (set value to true/false):

LT — *A* is lower triangular

UT — *A* is upper triangular

UHESS — *A* is upper Hessenberg (currently makes no difference)

SYM — *A* is symmetric or complex Hermitian (currently makes no difference)

POSDEF — *A* is positive definite

RECT — *A* is general rectangular (currently makes no difference)

TRANSA — Solve `A'*x = b` by `transpose (A) \ b`

The optional second output *R* is the inverse condition number of *A* (zero if matrix is singular).

See also: [mldivide], page 139, [matrix_type], page 438, [rcond], page 441.

`type = matrix_type (A)` [Built-in Function]
`type = matrix_type (A, "nocompute")` [Built-in Function]
`A = matrix_type (A, type)` [Built-in Function]
`A = matrix_type (A, "upper", perm)` [Built-in Function]
`A = matrix_type (A, "lower", perm)` [Built-in Function]
`A = matrix_type (A, "banded", nl, nu)` [Built-in Function]

Identify the matrix type or mark a matrix as a particular type. This allows more rapid solutions of linear equations involving *A* to be performed. Called with a single argument, `matrix_type` returns the type of the matrix and caches it for future use. Called with more than one argument, `matrix_type` allows the type of the matrix to be defined.

If the option `"nocompute"` is given, the function will not attempt to guess the type if it is still unknown. This is useful for debugging purposes.

The possible matrix types depend on whether the matrix is full or sparse, and can be one of the following

`"unknown"`
: Remove any previously cached matrix type, and mark type as unknown.

`"full"` Mark the matrix as full.

`"positive definite"`
: Probable full positive definite matrix.

`"diagonal"`
: Diagonal matrix. (Sparse matrices only)

`"permuted diagonal"`
: Permuted Diagonal matrix. The permutation does not need to be specifically indicated, as the structure of the matrix explicitly gives this. (Sparse matrices only)

`"upper"` Upper triangular. If the optional third argument *perm* is given, the matrix is assumed to be a permuted upper triangular with the permutations defined by the vector *perm*.

`"lower"` Lower triangular. If the optional third argument *perm* is given, the matrix is assumed to be a permuted lower triangular with the permutations defined by the vector *perm*.

`"banded"`
`"banded positive definite"`
: Banded matrix with the band size of *nl* below the diagonal and *nu* above it. If *nl* and *nu* are 1, then the matrix is tridiagonal and treated with specialized code. In addition the matrix can be marked as probably a positive definite. (Sparse matrices only)

`"singular"`
: The matrix is assumed to be singular and will be treated with a minimum norm solution.

Note that the matrix type will be discovered automatically on the first attempt to solve a linear equation involving A. Therefore `matrix_type` is only useful to give Octave hints of the matrix type. Incorrectly defining the matrix type will result in incorrect results from solutions of linear equations; it is entirely **the responsibility of the user** to correctly identify the matrix type.

Also, the test for positive definiteness is a low-cost test for a Hermitian matrix with a real positive diagonal. This does not guarantee that the matrix is positive definite, but only that it is a probable candidate. When such a matrix is factorized, a Cholesky factorization is first attempted, and if that fails the matrix is then treated with an LU factorization. Once the matrix has been factorized, `matrix_type` will return the correct classification of the matrix.

`norm (`*A*`)` [Built-in Function]
`norm (`*A*`, `*p*`)` [Built-in Function]
`norm (`*A*`, `*p*`, `*opt*`)` [Built-in Function]

Compute the p-norm of the matrix A. If the second argument is missing, `p = 2` is assumed.

If A is a matrix (or sparse matrix):

$p = 1$ 1-norm, the largest column sum of the absolute values of A.

$p = 2$ Largest singular value of A.

$p =$ `Inf` or `"inf"`
Infinity norm, the largest row sum of the absolute values of A.

$p =$ `"fro"`
Frobenius norm of A, `sqrt (sum (diag (A' * A)))`.

other p, `p > 1`
maximum `norm (A*x, p)` such that `norm (x, p) == 1`

If A is a vector or a scalar:

$p =$ `Inf` or `"inf"`
`max (abs (A))`.

$p =$ `-Inf` `min (abs (A))`.

$p =$ `"fro"`
Frobenius norm of A, `sqrt (sumsq (abs (A)))`.

$p = 0$ Hamming norm - the number of nonzero elements.

other p, `p > 1`
p-norm of A, `(sum (abs (A) .^ p)) ^ (1/p)`.

other p `p < 1`
the p-pseudonorm defined as above.

If *opt* is the value `"rows"`, treat each row as a vector and compute its norm. The result is returned as a column vector. Similarly, if *opt* is `"columns"` or `"cols"` then compute the norms of each column and return a row vector.

See also: [cond], page 436, [svd], page 450.

`null (A)` [Function File]
`null (A, tol)` [Function File]

Return an orthonormal basis of the null space of A.

The dimension of the null space is taken as the number of singular values of A not greater than *tol*. If the argument *tol* is missing, it is computed as

```
max (size (A)) * max (svd (A)) * eps
```

See also: [orth], page 440.

`orth (A)` [Function File]
`orth (A, tol)` [Function File]

Return an orthonormal basis of the range space of A.

The dimension of the range space is taken as the number of singular values of A greater than *tol*. If the argument *tol* is missing, it is computed as

```
max (size (A)) * max (svd (A)) * eps
```

See also: [null], page 440.

`[y, h] = mgorth (x, v)` [Built-in Function]

Orthogonalize a given column vector *x* with respect to a set of orthonormal vectors comprising the columns of *v* using the modified Gram-Schmidt method. On exit, *y* is a unit vector such that:

```
norm (y) = 1
v' * y = 0
x = [v, y]*h'
```

`pinv (x)` [Built-in Function]
`pinv (x, tol)` [Built-in Function]

Return the pseudoinverse of *x*. Singular values less than *tol* are ignored.

If the second argument is omitted, it is taken to be

```
tol = max (size (x)) * sigma_max (x) * eps,
```

where `sigma_max (x)` is the maximal singular value of *x*.

`rank (A)` [Function File]
`rank (A, tol)` [Function File]

Compute the rank of matrix *A*, using the singular value decomposition.

The rank is taken to be the number of singular values of *A* that are greater than the specified tolerance *tol*. If the second argument is omitted, it is taken to be

```
tol = max (size (A)) * sigma(1) * eps;
```

where `eps` is machine precision and `sigma(1)` is the largest singular value of *A*.

The rank of a matrix is the number of linearly independent rows or columns and determines how many particular solutions exist to a system of equations. Use `null` for finding the remaining homogenous solutions.

Example:

```
x = [1 2 3
     4 5 6
     7 8 9];
rank (x)
  ⇒ 2
```

The number of linearly independent rows is only 2 because the final row is a linear combination of -1*row1 + 2*row2.

See also: [null], page 440, [sprank], page 511, [svd], page 450.

`c = rcond (A)` [Built-in Function]

Compute the 1-norm estimate of the reciprocal condition number as returned by LAPACK. If the matrix is well-conditioned then *c* will be near 1 and if the matrix is poorly conditioned it will be close to zero.

The matrix *A* must not be sparse. If the matrix is sparse then `condest (A)` or `rcond (full (A))` should be used instead.

See also: [cond], page 436, [condest], page 510.

`trace (A)` [Function File]

Compute the trace of A, the sum of the elements along the main diagonal.

The implementation is straightforward: `sum (diag (A))`.

See also: [eig], page 437.

`rref (A)` [Function File]
`rref (A, tol)` [Function File]
`[r, k] = rref (...)` [Function File]

Return the reduced row echelon form of A. *tol* defaults to `eps * max (size (A)) * norm (A, inf)`.

Called with two return arguments, k returns the vector of "bound variables", which are those columns on which elimination has been performed.

18.3 Matrix Factorizations

`R = chol (A)` [Loadable Function]
`[R, p] = chol (A)` [Loadable Function]
`[R, p, Q] = chol (S)` [Loadable Function]
`[R, p, Q] = chol (S, "vector")` [Loadable Function]
`[L, ...] = chol (..., "lower")` [Loadable Function]
`[L, ...] = chol (..., "upper")` [Loadable Function]

Compute the Cholesky factor, R, of the symmetric positive definite matrix A, where $R^T R = A$.

Called with one output argument `chol` fails if A or S is not positive definite. With two or more output arguments p flags whether the matrix was positive definite and `chol` does not fail. A zero value indicated that the matrix was positive definite and the R gives the factorization, and p will have a positive value otherwise.

If called with 3 outputs then a sparsity preserving row/column permutation is applied to A prior to the factorization. That is R is the factorization of `A(Q,Q)` such that $R^T R = Q^T A Q$.

The sparsity preserving permutation is generally returned as a matrix. However, given the flag `"vector"`, Q will be returned as a vector such that $R^T R = A(Q, Q)$.

Called with either a sparse or full matrix and using the `"lower"` flag, `chol` returns the lower triangular factorization such that $LL^T = A$.

For full matrices, if the `"lower"` flag is set only the lower triangular part of the matrix is used for the factorization, otherwise the upper triangular part is used.

In general the lower triangular factorization is significantly faster for sparse matrices.

See also: [hess], page 444, [lu], page 444, [qr], page 446, [qz], page 448, [schur], page 450, [svd], page 450, [cholinv], page 442, [chol2inv], page 443, [cholupdate], page 443, [cholinsert], page 443, [choldelete], page 443, [cholshift], page 443.

`cholinv (A)` [Loadable Function]

Use the Cholesky factorization to compute the inverse of the symmetric positive definite matrix A.

See also: [chol], page 442, [chol2inv], page 443, [inv], page 437.

`chol2inv (U)` [Loadable Function]

Invert a symmetric, positive definite square matrix from its Cholesky decomposition, *U*. Note that *U* should be an upper-triangular matrix with positive diagonal elements. `chol2inv (U)` provides `inv (U'*U)` but it is much faster than using `inv`.

See also: [chol], page 442, [cholinv], page 442, [inv], page 437.

`[R1, info] = cholupdate (R, u, op)` [Loadable Function]

Update or downdate a Cholesky factorization. Given an upper triangular matrix *R* and a column vector *u*, attempt to determine another upper triangular matrix *R1* such that

- *R1*'**R1* = *R*'**R* + *u***u*' if *op* is `"+"`
- *R1*'**R1* = *R*'**R* - *u***u*' if *op* is `"-"`

If *op* is `"-"`, *info* is set to

- 0 if the downdate was successful,
- 1 if *R*'**R* - *u***u*' is not positive definite,
- 2 if *R* is singular.

If *info* is not present, an error message is printed in cases 1 and 2.

See also: [chol], page 442, [cholinsert], page 443, [choldelete], page 443, [cholshift], page 443.

`R1 = cholinsert (R, j, u)` [Loadable Function]
`[R1, info] = cholinsert (R, j, u)` [Loadable Function]

Given a Cholesky factorization of a real symmetric or complex Hermitian positive definite matrix *A* = *R*'**R*, *R* upper triangular, return the Cholesky factorization of *A1*, where A1(p,p) = A, A1(:,j) = A1(j,:)' = u and p = [1:j-1,j+1:n+1]. u(j) should be positive. On return, *info* is set to

- 0 if the insertion was successful,
- 1 if *A1* is not positive definite,
- 2 if *R* is singular.

If *info* is not present, an error message is printed in cases 1 and 2.

See also: [chol], page 442, [cholupdate], page 443, [choldelete], page 443, [cholshift], page 443.

`R1 = choldelete (R, j)` [Loadable Function]

Given a Cholesky factorization of a real symmetric or complex Hermitian positive definite matrix *A* = *R*'**R*, *R* upper triangular, return the Cholesky factorization of A(p,p), where p = [1:j-1,j+1:n+1].

See also: [chol], page 442, [cholupdate], page 443, [cholinsert], page 443, [cholshift], page 443.

`R1 = cholshift (R, i, j)` [Loadable Function]

Given a Cholesky factorization of a real symmetric or complex Hermitian positive definite matrix *A* = *R*'**R*, *R* upper triangular, return the Cholesky factorization of *A*(p,p), where p is the permutation

`p = [1:i-1, shift(i:j, 1), j+1:n]` if $i < j$
or
`p = [1:j-1, shift(j:i,-1), i+1:n]` if $j < i$.

See also: [chol], page 442, [cholupdate], page 443, [cholinsert], page 443, [choldelete], page 443.

`H = hess (A)` [Built-in Function]
`[P, H] = hess (A)` [Built-in Function]

Compute the Hessenberg decomposition of the matrix A.

The Hessenberg decomposition is

$$A = PHP^T$$

where P is a square unitary matrix ($P^T P = I$), and H is upper Hessenberg ($H_{i,j} = 0, \forall i \geq j + 1$).

The Hessenberg decomposition is usually used as the first step in an eigenvalue computation, but has other applications as well (see Golub, Nash, and Van Loan, IEEE Transactions on Automatic Control, 1979).

See also: [eig], page 437, [chol], page 442, [lu], page 444, [qr], page 446, [qz], page 448, [schur], page 450, [svd], page 450.

`[L, U] = lu (A)` [Built-in Function]
`[L, U, P] = lu (A)` [Built-in Function]
`[L, U, P, Q] = lu (S)` [Built-in Function]
`[L, U, P, Q, R] = lu (S)` [Built-in Function]
`[...] = lu (S, thres)` [Built-in Function]
`y = lu (...)` [Built-in Function]
`[...] = lu (..., "vector")` [Built-in Function]

Compute the LU decomposition of A. If A is full subroutines from LAPACK are used and if A is sparse then UMFPACK is used. The result is returned in a permuted form, according to the optional return value P. For example, given the matrix `a = [1, 2; 3, 4]`,

```
[l, u, p] = lu (a)
```

returns

```
l =

  1.00000  0.00000
  0.33333  1.00000

u =

  3.00000  4.00000
  0.00000  0.66667

p =

  0  1
  1  0
```

The matrix is not required to be square.

When called with two or three output arguments and a spare input matrix, `lu` does not attempt to perform sparsity preserving column permutations. Called with a fourth output argument, the sparsity preserving column transformation Q is returned, such that `P * A * Q = L * U`.

Called with a fifth output argument and a sparse input matrix, `lu` attempts to use a scaling factor R on the input matrix such that `P * (R \ A) * Q = L * U`. This typically leads to a sparser and more stable factorization.

An additional input argument *thres*, that defines the pivoting threshold can be given. *thres* can be a scalar, in which case it defines the UMFPACK pivoting tolerance for both symmetric and unsymmetric cases. If *thres* is a 2-element vector, then the first element defines the pivoting tolerance for the unsymmetric UMFPACK pivoting strategy and the second for the symmetric strategy. By default, the values defined by `spparms` are used ([0.1, 0.001]).

Given the string argument `"vector"`, `lu` returns the values of P and Q as vector values, such that for full matrix, `A (P,:) = L * U`, and `R(P,:) * A (:, Q) = L * U`.

With two output arguments, returns the permuted forms of the upper and lower triangular matrices, such that `A = L * U`. With one output argument *y*, then the matrix returned by the LAPACK routines is returned. If the input matrix is sparse then the matrix L is embedded into U to give a return value similar to the full case. For both full and sparse matrices, `lu` loses the permutation information.

See also: [luupdate], page 445, [chol], page 442, [hess], page 444, [qr], page 446, [qz], page 448, [schur], page 450, [svd], page 450.

`[L, U] = luupdate (L, U, x, y)` [Built-in Function]
`[L, U, P] = luupdate (L, U, P, x, y)` [Built-in Function]

Given an LU factorization of a real or complex matrix $A = L*U$, L lower unit trapezoidal and U upper trapezoidal, return the LU factorization of $A + x*y.'$, where x and y are column vectors (rank-1 update) or matrices with equal number of columns (rank-k update). Optionally, row-pivoted updating can be used by supplying a row permutation (pivoting) matrix P; in that case, an updated permutation matrix is returned. Note that if L, U, P is a pivoted LU factorization as obtained by `lu`:

```
[L, U, P] = lu (A);
```

then a factorization of `A+x*y.'` can be obtained either as

```
[L1, U1] = lu (L, U, P*x, y)
```

or

```
[L1, U1, P1] = lu (L, U, P, x, y)
```

The first form uses the unpivoted algorithm, which is faster, but less stable. The second form uses a slower pivoted algorithm, which is more stable.

The matrix case is done as a sequence of rank-1 updates; thus, for large enough k, it will be both faster and more accurate to recompute the factorization from scratch.

See also: [lu], page 444, [cholupdate], page 443, [qrupdate], page 447.

`[Q, R, P] = qr (A)` [Loadable Function]
`[Q, R, P] = qr (A, '0')` [Loadable Function]
`[C, R] = qr (A, B)` [Loadable Function]
`[C, R] = qr (A, B, '0')` [Loadable Function]

Compute the QR factorization of A, using standard LAPACK subroutines. For example, given the matrix `A = [1, 2; 3, 4]`,

```
[Q, R] = qr (A)
```

returns

```
Q =

  -0.31623  -0.94868
  -0.94868   0.31623

R =

  -3.16228  -4.42719
   0.00000  -0.63246
```

The `qr` factorization has applications in the solution of least squares problems

$$\min_x \|Ax - b\|_2$$

for overdetermined systems of equations (i.e., A is a tall, thin matrix). The QR factorization is $QR = A$ where Q is an orthogonal matrix and R is upper triangular.

If given a second argument of `'0'`, `qr` returns an economy-sized QR factorization, omitting zero rows of R and the corresponding columns of Q.

If the matrix A is full, the permuted QR factorization `[Q, R, P] = qr (A)` forms the QR factorization such that the diagonal entries of R are decreasing in magnitude order. For example, given the matrix `a = [1, 2; 3, 4]`,

```
[Q, R, P] = qr (A)
```

returns

```
Q =

  -0.44721  -0.89443
  -0.89443   0.44721

R =

  -4.47214  -3.13050
   0.00000   0.44721

P =

  0  1
  1  0
```

The permuted qr factorization `[Q, R, P] = qr (A)` factorization allows the construction of an orthogonal basis of `span (A)`.

If the matrix A is sparse, then compute the sparse QR factorization of A, using CSparse. As the matrix Q is in general a full matrix, this function returns the Q-less factorization R of A, such that `R = chol (A' * A)`.

If the final argument is the scalar `0` and the number of rows is larger than the number of columns, then an economy factorization is returned. That is R will have only `size (A,1)` rows.

If an additional matrix B is supplied, then `qr` returns C, where `C = Q' * B`. This allows the least squares approximation of `A \ B` to be calculated as

```
[C, R] = qr (A, B)
x = R \ C
```

See also: [chol], page 442, [hess], page 444, [lu], page 444, [qz], page 448, [schur], page 450, [svd], page 450, [qrupdate], page 447, [qrinsert], page 447, [qrdelete], page 448, [qrshift], page 448.

`[Q1, R1] = qrupdate (Q, R, u, v)` [Loadable Function]

Given a QR factorization of a real or complex matrix $A = Q*R$, Q unitary and R upper trapezoidal, return the QR factorization of $A + u*v'$, where u and v are column vectors (rank-1 update) or matrices with equal number of columns (rank-k update). Notice that the latter case is done as a sequence of rank-1 updates; thus, for k large enough, it will be both faster and more accurate to recompute the factorization from scratch.

The QR factorization supplied may be either full (Q is square) or economized (R is square).

See also: [qr], page 446, [qrinsert], page 447, [qrdelete], page 448, [qrshift], page 448.

`[Q1, R1] = qrinsert (Q, R, j, x, orient)` [Loadable Function]

Given a QR factorization of a real or complex matrix $A = Q*R$, Q unitary and R upper trapezoidal, return the QR factorization of [A(:,1:j-1) x A(:,j:n)], where u is a column vector to be inserted into A (if *orient* is `"col"`), or the QR factorization of [A(1:j-1,:);x;A(:,j:n)], where x is a row vector to be inserted into A (if *orient* is `"row"`).

The default value of *orient* is `"col"`. If *orient* is `"col"`, *u* may be a matrix and *j* an index vector resulting in the QR factorization of a matrix B such that B(:,*j*) gives *u* and B(:,*j*) = [] gives *A*. Notice that the latter case is done as a sequence of k insertions; thus, for k large enough, it will be both faster and more accurate to recompute the factorization from scratch.

If *orient* is `"col"`, the QR factorization supplied may be either full (Q is square) or economized (R is square).

If *orient* is `"row"`, full factorization is needed.

See also: [qr], page 446, [qrupdate], page 447, [qrdelete], page 448, [qrshift], page 448.

`[Q1, R1] = qrdelete (Q, R, j, orient)` [Loadable Function]

Given a QR factorization of a real or complex matrix $A = Q^*R$, Q unitary and R upper trapezoidal, return the QR factorization of [A(:,1:j-1) A(:,j+1:n)], i.e., A with one column deleted (if *orient* is `"col"`), or the QR factorization of [A(1:j-1,:);A(j+1:n,:)], i.e., A with one row deleted (if *orient* is `"row"`).

The default value of *orient* is `"col"`.

If *orient* is `"col"`, *j* may be an index vector resulting in the QR factorization of a matrix B such that A(:,*j*) = [] gives B. Notice that the latter case is done as a sequence of k deletions; thus, for k large enough, it will be both faster and more accurate to recompute the factorization from scratch.

If *orient* is `"col"`, the QR factorization supplied may be either full (Q is square) or economized (R is square).

If *orient* is `"row"`, full factorization is needed.

See also: [qr], page 446, [qrupdate], page 447, [qrinsert], page 447, [qrshift], page 448.

`[Q1, R1] = qrshift (Q, R, i, j)` [Loadable Function]

Given a QR factorization of a real or complex matrix $A = Q^*R$, Q unitary and R upper trapezoidal, return the QR factorization of A(:,p), where p is the permutation

`p = [1:i-1, shift(i:j, 1), j+1:n]` if *i* < *j*

or

`p = [1:j-1, shift(j:i,-1), i+1:n]` if *j* < *i*.

See also: [qr], page 446, [qrupdate], page 447, [qrinsert], page 447, [qrdelete], page 448.

`lambda = qz (A, B)` [Built-in Function]

`lambda = qz (A, B, opt)` [Built-in Function]

QZ decomposition of the generalized eigenvalue problem ($Ax = sBx$). There are three ways to call this function:

1. `lambda = qz (A, B)`

   Computes the generalized eigenvalues λ of $(A - sB)$.

2. `[AA, BB, Q, Z, V, W, lambda] = qz (A, B)`

   Computes QZ decomposition, generalized eigenvectors, and generalized eigenvalues of $(A - sB)$

$$AV = BV\text{diag}(\lambda)$$

$$W^T A = \text{diag}(\lambda) W^T B$$

$$AA = Q^T AZ, BB = Q^T BZ$$

   with Q and Z orthogonal (unitary)$= I$

3. `[AA,BB,Z{, lambda}] = qz (A, B, opt)`

   As in form [2], but allows ordering of generalized eigenpairs for (e.g.) solution of discrete time algebraic Riccati equations. Form 3 is not available for complex matrices, and does not compute the generalized eigenvectors V, W, nor the orthogonal matrix Q.

   opt for ordering eigenvalues of the GEP pencil. The leading block of the revised pencil contains all eigenvalues that satisfy:

"N"	= unordered (default)
"S"	= small: leading block has all |lambda| ≤ 1
"B"	= big: leading block has all |lambda| ≥ 1
"-"	= negative real part: leading block has all eigenvalues in the open left half-plane
"+"	= non-negative real part: leading block has all eigenvalues in the closed right half-plane

Note: `qz` performs permutation balancing, but not scaling (see [XREFbalance], page 435). The order of output arguments was selected for compatibility with MATLAB.

See also: [eig], page 437, [balance], page 435, [lu], page 444, [chol], page 442, [hess], page 444, [qr], page 446, [qzhess], page 449, [schur], page 450, [svd], page 450.

`[aa, bb, q, z] = qzhess (A, B)` [Function File]

Compute the Hessenberg-triangular decomposition of the matrix pencil `(A, B)`, returning `aa = q * A * z`, `bb = q * B * z`, with q and z orthogonal. For example:

```
[aa, bb, q, z] = qzhess ([1, 2; 3, 4], [5, 6; 7, 8])
     ⇒ aa = [ -3.02244, -4.41741;  0.92998,  0.69749 ]
     ⇒ bb = [ -8.60233, -9.99730;  0.00000, -0.23250 ]
     ⇒  q = [ -0.58124, -0.81373; -0.81373,  0.58124 ]
     ⇒  z = [ 1, 0; 0, 1 ]
```

The Hessenberg-triangular decomposition is the first step in Moler and Stewart's QZ decomposition algorithm.

Algorithm taken from Golub and Van Loan, *Matrix Computations, 2nd edition.*

See also: [lu], page 444, [chol], page 442, [hess], page 444, [qr], page 446, [qz], page 448, [schur], page 450, [svd], page 450.

`S = schur (A)` [Built-in Function]
`S = schur (A, "real")` [Built-in Function]
`S = schur (A, "complex")` [Built-in Function]
`S = schur (A, opt)` [Built-in Function]
`[U, S] = schur (A, ...)` [Built-in Function]

Compute the Schur decomposition of A

$$S = U^T A U$$

where U is a unitary matrix ($U^T U$ is identity) and S is upper triangular. The eigenvalues of A (and S) are the diagonal elements of S. If the matrix A is real, then the real Schur decomposition is computed, in which the matrix U is orthogonal and S is block upper triangular with blocks of size at most 2×2 along the diagonal. The diagonal elements of S (or the eigenvalues of the 2×2 blocks, when appropriate) are the eigenvalues of A and S.

The default for real matrices is a real Schur decomposition. A complex decomposition may be forced by passing the flag `"complex"`.

The eigenvalues are optionally ordered along the diagonal according to the value of *opt*. `opt = "a"` indicates that all eigenvalues with negative real parts should be moved to the leading block of S (used in `are`), `opt = "d"` indicates that all eigenvalues with magnitude less than one should be moved to the leading block of S (used in `dare`), and `opt = "u"`, the default, indicates that no ordering of eigenvalues should occur. The leading k columns of U always span the A-invariant subspace corresponding to the k leading eigenvalues of S.

The Schur decomposition is used to compute eigenvalues of a square matrix, and has applications in the solution of algebraic Riccati equations in control (see `are` and `dare`).

See also: [rsf2csf], page 450, [lu], page 444, [chol], page 442, [hess], page 444, [qr], page 446, [qz], page 448, [svd], page 450.

`[U, T] = rsf2csf (UR, TR)` [Function File]

Convert a real, upper quasi-triangular Schur form *TR* to a complex, upper triangular Schur form *T*.

Note that the following relations hold:

$UR \cdot TR \cdot UR^T = UTU^\dagger$ and $U^\dagger U$ is the identity matrix I.

Note also that U and T are not unique.

See also: [schur], page 450.

`angle = subspace (A, B)` [Function File]

Determine the largest principal angle between two subspaces spanned by the columns of matrices A and B.

`s = svd (A)` [Built-in Function]
`[U, S, V] = svd (A)` [Built-in Function]
`[U, S, V] = svd (A, econ)` [Built-in Function]

Compute the singular value decomposition of A

$$A = USV^\dagger$$

The function `svd` normally returns only the vector of singular values. When called with three return values, it computes U, S, and V. For example,

```
svd (hilb (3))
```

returns

```
ans =

  1.4083189
  0.1223271
  0.0026873
```

and

```
[u, s, v] = svd (hilb (3))
```

returns

```
u =

  -0.82704   0.54745   0.12766
  -0.45986  -0.52829  -0.71375
  -0.32330  -0.64901   0.68867

s =

  1.40832  0.00000  0.00000
  0.00000  0.12233  0.00000
  0.00000  0.00000  0.00269

v =

  -0.82704   0.54745   0.12766
  -0.45986  -0.52829  -0.71375
  -0.32330  -0.64901   0.68867
```

If given a second argument, `svd` returns an economy-sized decomposition, eliminating the unnecessary rows or columns of U or V.

See also: [svd_driver], page 451, [svds], page 515, [eig], page 437, [lu], page 444, [chol], page 442, [hess], page 444, [qr], page 446, [qz], page 448.

val = `svd_driver` () [Built-in Function]
old_val = `svd_driver` (*new_val*) [Built-in Function]
`svd_driver` (*new_val*, "*local*") [Built-in Function]

Query or set the underlying LAPACK driver used by `svd`. Currently recognized values are `"gesvd"` and `"gesdd"`. The default is `"gesvd"`.

When called from inside a function with the `"local"` option, the variable is changed locally for the function and any subroutines it calls. The original variable value is restored when exiting the function.

See also: [svd], page 450.

`[housv, beta, zer] = housh (x, j, z)` [Function File]

Compute Householder reflection vector *housv* to reflect x to be the j-th column of identity, i.e.,

```
(I - beta*housv*housv')x =  norm (x)*e(j) if x(j) < 0,
(I - beta*housv*housv')x = -norm (x)*e(j) if x(j) >= 0
```

Inputs

x — vector

j — index into vector

z — threshold for zero (usually should be the number 0)

Outputs (see Golub and Van Loan):

beta — If beta = 0, then no reflection need be applied (zer set to 0)

housv — householder vector

`[u, h, nu] = krylov (A, V, k, eps1, pflg)` [Function File]

Construct an orthogonal basis *u* of block Krylov subspace

```
[v a*v a^2*v ... a^(k+1)*v]
```

Using Householder reflections to guard against loss of orthogonality.

If *V* is a vector, then *h* contains the Hessenberg matrix such that `a*u == u*h+rk*ek'`, in which `rk = a*u(:,k)-u*h(:,k)`, and `ek'` is the vector `[0, 0, ..., 1]` of length `k`. Otherwise, *h* is meaningless.

If *V* is a vector and *k* is greater than `length (A) - 1`, then *h* contains the Hessenberg matrix such that `a*u == u*h`.

The value of *nu* is the dimension of the span of the Krylov subspace (based on *eps1*).

If *b* is a vector and *k* is greater than *m-1*, then *h* contains the Hessenberg decomposition of *A*.

The optional parameter *eps1* is the threshold for zero. The default value is 1e-12.

If the optional parameter *pflg* is nonzero, row pivoting is used to improve numerical behavior. The default value is 0.

Reference: A. Hodel, P. Misra, *Partial Pivoting in the Computation of Krylov Subspaces of Large Sparse Systems*, Proceedings of the 42nd IEEE Conference on Decision and Control, December 2003.

18.4 Functions of a Matrix

`expm (A)` [Function File]

Return the exponential of a matrix, defined as the infinite Taylor series

$$\exp(A) = I + A + \frac{A^2}{2!} + \frac{A^3}{3!} + \cdots$$

The Taylor series is *not* the way to compute the matrix exponential; see Moler and Van Loan, *Nineteen Dubious Ways to Compute the Exponential of a Matrix*, SIAM Review, 1978. This routine uses Ward's diagonal Padé approximation method with three step preconditioning (SIAM Journal on Numerical Analysis, 1977). Diagonal Padé approximations are rational polynomials of matrices $D_q(A)^{-1}N_q(A)$ whose Taylor series matches the first $2q+1$ terms of the Taylor series above; direct evaluation of the Taylor series (with the same preconditioning steps) may be desirable in lieu of the Padé approximation when $D_q(A)$ is ill-conditioned.

See also: [logm], page 453, [sqrtm], page 453.

`s = logm (A)` [Function File]
`s = logm (A, opt_iters)` [Function File]
`[s, iters] = logm (...)` [Function File]

Compute the matrix logarithm of the square matrix *A*. The implementation utilizes a Padé approximant and the identity

```
logm (A) = 2^k * logm (A^(1 / 2^k))
```

The optional argument *opt_iters* is the maximum number of square roots to compute and defaults to 100. The optional output *iters* is the number of square roots actually computed.

See also: [expm], page 452, [sqrtm], page 453.

`s = sqrtm (A)` [Built-in Function]
`[s, error_estimate] = sqrtm (A)` [Built-in Function]

Compute the matrix square root of the square matrix *A*.

Ref: N.J. Higham. *A New sqrtm for* MATLAB. Numerical Analysis Report No. 336, Manchester Centre for Computational Mathematics, Manchester, England, January 1999.

See also: [expm], page 452, [logm], page 453.

`kron (A, B)` [Built-in Function]
`kron (A1, A2, ...)` [Built-in Function]

Form the Kronecker product of two or more matrices, defined block by block as

```
x = [ a(i,j)*b ]
```

For example:

```
kron (1:4, ones (3, 1))
      ⇒  1  2  3  4
         1  2  3  4
         1  2  3  4
```

If there are more than two input arguments *A1*, *A2*, ..., *An* the Kronecker product is computed as

```
kron (kron (A1, A2), ..., An)
```

Since the Kronecker product is associative, this is well-defined.

`blkmm (A, B)` [Built-in Function]

Compute products of matrix blocks. The blocks are given as 2-dimensional subarrays of the arrays *A*, *B*. The size of *A* must have the form `[m,k,...]` and size of *B* must be `[k,n,...]`. The result is then of size `[m,n,...]` and is computed as follows:

```
for i = 1:prod (size (A)(3:end))
  C(:,:,i) = A(:,:,i) * B(:,:,i)
endfor
```

`x = syl (A, B, C)` [Built-in Function]

Solve the Sylvester equation

$$AX + XB + C = 0$$

using standard LAPACK subroutines. For example:

```
syl ([1, 2; 3, 4], [5, 6; 7, 8], [9, 10; 11, 12])
  ⇒ [ -0.50000, -0.66667; -0.66667, -0.50000 ]
```

18.5 Specialized Solvers

`x = bicg (A, b, rtol, maxit, M1, M2, x0)` [Function File]
`x = bicg (A, b, rtol, maxit, P)` [Function File]
`[x, flag, relres, iter, resvec] = bicg (A, b, ...)` [Function File]

Solve `A x = b` using the Bi-conjugate gradient iterative method.

- *rtol* is the relative tolerance, if not given or set to [] the default value 1e-6 is used.
- *maxit* the maximum number of outer iterations, if not given or set to [] the default value `min (20, numel (b))` is used.
- *x0* the initial guess, if not given or set to [] the default value `zeros (size (b))` is used.

A can be passed as a matrix or as a function handle or inline function `f` such that `f(x, "notransp") = A*x` and `f(x, "transp") = A'*x`.

The preconditioner *P* is given as `P = M1 * M2`. Both *M1* and *M2* can be passed as a matrix or as a function handle or inline function `g` such that `g(x, "notransp") = M1 \ x` or `g(x, "notransp") = M2 \ x` and `g(x, "transp") = M1' \ x` or `g(x, "transp") = M2' \ x`.

If called with more than one output parameter

- *flag* indicates the exit status:
  - 0: iteration converged to the within the chosen tolerance
  - 1: the maximum number of iterations was reached before convergence
  - 3: the algorithm reached stagnation

  (the value 2 is unused but skipped for compatibility).
- *relres* is the final value of the relative residual.
- *iter* is the number of iterations performed.
- *resvec* is a vector containing the relative residual at each iteration.

See also: [bicgstab], page 454, [cgs], page 455, [gmres], page 456, [pcg], page 516.

`x = bicgstab (A, b, rtol, maxit, M1, M2, x0)` [Function File]
`x = bicgstab (A, b, rtol, maxit, P)` [Function File]
`[x, flag, relres, iter, resvec] = bicgstab (A, b, ...)` [Function File]

Solve `A x = b` using the stabilizied Bi-conjugate gradient iterative method.

- *rtol* is the relative tolerance, if not given or set to [] the default value 1e-6 is used.
- *maxit* the maximum number of outer iterations, if not given or set to [] the default value `min (20, numel (b))` is used.
- *x0* the initial guess, if not given or set to [] the default value `zeros (size (b))` is used.

A can be passed as a matrix or as a function handle or inline function `f` such that `f(x) = A*x`.

The preconditioner *P* is given as `P = M1 * M2`. Both *M1* and *M2* can be passed as a matrix or as a function handle or inline function `g` such that `g(x) = M1 \ x` or `g(x) = M2 \ x`.

If called with more than one output parameter

- *flag* indicates the exit status:
  - 0: iteration converged to the within the chosen tolerance
  - 1: the maximum number of iterations was reached before convergence
  - 3: the algorithm reached stagnation

  (the value 2 is unused but skipped for compatibility).
- *relres* is the final value of the relative residual.
- *iter* is the number of iterations performed.
- *resvec* is a vector containing the relative residual at each iteration.

See also: [bicg], page 454, [cgs], page 455, [gmres], page 456, [pcg], page 516.

x = `cgs` (*A*, *b*, *rtol*, *maxit*, *M1*, *M2*, *x0*) [Function File]
x = `cgs` (*A*, *b*, *rtol*, *maxit*, *P*) [Function File]
[*x*, *flag*, *relres*, *iter*, *resvec*] = `cgs` (*A*, *b*, ...) [Function File]

Solve `A x = b`, where *A* is a square matrix, using the Conjugate Gradients Squared method.

- *rtol* is the relative tolerance, if not given or set to [] the default value 1e-6 is used.
- *maxit* the maximum number of outer iterations, if not given or set to [] the default value `min (20, numel (b))` is used.
- *x0* the initial guess, if not given or set to [] the default value `zeros (size (b))` is used.

A can be passed as a matrix or as a function handle or inline function `f` such that `f(x) = A*x`.

The preconditioner *P* is given as `P = M1 * M2`. Both *M1* and *M2* can be passed as a matrix or as a function handle or inline function `g` such that `g(x) = M1 \ x` or `g(x) = M2 \ x`.

If called with more than one output parameter

- *flag* indicates the exit status:
  - 0: iteration converged to the within the chosen tolerance
  - 1: the maximum number of iterations was reached before convergence
  - 3: the algorithm reached stagnation

(the value 2 is unused but skipped for compatibility).

- *relres* is the final value of the relative residual.
- *iter* is the number of iterations performed.
- *resvec* is a vector containing the relative residual at each iteration.

See also: [pcg], page 516, [bicgstab], page 454, [bicg], page 454, [gmres], page 456.

`x = gmres (A, b, m, rtol, maxit, M1, M2, x0)` [Function File]
`x = gmres (A, b, m, rtol, maxit, P)` [Function File]
`[x, flag, relres, iter, resvec] = gmres (...)` [Function File]

Solve `A x = b` using the Preconditioned GMRES iterative method with restart, a.k.a. PGMRES(m).

- *rtol* is the relative tolerance, if not given or set to [] the default value 1e-6 is used.
- *maxit* is the maximum number of outer iterations, if not given or set to [] the default value `min (10, numel (b) / restart)` is used.
- *x0* is the initial guess, if not given or set to [] the default value `zeros (size (b))` is used.
- *m* is the restart parameter, if not given or set to [] the default value `numel (b)` is used.

Argument *A* can be passed as a matrix, function handle, or inline function `f` such that `f(x) = A*x`.

The preconditioner *P* is given as `P = M1 * M2`. Both *M1* and *M2* can be passed as a matrix, function handle, or inline function `g` such that `g(x) = M1\x` or `g(x) = M2\x`.

Besides the vector *x*, additional outputs are:

- *flag* indicates the exit status:

  0 : iteration converged to within the specified tolerance
  1 : maximum number of iterations exceeded
  2 : unused, but skipped for compatibility
  3 : algorithm reached stagnation (no change between iterations)
- *relres* is the final value of the relative residual.
- *iter* is a vector containing the number of outer iterations and total iterations performed.
- *resvec* is a vector containing the relative residual at each iteration.

See also: [bicg], page 454, [bicgstab], page 454, [cgs], page 455, [pcg], page 516.

19 Vectorization and Faster Code Execution

Vectorization is a programming technique that uses vector operations instead of element-by-element loop-based operations. Besides frequently producing more succinct Octave code, vectorization also allows for better optimization in the subsequent implementation. The optimizations may occur either in Octave's own Fortran, C, or C++ internal implementation, or even at a lower level depending on the compiler and external numerical libraries used to build Octave. The ultimate goal is to make use of your hardware's vector instructions if possible or to perform other optimizations in software.

Vectorization is not a concept unique to Octave, but it is particularly important because Octave is a matrix-oriented language. Vectorized Octave code will see a dramatic speed up (10X–100X) in most cases.

This chapter discusses vectorization and other techniques for writing faster code.

19.1 Basic Vectorization

To a very good first approximation, the goal in vectorization is to write code that avoids loops and uses whole-array operations. As a trivial example, consider

```
for i = 1:n
  for j = 1:m
    c(i,j) = a(i,j) + b(i,j);
  endfor
endfor
```

compared to the much simpler

```
c = a + b;
```

This isn't merely easier to write; it is also internally much easier to optimize. Octave delegates this operation to an underlying implementation which, among other optimizations, may use special vector hardware instructions or could conceivably even perform the additions in parallel. In general, if the code is vectorized, the underlying implementation has more freedom about the assumptions it can make in order to achieve faster execution.

This is especially important for loops with "cheap" bodies. Often it suffices to vectorize just the innermost loop to get acceptable performance. A general rule of thumb is that the "order" of the vectorized body should be greater or equal to the "order" of the enclosing loop.

As a less trivial example, instead of

```
for i = 1:n-1
  a(i) = b(i+1) - b(i);
endfor
```

write

```
a = b(2:n) - b(1:n-1);
```

This shows an important general concept about using arrays for indexing instead of looping over an index variable. See Section 8.1 [Index Expressions], page 131. Also use boolean indexing generously. If a condition needs to be tested, this condition can also be written as a boolean index. For instance, instead of

```
for i = 1:n
  if (a(i) > 5)
    a(i) -= 20
  endif
endfor
```

write

```
a(a>5) -= 20;
```

which exploits the fact that `a > 5` produces a boolean index.

Use elementwise vector operators whenever possible to avoid looping (operators like `.*` and `.^`). See Section 8.3 [Arithmetic Ops], page 137. For simple inline functions, the `vectorize` function can do this automatically.

`vectorize (`*fun*`)` [Built-in Function]

Create a vectorized version of the inline function *fun* by replacing all occurrences of `*`, `/`, etc., with `.*`, `./`, etc.

This may be useful, for example, when using inline functions with numerical integration or optimization where a vector-valued function is expected.

```
fcn = vectorize (inline ("x^2 - 1"))
   ⇒ fcn = f(x) = x.^2 - 1
quadv (fcn, 0, 3)
   ⇒ 6
```

See also: [inline], page 192, [formula], page 193, [argnames], page 193.

Also exploit broadcasting in these elementwise operators both to avoid looping and unnecessary intermediate memory allocations. See Section 19.2 [Broadcasting], page 459.

Use built-in and library functions if possible. Built-in and compiled functions are very fast. Even with an m-file library function, chances are good that it is already optimized, or will be optimized more in a future release.

For instance, even better than

```
a = b(2:n) - b(1:n-1);
```

is

```
a = diff (b);
```

Most Octave functions are written with vector and array arguments in mind. If you find yourself writing a loop with a very simple operation, chances are that such a function already exists. The following functions occur frequently in vectorized code:

- Index manipulation
  - find
  - sub2ind
  - ind2sub
  - sort
  - unique
  - lookup
  - ifelse / merge

- Repetition
  - repmat
  - repelems
- Vectorized arithmetic
  - sum
  - prod
  - cumsum
  - cumprod
  - sumsq
  - diff
  - dot
  - cummax
  - cummin
- Shape of higher dimensional arrays
  - reshape
  - resize
  - permute
  - squeeze
  - deal

19.2 Broadcasting

Broadcasting refers to how Octave binary operators and functions behave when their matrix or array operands or arguments differ in size. Since version 3.6.0, Octave now automatically broadcasts vectors, matrices, and arrays when using elementwise binary operators and functions. Broadly speaking, smaller arrays are "broadcast" across the larger one, until they have a compatible shape. The rule is that corresponding array dimensions must either

1. be equal, or
2. one of them must be 1.

In case all dimensions are equal, no broadcasting occurs and ordinary element-by-element arithmetic takes place. For arrays of higher dimensions, if the number of dimensions isn't the same, then missing trailing dimensions are treated as 1. When one of the dimensions is 1, the array with that singleton dimension gets copied along that dimension until it matches the dimension of the other array. For example, consider

```
x = [1 2 3;
     4 5 6;
     7 8 9];

y = [10 20 30];

x + y
```

Without broadcasting, x + y would be an error because the dimensions do not agree. However, with broadcasting it is as if the following operation were performed:

```
x = [1 2 3
     4 5 6
     7 8 9];

y = [10 20 30
     10 20 30
     10 20 30];

x + y
⇒    11   22   33
     14   25   36
     17   28   39
```

That is, the smaller array of size `[1 3]` gets copied along the singleton dimension (the number of rows) until it is `[3 3]`. No actual copying takes place, however. The internal implementation reuses elements along the necessary dimension in order to achieve the desired effect without copying in memory.

Both arrays can be broadcast across each other, for example, all pairwise differences of the elements of a vector with itself:

```
y - y'
⇒    0   10   20
   -10    0   10
   -20  -10    0
```

Here the vectors of size `[1 3]` and `[3 1]` both get broadcast into matrices of size `[3 3]` before ordinary matrix subtraction takes place.

A special case of broadcasting that may be familiar is when all dimensions of the array being broadcast are 1, i.e., the array is a scalar. Thus for example, operations like `x - 42` and `max (x, 2)` are basic examples of broadcasting.

For a higher-dimensional example, suppose `img` is an RGB image of size `[m n 3]` and we wish to multiply each color by a different scalar. The following code accomplishes this with broadcasting,

```
img .*= permute ([0.8, 0.9, 1.2], [1, 3, 2]);
```

Note the usage of permute to match the dimensions of the `[0.8, 0.9, 1.2]` vector with `img`.

For functions that are not written with broadcasting semantics, `bsxfun` can be useful for coercing them to broadcast.

`bsxfun` (*f*, *A*, *B*) [Built-in Function]

The binary singleton expansion function applier performs broadcasting, that is, applies a binary function *f* element-by-element to two array arguments *A* and *B*, and expands as necessary singleton dimensions in either input argument. *f* is a function handle, inline function, or string containing the name of the function to evaluate. The function *f* must be capable of accepting two column-vector arguments of equal length, or one column vector argument and a scalar.

The dimensions of *A* and *B* must be equal or singleton. The singleton dimensions of the arrays will be expanded to the same dimensionality as the other array.

See also: [arrayfun], page 462, [cellfun], page 464.

Broadcasting is only applied if either of the two broadcasting conditions hold. As usual, however, broadcasting does not apply when two dimensions differ and neither is 1:

```
x = [1 2 3
     4 5 6];
y = [10 20
     30 40];
x + y
```

This will produce an error about nonconformant arguments.

Besides common arithmetic operations, several functions of two arguments also broadcast. The full list of functions and operators that broadcast is

```
plus      +   .+
minus     -   .-
times     .*
rdivide   ./
ldivide   .\
power     .^   .**
lt        <
le        <=
eq        ==
gt        >
ge        >=
ne        !=   ~=
and       &
or        |
atan2
hypot
max
min
mod
rem
xor

+=   -=   .+=   .-=   .*=   ./=   .\=   .^=   .**=   &=   |=
```

Beware of resorting to broadcasting if a simpler operation will suffice. For matrices *a* and *b*, consider the following:

```
c = sum (permute (a, [1, 3, 2]) .* permute (b, [3, 2, 1]), 3);
```

This operation broadcasts the two matrices with permuted dimensions across each other during elementwise multiplication in order to obtain a larger 3-D array, and this array is then summed along the third dimension. A moment of thought will prove that this operation is simply the much faster ordinary matrix multiplication, `c = a*b;`.

A note on terminology: "broadcasting" is the term popularized by the Numpy numerical environment in the Python programming language. In other programming languages and environments, broadcasting may also be known as *binary singleton expansion* (BSX, in

MATLAB, and the origin of the name of the **bsxfun** function), *recycling* (R programming language), *single-instruction multiple data* (SIMD), or *replication*.

19.2.1 Broadcasting and Legacy Code

The new broadcasting semantics almost never affect code that worked in previous versions of Octave. Consequently, all code inherited from MATLAB that worked in previous versions of Octave should still work without change in Octave. The only exception is code such as

```
try
  c = a.*b;
catch
  c = a.*a;
end_try_catch
```

that may have relied on matrices of different size producing an error. Due to how broadcasting changes semantics with older versions of Octave, by default Octave warns if a broadcasting operation is performed. To disable this warning, refer to its ID (see [warning_ids], page 205):

```
warning ("off", "Octave:broadcast");
```

If you want to recover the old behavior and produce an error, turn this warning into an error:

```
warning ("error", "Octave:broadcast");
```

For broadcasting on scalars that worked in previous versions of Octave, this warning will not be emitted.

19.3 Function Application

As a general rule, functions should already be written with matrix arguments in mind and should consider whole matrix operations in a vectorized manner. Sometimes, writing functions in this way appears difficult or impossible for various reasons. For those situations, Octave provides facilities for applying a function to each element of an array, cell, or struct.

arrayfun (*func*, *A*) [Function File]
x = arrayfun (*func*, *A*) [Function File]
x = arrayfun (*func*, *A*, *b*, ...) [Function File]
[*x*, *y*, ...] = arrayfun (*func*, *A*, ...) [Function File]
arrayfun (..., "*UniformOutput*", *val*) [Function File]
arrayfun (..., "*ErrorHandler*", *errfunc*) [Function File]

Execute a function on each element of an array. This is useful for functions that do not accept array arguments. If the function does accept array arguments it is better to call the function directly.

The first input argument *func* can be a string, a function handle, an inline function, or an anonymous function. The input argument *A* can be a logic array, a numeric array, a string array, a structure array, or a cell array. By a call of the function **arrayfun** all elements of *A* are passed on to the named function *func* individually.

The named function can also take more than two input arguments, with the input arguments given as third input argument *b*, fourth input argument *c*, ... If given more than one array input argument then all input arguments must have the same sizes, for example:

```
arrayfun (@atan2, [1, 0], [0, 1])
     ⇒ [ 1.5708   0.0000 ]
```

If the parameter *val* after a further string input argument `"UniformOutput"` is set `true` (the default), then the named function *func* must return a single element which then will be concatenated into the return value and is of type matrix. Otherwise, if that parameter is set to `false`, then the outputs are concatenated in a cell array. For example:

```
arrayfun (@(x,y) x:y, "abc", "def", "UniformOutput", false)
⇒
   {
     [1,1] = abcd
     [1,2] = bcde
     [1,3] = cdef
   }
```

If more than one output arguments are given then the named function must return the number of return values that also are expected, for example:

```
[A, B, C] = arrayfun (@find, [10; 0], "UniformOutput", false)
⇒
A =
{
   [1,1] =  1
   [2,1] = [](0x0)
}
B =
{
   [1,1] =  1
   [2,1] = [](0x0)
}
C =
{
   [1,1] =  10
   [2,1] = [](0x0)
}
```

If the parameter *errfunc* after a further string input argument `"ErrorHandler"` is another string, a function handle, an inline function, or an anonymous function, then *errfunc* defines a function to call in the case that *func* generates an error. The definition of the function must be of the form

```
function [...] = errfunc (s, ...)
```

where there is an additional input argument to *errfunc* relative to *func*, given by *s*. This is a structure with the elements `"identifier"`, `"message"`, and `"index"` giving, respectively, the error identifier, the error message, and the index of the array elements that caused the error. The size of the output argument of *errfunc* must have the same size as the output argument of *func*, otherwise a real error is thrown. For example:

```
function y = ferr (s, x), y = "MyString"; endfunction
arrayfun (@str2num, [1234],
          "UniformOutput", false, "ErrorHandler", @ferr)
⇒
   {
     [1,1] = MyString
   }
```

See also: [spfun], page 464, [cellfun], page 464, [structfun], page 466.

y = `spfun` (*f*, *S*) [Function File]

Compute `f(S)` for the non-zero values of *S*. This results in a sparse matrix with the same structure as *S*. The function *f* can be passed as a string, a function handle, or an inline function.

See also: [arrayfun], page 462, [cellfun], page 464, [structfun], page 466.

`cellfun` (*name*, *C*) [Built-in Function]
`cellfun` ("*size*", *C*, *k*) [Built-in Function]
`cellfun` ("*isclass*", *C*, *class*) [Built-in Function]
`cellfun` (*func*, *C*) [Built-in Function]
`cellfun` (*func*, *C*, *D*) [Built-in Function]
`[a, ...] = cellfun (...)` [Built-in Function]
`cellfun` (..., "*ErrorHandler*", *errfunc*) [Built-in Function]
`cellfun` (..., "*UniformOutput*", *val*) [Built-in Function]

Evaluate the function named *name* on the elements of the cell array *C*. Elements in *C* are passed on to the named function individually. The function *name* can be one of the functions

`isempty` Return 1 for empty elements.

`islogical`
Return 1 for logical elements.

`isnumeric`
Return 1 for numeric elements.

`isreal` Return 1 for real elements.

`length` Return a vector of the lengths of cell elements.

`ndims` Return the number of dimensions of each element.

`numel`
`prodofsize`
Return the number of elements contained within each cell element. The number is the product of the dimensions of the object at each cell element.

`size` Return the size along the *k*-th dimension.

`isclass` Return 1 for elements of *class*.

Additionally, `cellfun` accepts an arbitrary function *func* in the form of an inline function, function handle, or the name of a function (in a character string). The function can take one or more arguments, with the inputs arguments given by *C*, *D*, etc. Equally the function can return one or more output arguments. For example:

```
cellfun ("atan2", {1, 0}, {0, 1})
     ⇒ [ 1.57080   0.00000 ]
```

The number of output arguments of `cellfun` matches the number of output arguments of the function. The outputs of the function will be collected into the output arguments of `cellfun` like this:

```
function [a, b] = twoouts (x)
  a = x;
  b = x*x;
endfunction
[aa, bb] = cellfun (@twoouts, {1, 2, 3})
     ⇒
        aa =
           1 2 3
        bb =
           1 4 9
```

Note that per default the output argument(s) are arrays of the same size as the input arguments. Input arguments that are singleton (1x1) cells will be automatically expanded to the size of the other arguments.

If the parameter `"UniformOutput"` is set to true (the default), then the function must return scalars which will be concatenated into the return array(s). If `"UniformOutput"` is false, the outputs are concatenated into a cell array (or cell arrays). For example:

```
cellfun ("tolower", {"Foo", "Bar", "FooBar"},
          "UniformOutput", false)
⇒ {"foo", "bar", "foobar"}
```

Given the parameter `"ErrorHandler"`, then *errfunc* defines a function to call in case *func* generates an error. The form of the function is

```
function [...] = errfunc (s, ...)
```

where there is an additional input argument to *errfunc* relative to *func*, given by *s*. This is a structure with the elements `"identifier"`, `"message"` and `"index"`, giving respectively the error identifier, the error message, and the index into the input arguments of the element that caused the error. For example:

```
function y = foo (s, x), y = NaN; endfunction
cellfun ("factorial", {-1,2}, "ErrorHandler", @foo)
⇒ [NaN 2]
```

Use `cellfun` intelligently. The `cellfun` function is a useful tool for avoiding loops. It is often used with anonymous function handles; however, calling an anonymous function involves an overhead quite comparable to the overhead of an m-file function. Passing a handle to a built-in function is faster, because the interpreter is not involved in the internal loop. For example:

```
a = {...}
v = cellfun (@(x) det (x), a); # compute determinants
v = cellfun (@det, a); # faster
```

See also: [arrayfun], page 462, [structfun], page 466, [spfun], page 464.

`structfun` (*func*, *S*) [Function File]
`[A, ...] = structfun (...)` [Function File]
`structfun (..., "`*ErrorHandler*`", errfunc)` [Function File]
`structfun (..., "`*UniformOutput*`", val)` [Function File]

Evaluate the function named *name* on the fields of the structure *S*. The fields of *S* are passed to the function *func* individually.

`structfun` accepts an arbitrary function *func* in the form of an inline function, function handle, or the name of a function (in a character string). In the case of a character string argument, the function must accept a single argument named *x*, and it must return a string value. If the function returns more than one argument, they are returned as separate output variables.

If the parameter `"UniformOutput"` is set to true (the default), then the function must return a single element which will be concatenated into the return value. If `"UniformOutput"` is false, the outputs are placed into a structure with the same fieldnames as the input structure.

```
s.name1 = "John Smith";
s.name2 = "Jill Jones";
structfun (@(x) regexp (x, '(\w+)$', "matches"){1}, s,
           "UniformOutput", false)
⇒
  {
    name1 = Smith
    name2 = Jones
  }
```

Given the parameter `"ErrorHandler"`, *errfunc* defines a function to call in case *func* generates an error. The form of the function is

```
function [...] = errfunc (se, ...)
```

where there is an additional input argument to *errfunc* relative to *func*, given by *se*. This is a structure with the elements `"identifier"`, `"message"` and `"index"`, giving respectively the error identifier, the error message, and the index into the input arguments of the element that caused the error. For an example on how to use an error handler, see [cellfun], page 464.

See also: [cellfun], page 464, [arrayfun], page 462, [spfun], page 464.

19.4 Accumulation

Whenever it's possible to categorize according to indices the elements of an array when performing a computation, accumulation functions can be useful.

`accumarray` (*subs*, *vals*, *sz*, *func*, *fillval*, *issparse*) [Function File]
`accumarray` (*subs*, *vals*, ...) [Function File]

Create an array by accumulating the elements of a vector into the positions defined by their subscripts. The subscripts are defined by the rows of the matrix *subs* and the values by *vals*. Each row of *subs* corresponds to one of the values in *vals*. If *vals* is a scalar, it will be used for each of the row of *subs*. If *subs* is a cell array of vectors, all vectors must be of the same length, and the subscripts in the *k*th vector must correspond to the *k*th dimension of the result.

The size of the matrix will be determined by the subscripts themselves. However, if *sz* is defined it determines the matrix size. The length of *sz* must correspond to the number of columns in *subs*. An exception is if *subs* has only one column, in which case *sz* may be the dimensions of a vector and the subscripts of *subs* are taken as the indices into it.

The default action of `accumarray` is to sum the elements with the same subscripts. This behavior can be modified by defining the *func* function. This should be a function or function handle that accepts a column vector and returns a scalar. The result of the function should not depend on the order of the subscripts.

The elements of the returned array that have no subscripts associated with them are set to zero. Defining *fillval* to some other value allows these values to be defined. This behavior changes, however, for certain values of *func*. If *func* is `min` (respectively, `max`) then the result will be filled with the minimum (respectively, maximum) integer if *vals* is of integral type, logical false (respectively, logical true) if *vals* is of logical type, zero if *fillval* is zero and all values are non-positive (respectively, non-negative), and NaN otherwise.

By default `accumarray` returns a full matrix. If *issparse* is logically true, then a sparse matrix is returned instead.

The following `accumarray` example constructs a frequency table that in the first column counts how many occurrences each number in the second column has, taken from the vector *x*. Note the usage of `unique` for assigning to all repeated elements of *x* the same index (see [unique], page 597).

```
x = [91, 92, 90, 92, 90, 89, 91, 89, 90, 100, 100, 100];
[u, ~, j] = unique (x);
[accumarray(j', 1), u']
  ⇒ 2    89
    3    90
    2    91
    2    92
    3   100
```

Another example, where the result is a multi-dimensional 3-D array and the default value (zero) appears in the output:

```
accumarray ([1, 1, 1;
             2, 1, 2;
             2, 3, 2;
             2, 1, 2;
             2, 3, 2], 101:105)
⇒ ans(:,:,1) = [101, 0, 0; 0, 0, 0]
⇒ ans(:,:,2) = [0, 0, 0; 206, 0, 208]
```

The sparse option can be used as an alternative to the `sparse` constructor (see [sparse], page 492). Thus

```
sparse (i, j, sv)
```

can be written with `accumarray` as

```
accumarray ([i, j], sv', [], [], 0, true)
```

For repeated indices, `sparse` adds the corresponding value. To take the minimum instead, use `min` as an accumulator function:

```
accumarray ([i, j], sv', [], @min, 0, true)
```

The complexity of accumarray in general for the non-sparse case is generally O(M+N), where N is the number of subscripts and M is the maximum subscript (linearized in multi-dimensional case). If *func* is one of `@sum` (default), `@max`, `@min` or `@(x) {x}`, an optimized code path is used. Note that for general reduction function the interpreter overhead can play a major part and it may be more efficient to do multiple accumarray calls and compute the results in a vectorized manner.

See also: [accumdim], page 468, [unique], page 597, [sparse], page 492.

`accumdim (`*`subs, vals, dim, n, func, fillval`*`)` [Function File]

Create an array by accumulating the slices of an array into the positions defined by their subscripts along a specified dimension. The subscripts are defined by the index vector *subs*. The dimension is specified by *dim*. If not given, it defaults to the first non-singleton dimension. The length of *subs* must be equal to `size (vals, dim)`.

The extent of the result matrix in the working dimension will be determined by the subscripts themselves. However, if *n* is defined it determines this extent.

The default action of `accumdim` is to sum the subarrays with the same subscripts. This behavior can be modified by defining the *func* function. This should be a function or function handle that accepts an array and a dimension, and reduces the array along this dimension. As a special exception, the built-in `min` and `max` functions can be used directly, and `accumdim` accounts for the middle empty argument that is used in their calling.

The slices of the returned array that have no subscripts associated with them are set to zero. Defining *fillval* to some other value allows these values to be defined.

An example of the use of `accumdim` is:

```
accumdim ([1, 2, 1, 2, 1], [ 7, -10,   4;
                             -5, -12,   8;
                            -12,   2,   8;
                            -10,   9,  -3;
                             -5,  -3, -13])
⇒ [-10,-11,-1;-15,-3,5]
```

See also: [accumarray], page 466.

19.5 JIT Compiler

Vectorization is the preferred technique for eliminating loops and speeding up code. Nevertheless, it is not always possible to replace every loop. In such situations it may be worth trying Octave's **experimental** Just-In-Time (JIT) compiler.

A JIT compiler works by analyzing the body of a loop, translating the Octave statements into another language, compiling the new code segment into an executable, and then running the executable and collecting any results. The process is not simple and there is a significant amount of work to perform for each step. It can still make sense, however, if the number of loop iterations is large. Because Octave is an interpreted language every time through a loop Octave must parse the statements in the loop body before executing them. With a JIT compiler this is done just once when the body is translated to another language.

The JIT compiler is a very new feature in Octave and not all valid Octave statements can currently be accelerated. However, if no other technique is available it may be worth benchmarking the code with JIT enabled. The function `jit_enable` is used to turn compilation on or off. The function `jit_startcnt` sets the threshold for acceleration. Loops with iteration counts above `jit_startcnt` will be accelerated. The function `debug_jit` is not likely to be of use to anyone not working directly on the implementation of the JIT compiler.

val = **jit_enable** () [Built-in Function]
old_val = **jit_enable** (*new_val*) [Built-in Function]
jit_enable (*new_val*, "*local*") [Built-in Function]

Query or set the internal variable that enables Octave's JIT compiler.

When called from inside a function with the `"local"` option, the variable is changed locally for the function and any subroutines it calls. The original variable value is restored when exiting the function.

See also: [jit_startcnt], page 469, [debug_jit], page 469.

val = **jit_startcnt** () [Built-in Function]
old_val = **jit_startcnt** (*new_val*) [Built-in Function]
jit_startcnt (*new_val*, "*local*") [Built-in Function]

Query or set the internal variable that determines whether JIT compilation will take place for a specific loop. Because compilation is a costly operation it does not make sense to employ JIT when the loop count is low. By default only loops with greater than 1000 iterations will be accelerated.

When called from inside a function with the `"local"` option, the variable is changed locally for the function and any subroutines it calls. The original variable value is restored when exiting the function.

See also: [jit_enable], page 469, [debug_jit], page 469.

val = **debug_jit** () [Built-in Function]
old_val = **debug_jit** (*new_val*) [Built-in Function]

`debug_jit (new_val, "local")` [Built-in Function]

Query or set the internal variable that determines whether debugging/tracing is enabled for Octave's JIT compiler.

When called from inside a function with the `"local"` option, the variable is changed locally for the function and any subroutines it calls. The original variable value is restored when exiting the function.

See also: [jit_enable], page 469, [jit_startcnt], page 469.

19.6 Miscellaneous Techniques

Here are some other ways of improving the execution speed of Octave programs.

- Avoid computing costly intermediate results multiple times. Octave currently does not eliminate common subexpressions. Also, certain internal computation results are cached for variables. For instance, if a matrix variable is used multiple times as an index, checking the indices (and internal conversion to integers) is only done once.
- Be aware of lazy copies (copy-on-write). When a copy of an object is created, the data is not immediately copied, but rather shared. The actual copying is postponed until the copied data needs to be modified. For example:

  ```
  a = zeros (1000); # create a 1000x1000 matrix
  b = a; # no copying done here
  b(1) = 1; # copying done here
  ```

  Lazy copying applies to whole Octave objects such as matrices, cells, struct, and also individual cell or struct elements (not array elements).

  Additionally, index expressions also use lazy copying when Octave can determine that the indexed portion is contiguous in memory. For example:

  ```
  a = zeros (1000); # create a 1000x1000 matrix
  b = a(:,10:100);  # no copying done here
  b = a(10:100,:);  # copying done here
  ```

  This applies to arrays (matrices), cell arrays, and structs indexed using '`()`'. Index expressions generating comma-separated lists can also benefit from shallow copying in some cases. In particular, when *a* is a struct array, expressions like `{a.x}, {a(:,2).x}` will use lazy copying, so that data can be shared between a struct array and a cell array.

  Most indexing expressions do not live longer than their parent objects. In rare cases, however, a lazily copied slice outlasts its parent, in which case it becomes orphaned, still occupying unnecessarily more memory than needed. To provide a remedy working in most real cases, Octave checks for orphaned lazy slices at certain situations, when a value is stored into a "permanent" location, such as a named variable or cell or struct element, and possibly economizes them. For example:

  ```
  a = zeros (1000); # create a 1000x1000 matrix
  b = a(:,10:100);  # lazy slice
  a = []; # the original "a" array is still allocated
  c{1} = b; # b is reallocated at this point
  ```

- Avoid deep recursion. Function calls to m-file functions carry a relatively significant overhead, so rewriting a recursion as a loop often helps. Also, note that the maximum level of recursion is limited.

- Avoid resizing matrices unnecessarily. When building a single result matrix from a series of calculations, set the size of the result matrix first, then insert values into it. Write

```
result = zeros (big_n, big_m)
for i = over:and_over
  ridx = ...
  cidx = ...
  result(ridx, cidx) = new_value ();
endfor
```

  instead of

```
result = [];
for i = ever:and_ever
  result = [ result, new_value() ];
endfor
```

  Sometimes the number of items can not be computed in advance, and stack-like operations are needed. When elements are being repeatedly inserted or removed from the end of an array, Octave detects it as stack usage and attempts to use a smarter memory management strategy by pre-allocating the array in bigger chunks. This strategy is also applied to cell and struct arrays.

```
a = [];
while (condition)
  ...
  a(end+1) = value; # "push" operation
  ...
  a(end) = []; # "pop" operation
  ...
endwhile
```

- Avoid calling `eval` or `feval` excessively. Parsing input or looking up the name of a function in the symbol table are relatively expensive operations.

  If you are using `eval` merely as an exception handling mechanism, and not because you need to execute some arbitrary text, use the `try` statement instead. See Section 10.9 [The try Statement], page 164.

- Use `ignore_function_time_stamp` when appropriate. If you are calling lots of functions, and none of them will need to change during your run, set the variable `ignore_function_time_stamp` to `"all"`. This will stop Octave from checking the time stamp of a function file to see if it has been updated while the program is being run.

19.7 Examples

The following are examples of vectorization questions asked by actual users of Octave and their solutions.

- For a vector `A`, the following loop

```
n = length (A);
B = zeros (n, 2);
for i = 1:length (A)
  ## this will be two columns, the first is the difference and
  ## the second the mean of the two elements used for the diff.
  B(i,:) = [A(i+1)-A(i), (A(i+1) + A(i))/2];
endfor
```

can be turned into the following one-liner:

```
B = [diff(A)(:), 0.5*(A(1:end-1)+A(2:end))(:)]
```

Note the usage of colon indexing to flatten an intermediate result into a column vector. This is a common vectorization trick.

20 Nonlinear Equations

20.1 Solvers

Octave can solve sets of nonlinear equations of the form

$$f(x) = 0$$

using the function `fsolve`, which is based on the MINPACK subroutine `hybrd`. This is an iterative technique so a starting point must be provided. This also has the consequence that convergence is not guaranteed even if a solution exists.

`fsolve (`*fcn*`, `*x0*`, `*options*`)` [Function File]
`[`*x*`, `*fvec*`, `*info*`, `*output*`, `*fjac*`] = fsolve (`*fcn*`, ...)` [Function File]

Solve a system of nonlinear equations defined by the function *fcn*. *fcn* should accept a vector (array) defining the unknown variables, and return a vector of left-hand sides of the equations. Right-hand sides are defined to be zeros. In other words, this function attempts to determine a vector *x* such that `fcn (x)` gives (approximately) all zeros. *x0* determines a starting guess. The shape of *x0* is preserved in all calls to *fcn*, but otherwise it is treated as a column vector. *options* is a structure specifying additional options. Currently, `fsolve` recognizes these options: `"FunValCheck"`, `"OutputFcn"`, `"TolX"`, `"TolFun"`, `"MaxIter"`, `"MaxFunEvals"`, `"Jacobian"`, `"Updating"`, `"ComplexEqn"` `"TypicalX"`, `"AutoScaling"` and `"FinDiffType"`.

If `"Jacobian"` is `"on"`, it specifies that *fcn*, called with 2 output arguments, also returns the Jacobian matrix of right-hand sides at the requested point. `"TolX"` specifies the termination tolerance in the unknown variables, while `"TolFun"` is a tolerance for equations. Default is `1e-7` for both `"TolX"` and `"TolFun"`.

If `"AutoScaling"` is on, the variables will be automatically scaled according to the column norms of the (estimated) Jacobian. As a result, TolF becomes scaling-independent. By default, this option is off, because it may sometimes deliver unexpected (though mathematically correct) results.

If `"Updating"` is `"on"`, the function will attempt to use Broyden updates to update the Jacobian, in order to reduce the amount of Jacobian calculations. If your user function always calculates the Jacobian (regardless of number of output arguments), this option provides no advantage and should be set to false.

`"ComplexEqn"` is `"on"`, `fsolve` will attempt to solve complex equations in complex variables, assuming that the equations possess a complex derivative (i.e., are holomorphic). If this is not what you want, should unpack the real and imaginary parts of the system to get a real system.

For description of the other options, see `optimset`.

On return, *fval* contains the value of the function *fcn* evaluated at *x*, and *info* may be one of the following values:

1 Converged to a solution point. Relative residual error is less than specified by TolFun.

2 Last relative step size was less that TolX.

3 Last relative decrease in residual was less than TolF.

0 Iteration limit exceeded.

-3 The trust region radius became excessively small.

Note: If you only have a single nonlinear equation of one variable, using **fzero** is usually a much better idea.

Note about user-supplied Jacobians: As an inherent property of the algorithm, Jacobian is always requested for a solution vector whose residual vector is already known, and it is the last accepted successful step. Often this will be one of the last two calls, but not always. If the savings by reusing intermediate results from residual calculation in Jacobian calculation are significant, the best strategy is to employ OutputFcn: After a vector is evaluated for residuals, if OutputFcn is called with that vector, then the intermediate results should be saved for future Jacobian evaluation, and should be kept until a Jacobian evaluation is requested or until OutputFcn is called with a different vector, in which case they should be dropped in favor of this most recent vector. A short example how this can be achieved follows:

```
function [fvec, fjac] = user_func (x, optimvalues, state)
persistent sav = [], sav0 = [];
if (nargin == 1)
  ## evaluation call
  if (nargout == 1)
    sav0.x = x; # mark saved vector
    ## calculate fvec, save results to sav0.
  elseif (nargout == 2)
    ## calculate fjac using sav.
  endif
else
  ## outputfcn call.
  if (all (x == sav0.x))
    sav = sav0;
  endif
  ## maybe output iteration status, etc.
endif
endfunction

## ...

fsolve (@user_func, x0, optimset ("OutputFcn", @user_func, ...))
```

See also: [fzero], page 475, [optimset], page 558.

The following is a complete example. To solve the set of equations

$$-2x^2 + 3xy + 4\sin(y) - 6 = 0$$
$$3x^2 - 2xy^2 + 3\cos(x) + 4 = 0$$

you first need to write a function to compute the value of the given function. For example:

```
function y = f (x)
  y = zeros (2, 1);
  y(1) = -2*x(1)^2 + 3*x(1)*x(2)   + 4*sin(x(2)) - 6;
  y(2) =  3*x(1)^2 - 2*x(1)*x(2)^2 + 3*cos(x(1)) + 4;
endfunction
```

Then, call `fsolve` with a specified initial condition to find the roots of the system of equations. For example, given the function `f` defined above,

```
[x, fval, info] = fsolve (@f, [1; 2])
```

results in the solution

```
x =

  0.57983
  2.54621

fval =

  -5.7184e-10
   5.5460e-10

info = 1
```

A value of `info = 1` indicates that the solution has converged.

When no Jacobian is supplied (as in the example above) it is approximated numerically. This requires more function evaluations, and hence is less efficient. In the example above we could compute the Jacobian analytically as

$$\begin{bmatrix} \frac{\partial f_1}{\partial x_1} & \frac{\partial f_1}{\partial x_2} \\ \frac{\partial f_2}{\partial x_1} & \frac{\partial f_2}{\partial x_2} \end{bmatrix} = \begin{bmatrix} 3x_2 - 4x_1 & 4\cos(x_2) + 3x_1 \\ -2x_2^2 - 3\sin(x_1) + 6x_1 & -4x_1x_2 \end{bmatrix}$$

and compute it with the following Octave function

```
function [y, jac] = f (x)
  y = zeros (2, 1);
  y(1) = -2*x(1)^2 + 3*x(1)*x(2)   + 4*sin(x(2)) - 6;
  y(2) =  3*x(1)^2 - 2*x(1)*x(2)^2 + 3*cos(x(1)) + 4;
  if (nargout == 2)
    jac = zeros (2, 2);
    jac(1,1) =  3*x(2) - 4*x(1);
    jac(1,2) =  4*cos(x(2)) + 3*x(1);
    jac(2,1) = -2*x(2)^2 - 3*sin(x(1)) + 6*x(1);
    jac(2,2) = -4*x(1)*x(2);
  endif
endfunction
```

The Jacobian can then be used with the following call to `fsolve`:

```
[x, fval, info] = fsolve (@f, [1; 2], optimset ("jacobian", "on"));
```

which gives the same solution as before.

`fzero (fun, x0)` [Function File]
`fzero (fun, x0, options)` [Function File]
`[x, fval, info, output] = fzero (...)` [Function File]

Find a zero of a univariate function.

fun is a function handle, inline function, or string containing the name of the function to evaluate. *x0* should be a two-element vector specifying two points which bracket a zero. In other words, there must be a change in sign of the function between *x0*(1) and *x0*(2). More mathematically, the following must hold

```
sign (fun(x0(1))) * sign (fun(x0(2))) <= 0
```

If *x0* is a single scalar then several nearby and distant values are probed in an attempt to obtain a valid bracketing. If this is not successful, the function fails. *options* is a structure specifying additional options. Currently, `fzero` recognizes these options: `"FunValCheck"`, `"OutputFcn"`, `"TolX"`, `"MaxIter"`, `"MaxFunEvals"`. For a description of these options, see [optimset], page 558.

On exit, the function returns *x*, the approximate zero point and *fval*, the function value thereof. *info* is an exit flag that can have these values:

- 1 The algorithm converged to a solution.
- 0 Maximum number of iterations or function evaluations has been reached.
- -1 The algorithm has been terminated from user output function.
- -5 The algorithm may have converged to a singular point.

output is a structure containing runtime information about the `fzero` algorithm. Fields in the structure are:

- iterations Number of iterations through loop.
- nfev Number of function evaluations.
- bracketx A two-element vector with the final bracketing of the zero along the x-axis.
- brackety A two-element vector with the final bracketing of the zero along the y-axis.

See also: [optimset], page 558, [fsolve], page 473.

20.2 Minimizers

Often it is useful to find the minimum value of a function rather than just the zeroes where it crosses the x-axis. `fminbnd` is designed for the simpler, but very common, case of a univariate function where the interval to search is bounded. For unbounded minimization of a function with potentially many variables use `fminunc` or `fminsearch`. The two functions use different internal algorithms and some knowledge of the objective function is required. For functions which can be differentiated, `fminunc` is appropriate. For functions with discontinuities, or for which a gradient search would fail, use `fminsearch`. See Chapter 25 [Optimization], page 547, for minimization with the presence of constraint functions. Note that searches can be made for maxima by simply inverting the objective function ($F_{max} = -F_{min}$).

`[x, fval, info, output] = fminbnd (fun, a, b, options)` [Function File]

Find a minimum point of a univariate function.

fun should be a function handle or name. *a*, *b* specify a starting interval. *options* is a structure specifying additional options. Currently, `fminbnd` recognizes these options: `"FunValCheck"`, `"OutputFcn"`, `"TolX"`, `"MaxIter"`, `"MaxFunEvals"`. For a description of these options, see [optimset], page 558.

On exit, the function returns *x*, the approximate minimum point and *fval*, the function value thereof. *info* is an exit flag that can have these values:

- 1 The algorithm converged to a solution.
- 0 Maximum number of iterations or function evaluations has been exhausted.
- -1 The algorithm has been terminated from user output function.

Notes: The search for a minimum is restricted to be in the interval bound by *a* and *b*. If you only have an initial point to begin searching from you will need to use an unconstrained minimization algorithm such as `fminunc` or `fminsearch`. `fminbnd` internally uses a Golden Section search strategy.

See also: [fzero], page 475, [fminunc], page 477, [fminsearch], page 478, [optimset], page 558.

`fminunc (fcn, x0)` [Function File]
`fminunc (fcn, x0, options)` [Function File]
`[x, fval, info, output, grad, hess] = fminunc (fcn, ...)` [Function File]

Solve an unconstrained optimization problem defined by the function *fcn*.

fcn should accept a vector (array) defining the unknown variables, and return the objective function value, optionally with gradient. `fminunc` attempts to determine a vector *x* such that `fcn (x)` is a local minimum. *x0* determines a starting guess. The shape of *x0* is preserved in all calls to *fcn*, but otherwise is treated as a column vector. *options* is a structure specifying additional options. Currently, `fminunc` recognizes these options: `"FunValCheck"`, `"OutputFcn"`, `"TolX"`, `"TolFun"`, `"MaxIter"`, `"MaxFunEvals"`, `"GradObj"`, `"FinDiffType"`, `"TypicalX"`, `"AutoScaling"`.

If `"GradObj"` is `"on"`, it specifies that *fcn*, when called with 2 output arguments, also returns the Jacobian matrix of partial first derivatives at the requested point. `TolX` specifies the termination tolerance for the unknown variables *x*, while `TolFun` is a tolerance for the objective function value *fval*. The default is `1e-7` for both options.

For a description of the other options, see `optimset`.

On return, *x* is the location of the minimum and *fval* contains the value of the objective function at *x*. *info* may be one of the following values:

1 Converged to a solution point. Relative gradient error is less than specified by `TolFun`.

2 Last relative step size was less than `TolX`.

3 Last relative change in function value was less than `TolFun`.

0 Iteration limit exceeded—either maximum numer of algorithm iterations `MaxIter` or maximum number of function evaluations `MaxFunEvals`.

-1 Alogrithm terminated by `OutputFcn`.

-3 The trust region radius became excessively small.

Optionally, `fminunc` can return a structure with convergence statistics (*output*), the output gradient (*grad*) at the solution *x*, and approximate Hessian (*hess*) at the solution *x*.

Notes: If have only a single nonlinear equation of one variable then using `fminbnd` is usually a much better idea. The algorithm used is a gradient search which depends on the objective function being differentiable. If the function has discontinuities it may be better to use a derivative-free algorithm such as `fminsearch`.

See also: [fminbnd], page 476, [fminsearch], page 478, [optimset], page 558.

`x = fminsearch (fun, x0)` [Function File]
`x = fminsearch (fun, x0, options)` [Function File]
`[x, fval] = fminsearch (...)` [Function File]

Find a value of *x* which minimizes the function *fun*. The search begins at the point *x0* and iterates using the Nelder & Mead Simplex algorithm (a derivative-free method). This algorithm is better-suited to functions which have discontinuities or for which a gradient-based search such as `fminunc` fails.

Options for the search are provided in the parameter *options* using the function `optimset`. Currently, `fminsearch` accepts the options: `"TolX"`, `"MaxFunEvals"`, `"MaxIter"`, `"Display"`. For a description of these options, see `optimset`.

On exit, the function returns *x*, the minimum point, and *fval*, the function value thereof.

Example usages:

```
fminsearch (@(x) (x(1)-5).^2+(x(2)-8).^4, [0;0])

fminsearch (inline ("(x(1)-5).^2+(x(2)-8).^4", "x"), [0;0])
```

See also: [fminbnd], page 476, [fminunc], page 477, [optimset], page 558.

21 Diagonal and Permutation Matrices

21.1 Creating and Manipulating Diagonal/Permutation Matrices

A diagonal matrix is defined as a matrix that has zero entries outside the main diagonal; that is, $D_{ij} = 0$ if $i \neq j$ Most often, square diagonal matrices are considered; however, the definition can equally be applied to non-square matrices, in which case we usually speak of a rectangular diagonal matrix.

A permutation matrix is defined as a square matrix that has a single element equal to unity in each row and each column; all other elements are zero. That is, there exists a permutation (vector) p such that $P_{ij} = 1$ if $j = p_i$ and $P_{ij} = 0$ otherwise.

Octave provides special treatment of real and complex rectangular diagonal matrices, as well as permutation matrices. They are stored as special objects, using efficient storage and algorithms, facilitating writing both readable and efficient matrix algebra expressions in the Octave language.

21.1.1 Creating Diagonal Matrices

The most common and easiest way to create a diagonal matrix is using the built-in function *diag*. The expression `diag (v)`, with *v* a vector, will create a square diagonal matrix with elements on the main diagonal given by the elements of *v*, and size equal to the length of *v*. `diag (v, m, n)` can be used to construct a rectangular diagonal matrix. The result of these expressions will be a special diagonal matrix object, rather than a general matrix object.

Diagonal matrix with unit elements can be created using *eye*. Some other built-in functions can also return diagonal matrices. Examples include *balance* or *inv*.

Example:

```
  diag (1:4)
⇒
Diagonal Matrix

   1   0   0   0
   0   2   0   0
   0   0   3   0
   0   0   0   4

  diag (1:3,5,3)

⇒
Diagonal Matrix

   1   0   0
   0   2   0
   0   0   3
   0   0   0
   0   0   0
```

21.1.2 Creating Permutation Matrices

For creating permutation matrices, Octave does not introduce a new function, but rather overrides an existing syntax: permutation matrices can be conveniently created by indexing an identity matrix by permutation vectors. That is, if q is a permutation vector of length n, the expression

```
P = eye (n) (:, q);
```

will create a permutation matrix - a special matrix object.

```
eye (n) (q, :)
```

will also work (and create a row permutation matrix), as well as

```
eye (n) (q1, q2).
```

For example:

```
  eye (4) ([1,3,2,4],:)
⇒
Permutation Matrix

   1   0   0   0
   0   0   1   0
   0   1   0   0
   0   0   0   1

  eye (4) (:,[1,3,2,4])
⇒
Permutation Matrix

   1   0   0   0
   0   0   1   0
   0   1   0   0
   0   0   0   1
```

Mathematically, an identity matrix is both diagonal and permutation matrix. In Octave, `eye (n)` returns a diagonal matrix, because a matrix can only have one class. You can convert this diagonal matrix to a permutation matrix by indexing it by an identity permutation, as shown below. This is a special property of the identity matrix; indexing other diagonal matrices generally produces a full matrix.

```
  eye (3)
⇒
Diagonal Matrix

   1   0   0
   0   1   0
   0   0   1

  eye(3)(1:3,:)
⇒
Permutation Matrix

   1   0   0
   0   1   0
   0   0   1
```

Some other built-in functions can also return permutation matrices. Examples include *inv* or *lu*.

21.1.3 Explicit and Implicit Conversions

The diagonal and permutation matrices are special objects in their own right. A number of operations and built-in functions are defined for these matrices to use special, more efficient code than would be used for a full matrix in the same place. Examples are given in further sections.

To facilitate smooth mixing with full matrices, backward compatibility, and compatibility with MATLAB, the diagonal and permutation matrices should allow any operation that works on full matrices, and will either treat it specially, or implicitly convert themselves to full matrices.

Instances include matrix indexing, except for extracting a single element or a leading submatrix, indexed assignment, or applying most mapper functions, such as *exp*.

An explicit conversion to a full matrix can be requested using the built-in function *full*. It should also be noted that the diagonal and permutation matrix objects will cache the result of the conversion after it is first requested (explicitly or implicitly), so that subsequent conversions will be very cheap.

21.2 Linear Algebra with Diagonal/Permutation Matrices

As has been already said, diagonal and permutation matrices make it possible to use efficient algorithms while preserving natural linear algebra syntax. This section describes in detail the operations that are treated specially when performed on these special matrix objects.

21.2.1 Expressions Involving Diagonal Matrices

Assume D is a diagonal matrix. If M is a full matrix, then `D*M` will scale the rows of M. That means, if `S = D*M`, then for each pair of indices i,j it holds

$$S_{ij} = D_{ii} M_{ij}$$

Similarly, `M*D` will do a column scaling.

The matrix D may also be rectangular, m-by-n where `m != n`. If `m < n`, then the expression `D*M` is equivalent to

```
D(:,1:m) * M(1:m,:),
```

i.e., trailing `n-m` rows of M are ignored. If `m > n`, then `D*M` is equivalent to

```
[D(1:n,n) * M; zeros(m-n, columns (M))],
```

i.e., null rows are appended to the result. The situation for right-multiplication `M*D` is analogous.

The expressions `D \ M` and `M / D` perform inverse scaling. They are equivalent to solving a diagonal (or rectangular diagonal) in a least-squares minimum-norm sense. In exact arithmetic, this is equivalent to multiplying by a pseudoinverse. The pseudoinverse of a rectangular diagonal matrix is again a rectangular diagonal matrix with swapped dimensions, where each nonzero diagonal element is replaced by its reciprocal. The matrix division algorithms do, in fact, use division rather than multiplication by reciprocals for better numerical accuracy; otherwise, they honor the above definition. Note that a diagonal matrix is never truncated due to ill-conditioning; otherwise, it would not be of much use for scaling. This is typically consistent with linear algebra needs. A full matrix that only happens to be diagonal (and is thus not a special object) is of course treated normally.

Multiplication and division by diagonal matrices work efficiently also when combined with sparse matrices, i.e., `D*S`, where D is a diagonal matrix and S is a sparse matrix scales the rows of the sparse matrix and returns a sparse matrix. The expressions `S*D`, `D\S`, `S/D` work analogically.

If *D1* and *D2* are both diagonal matrices, then the expressions

```
D1 + D2
D1 - D2
D1 * D2
D1 / D2
D1 \ D2
```

again produce diagonal matrices, provided that normal dimension matching rules are obeyed. The relations used are same as described above.

Also, a diagonal matrix D can be multiplied or divided by a scalar, or raised to a scalar power if it is square, producing diagonal matrix result in all cases.

A diagonal matrix can also be transposed or conjugate-transposed, giving the expected result. Extracting a leading submatrix of a diagonal matrix, i.e., `D(1:m,1:n)`, will produce a diagonal matrix, other indexing expressions will implicitly convert to full matrix.

Adding a diagonal matrix to a full matrix only operates on the diagonal elements. Thus,

```
A = A + eps * eye (n)
```

is an efficient method of augmenting the diagonal of a matrix. Subtraction works analogically.

When involved in expressions with other element-by-element operators, `.*`, `./`, `.\` or `.^`, an implicit conversion to full matrix will take place. This is not always strictly necessary but chosen to facilitate better consistency with MATLAB.

21.2.2 Expressions Involving Permutation Matrices

If P is a permutation matrix and M a matrix, the expression `P*M` will permute the rows of M. Similarly, `M*P` will yield a column permutation. Matrix division `P\M` and `M/P` can be used to do inverse permutation.

The previously described syntax for creating permutation matrices can actually help an user to understand the connection between a permutation matrix and a permuting vector. Namely, the following holds, where `I = eye (n)` is an identity matrix:

```
I(p,:) * M = (I*M) (p,:) = M(p,:)
```

Similarly,

```
M * I(:,p) = (M*I) (:,p) = M(:,p)
```

The expressions `I(p,:)` and `I(:,p)` are permutation matrices.

A permutation matrix can be transposed (or conjugate-transposed, which is the same, because a permutation matrix is never complex), inverting the permutation, or equivalently, turning a row-permutation matrix into a column-permutation one. For permutation matrices, transpose is equivalent to inversion, thus `P\M` is equivalent to `P'*M`. Transpose of a permutation matrix (or inverse) is a constant-time operation, flipping only a flag internally, and thus the choice between the two above equivalent expressions for inverse permuting is completely up to the user's taste.

Multiplication and division by permutation matrices works efficiently also when combined with sparse matrices, i.e., `P*S`, where P is a permutation matrix and S is a sparse matrix permutes the rows of the sparse matrix and returns a sparse matrix. The expressions `S*P`, `P\S`, `S/P` work analogically.

Two permutation matrices can be multiplied or divided (if their sizes match), performing a composition of permutations. Also a permutation matrix can be indexed by a permutation vector (or two vectors), giving again a permutation matrix. Any other operations do not generally yield a permutation matrix and will thus trigger the implicit conversion.

21.3 Functions That Are Aware of These Matrices

This section lists the built-in functions that are aware of diagonal and permutation matrices on input, or can return them as output. Passed to other functions, these matrices will in general trigger an implicit conversion. (Of course, user-defined dynamically linked functions may also work with diagonal or permutation matrices).

21.3.1 Diagonal Matrix Functions

inv and *pinv* can be applied to a diagonal matrix, yielding again a diagonal matrix. *det* will use an efficient straightforward calculation when given a diagonal matrix, as well as *cond*. The following mapper functions can be applied to a diagonal matrix without converting it to a full one: *abs*, *real*, *imag*, *conj*, *sqrt*. A diagonal matrix can also be returned from the *balance* and *svd* functions. The *sparse* function will convert a diagonal matrix efficiently to a sparse matrix.

21.3.2 Permutation Matrix Functions

inv and *pinv* will invert a permutation matrix, preserving its specialness. *det* can be applied to a permutation matrix, efficiently calculating the sign of the permutation (which is equal to the determinant).

A permutation matrix can also be returned from the built-in functions *lu* and *qr*, if a pivoted factorization is requested.

The *sparse* function will convert a permutation matrix efficiently to a sparse matrix. The *find* function will also work efficiently with a permutation matrix, making it possible to conveniently obtain the permutation indices.

21.4 Examples of Usage

The following can be used to solve a linear system `A*x = b` using the pivoted LU factorization:

```
[L, U, P] = lu (A); ## now L*U = P*A
x = U \ L \ P*b;
```

This is one way to normalize columns of a matrix X to unit norm:

```
s = norm (X, "columns");
X /= diag (s);
```

The same can also be accomplished with broadcasting (see Section 19.2 [Broadcasting], page 459):

```
s = norm (X, "columns");
X ./= s;
```

The following expression is a way to efficiently calculate the sign of a permutation, given by a permutation vector *p*. It will also work in earlier versions of Octave, but slowly.

```
det (eye (length (p))(p, :))
```

Finally, here's how to solve a linear system `A*x = b` with Tikhonov regularization (ridge regression) using SVD (a skeleton only):

```
m = rows (A); n = columns (A);
[U, S, V] = svd (A);
## determine the regularization factor alpha
## alpha = ...
## transform to orthogonal basis
b = U'*b;
## Use the standard formula, replacing A with S.
## S is diagonal, so the following will be very fast and accurate.
x = (S'*S + alpha^2 * eye (n)) \ (S' * b);
## transform to solution basis
x = V*x;
```

21.5 Differences in Treatment of Zero Elements

Making diagonal and permutation matrices special matrix objects in their own right and the consequent usage of smarter algorithms for certain operations implies, as a side effect, small differences in treating zeros. The contents of this section apply also to sparse matrices, discussed in the following chapter. (see Chapter 22 [Sparse Matrices], page 487)

The IEEE floating point standard defines the result of the expressions `0*Inf` and `0*NaN` as `NaN`. This is widely agreed to be a good compromise. Numerical software dealing with structured and sparse matrices (including Octave) however, almost always makes a distinction between a "numerical zero" and an "assumed zero". A "numerical zero" is a zero value occurring in a place where any floating-point value could occur. It is normally stored somewhere in memory as an explicit value. An "assumed zero", on the contrary, is a zero matrix element implied by the matrix structure (diagonal, triangular) or a sparsity pattern; its value is usually not stored explicitly anywhere, but is implied by the underlying data structure.

The primary distinction is that an assumed zero, when multiplied by any number, or divided by any nonzero number, yields *always* a zero, even when, e.g., multiplied by `Inf` or divided by `NaN`. The reason for this behavior is that the numerical multiplication is not actually performed anywhere by the underlying algorithm; the result is just assumed to be zero. Equivalently, one can say that the part of the computation involving assumed zeros is performed symbolically, not numerically.

This behavior not only facilitates the most straightforward and efficient implementation of algorithms, but also preserves certain useful invariants, like:

- scalar * diagonal matrix is a diagonal matrix
- sparse matrix / scalar preserves the sparsity pattern
- permutation matrix * matrix is equivalent to permuting rows

all of these natural mathematical truths would be invalidated by treating assumed zeros as numerical ones.

Note that MATLAB does not strictly follow this principle and converts assumed zeros to numerical zeros in certain cases, while not doing so in other cases. As of today, there are no intentions to mimic such behavior in Octave.

Examples of effects of assumed zeros vs. numerical zeros:

```
Inf * eye (3)
⇒
   Inf     0     0
     0   Inf     0
     0     0   Inf

Inf * speye (3)
⇒
Compressed Column Sparse (rows = 3, cols = 3, nnz = 3 [33%])

  (1, 1) -> Inf
  (2, 2) -> Inf
  (3, 3) -> Inf

Inf * full (eye (3))
⇒
   Inf   NaN   NaN
   NaN   Inf   NaN
   NaN   NaN   Inf
```

```
diag (1:3) * [NaN; 1; 1]
⇒
   NaN
     2
     3

sparse (1:3,1:3,1:3) * [NaN; 1; 1]
⇒
   NaN
     2
     3
[1,0,0;0,2,0;0,0,3] * [NaN; 1; 1]
⇒
   NaN
   NaN
   NaN
```

22 Sparse Matrices

22.1 Creation and Manipulation of Sparse Matrices

The size of mathematical problems that can be treated at any particular time is generally limited by the available computing resources. Both, the speed of the computer and its available memory place limitation on the problem size.

There are many classes of mathematical problems which give rise to matrices, where a large number of the elements are zero. In this case it makes sense to have a special matrix type to handle this class of problems where only the non-zero elements of the matrix are stored. Not only does this reduce the amount of memory to store the matrix, but it also means that operations on this type of matrix can take advantage of the a priori knowledge of the positions of the non-zero elements to accelerate their calculations.

A matrix type that stores only the non-zero elements is generally called sparse. It is the purpose of this document to discuss the basics of the storage and creation of sparse matrices and the fundamental operations on them.

22.1.1 Storage of Sparse Matrices

It is not strictly speaking necessary for the user to understand how sparse matrices are stored. However, such an understanding will help to get an understanding of the size of sparse matrices. Understanding the storage technique is also necessary for those users wishing to create their own oct-files.

There are many different means of storing sparse matrix data. What all of the methods have in common is that they attempt to reduce the complexity and storage given a priori knowledge of the particular class of problems that will be solved. A good summary of the available techniques for storing sparse matrix is given by Saad[1]. With full matrices, knowledge of the point of an element of the matrix within the matrix is implied by its position in the computers memory. However, this is not the case for sparse matrices, and so the positions of the non-zero elements of the matrix must equally be stored.

An obvious way to do this is by storing the elements of the matrix as triplets, with two elements being their position in the array (rows and column) and the third being the data itself. This is conceptually easy to grasp, but requires more storage than is strictly needed.

The storage technique used within Octave is the compressed column format. It is similar to the Yale format.[2] In this format the position of each element in a row and the data are stored as previously. However, if we assume that all elements in the same column are stored adjacent in the computers memory, then we only need to store information on the number of non-zero elements in each column, rather than their positions. Thus assuming that the matrix has more non-zero elements than there are columns in the matrix, we win in terms of the amount of memory used.

In fact, the column index contains one more element than the number of columns, with the first element always being zero. The advantage of this is a simplification in the code, in that there is no special case for the first or last columns. A short example, demonstrating this in C is.

[1] Y. Saad "SPARSKIT: A basic toolkit for sparse matrix computation", 1994, `http://www-users.cs.umn.edu/~saad/software/`

[2] `http://en.wikipedia.org/wiki/Sparse_matrix#Yale_format`

```
for (j = 0; j < nc; j++)
  for (i = cidx(j); i < cidx(j+1); i++)
     printf ("non-zero element (%i,%i) is %d\n",
         ridx(i), j, data(i));
```

A clear understanding might be had by considering an example of how the above applies to an example matrix. Consider the matrix

```
1   2   0  0
0   0   0  3
0   0   0  4
```

The non-zero elements of this matrix are

```
(1, 1)  ⇒ 1
(1, 2)  ⇒ 2
(2, 4)  ⇒ 3
(3, 4)  ⇒ 4
```

This will be stored as three vectors *cidx*, *ridx* and *data*, representing the column indexing, row indexing and data respectively. The contents of these three vectors for the above matrix will be

```
cidx = [0, 1, 2, 2, 4]
ridx = [0, 0, 1, 2]
data = [1, 2, 3, 4]
```

Note that this is the representation of these elements with the first row and column assumed to start at zero, while in Octave itself the row and column indexing starts at one. Thus the number of elements in the *i*-th column is given by `cidx (i + 1) - cidx (i)`.

Although Octave uses a compressed column format, it should be noted that compressed row formats are equally possible. However, in the context of mixed operations between mixed sparse and dense matrices, it makes sense that the elements of the sparse matrices are in the same order as the dense matrices. Octave stores dense matrices in column major ordering, and so sparse matrices are equally stored in this manner.

A further constraint on the sparse matrix storage used by Octave is that all elements in the rows are stored in increasing order of their row index, which makes certain operations faster. However, it imposes the need to sort the elements on the creation of sparse matrices. Having disordered elements is potentially an advantage in that it makes operations such as concatenating two sparse matrices together easier and faster, however it adds complexity and speed problems elsewhere.

22.1.2 Creating Sparse Matrices

There are several means to create sparse matrix.

Returned from a function

: There are many functions that directly return sparse matrices. These include *speye*, *sprand*, *diag*, etc.

Constructed from matrices or vectors

: The function *sparse* allows a sparse matrix to be constructed from three vectors representing the row, column and data. Alternatively, the function *spconvert* uses a three column matrix format to allow easy importation of data from elsewhere.

Created and then filled
: The function *sparse* or *spalloc* can be used to create an empty matrix that is then filled by the user

From a user binary program
: The user can directly create the sparse matrix within an oct-file.

There are several basic functions to return specific sparse matrices. For example the sparse identity matrix, is a matrix that is often needed. It therefore has its own function to create it as `speye (n)` or `speye (r, c)`, which creates an n-by-n or r-by-c sparse identity matrix.

Another typical sparse matrix that is often needed is a random distribution of random elements. The functions *sprand* and *sprandn* perform this for uniform and normal random distributions of elements. They have exactly the same calling convention, where `sprand (r, c, d)`, creates an r-by-c sparse matrix with a density of filled elements of d.

Other functions of interest that directly create sparse matrices, are *diag* or its generalization *spdiags*, that can take the definition of the diagonals of the matrix and create the sparse matrix that corresponds to this. For example,

```
s = diag (sparse (randn (1,n)), -1);
```

creates a sparse (n+1)-by-(n+1) sparse matrix with a single diagonal defined.

`[b, c] = spdiags (A)` [Function File]
`b = spdiags (A, c)` [Function File]
`b = spdiags (v, c, A)` [Function File]
`b = spdiags (v, c, m, n)` [Function File]

A generalization of the function `diag`. Called with a single input argument, the non-zero diagonals c of A are extracted. With two arguments the diagonals to extract are given by the vector c.

The other two forms of `spdiags` modify the input matrix by replacing the diagonals. They use the columns of v to replace the columns represented by the vector c. If the sparse matrix A is defined then the diagonals of this matrix are replaced. Otherwise a matrix of m by n is created with the diagonals given by v.

Negative values of c represent diagonals below the main diagonal, and positive values of c diagonals above the main diagonal.

For example:

```
spdiags (reshape (1:12, 4, 3), [-1 0 1], 5, 4)
   ⇒ 5 10  0  0
     1  6 11  0
     0  2  7 12
     0  0  3  8
     0  0  0  4
```

`s = speye (m, n)` [Function File]
`s = speye (m)` [Function File]

`s = speye (sz)` [Function File]

Return a sparse identity matrix of size *mxn*.

The implementation is significantly more efficient than `sparse (eye (m))` as the full matrix is not constructed.

Called with a single argument a square matrix of size *m*-by-*m* is created. If called with a single vector argument *sz*, this argument is taken to be the size of the matrix to create.

See also: [sparse], page 492, [spdiags], page 489, [eye], page 391.

`r = spones (S)` [Function File]

Replace the non-zero entries of *S* with ones. This creates a sparse matrix with the same structure as *S*.

See also: [sparse], page 492, [sprand], page 490, [sprandn], page 490, [sprandsym], page 490, [spfun], page 464, [spy], page 496.

`sprand (m, n, d)` [Function File]
`sprand (s)` [Function File]

Generate a random sparse matrix. The size of the matrix will be *mxn*, with a density of values given by *d*. *d* must be between 0 and 1 inclusive. Values will be uniformly distributed between 0 and 1.

If called with a single matrix argument, a random sparse matrix is generated wherever the matrix *S* is non-zero.

See also: [sprandn], page 490, [sprandsym], page 490, [spones], page 490, [sparse], page 492.

`sprandn (m, n, d)` [Function File]
`sprandn (s)` [Function File]

Generate a random sparse matrix. The size of the matrix will be *mxn*, with a density of values given by *d*. *d* must be between 0 and 1 inclusive. Values will be normally distributed with a mean of zero and a variance of 1.

If called with a single matrix argument, a random sparse matrix is generated wherever the matrix *S* is non-zero.

See also: [sprand], page 490, [sprandsym], page 490, [spones], page 490, [sparse], page 492.

`sprandsym (n, d)` [Function File]
`sprandsym (s)` [Function File]

Generate a symmetric random sparse matrix.

The size of the matrix will be *nxn*, with a density of values given by *d*. *d* must be between 0 and 1 inclusive. Values will be normally distributed with a mean of zero and a variance of 1.

If called with a single matrix argument, a random sparse matrix is generated wherever the matrix *S* is non-zero in its lower triangular part.

See also: [sprand], page 490, [sprandn], page 490, [spones], page 490, [sparse], page 492.

The recommended way for the user to create a sparse matrix, is to create two vectors containing the row and column index of the data and a third vector of the same size containing the data to be stored. For example,

```
ri = ci = d = [];
for j = 1:c
  ri = [ri; randperm(r,n)'];
  ci = [ci; j*ones(n,1)];
  d = [d; rand(n,1)];
endfor
s = sparse (ri, ci, d, r, c);
```

creates an *r*-by-*c* sparse matrix with a random distribution of *n* (<*r*) elements per column. The elements of the vectors do not need to be sorted in any particular order as Octave will sort them prior to storing the data. However, pre-sorting the data will make the creation of the sparse matrix faster.

The function *spconvert* takes a three or four column real matrix. The first two columns represent the row and column index respectively and the third and four columns, the real and imaginary parts of the sparse matrix. The matrix can contain zero elements and the elements can be sorted in any order. Adding zero elements is a convenient way to define the size of the sparse matrix. For example:

```
s = spconvert ([1 2 3 4; 1 3 4 4; 1 2 3 0]')
⇒ Compressed Column Sparse (rows=4, cols=4, nnz=3)
     (1 , 1) -> 1
     (2 , 3) -> 2
     (3 , 4) -> 3
```

An example of creating and filling a matrix might be

```
k = 5;
nz = r * k;
s = spalloc (r, c, nz)
for j = 1:c
  idx = randperm (r);
  s (:, j) = [zeros(r - k, 1); ...
        rand(k, 1)] (idx);
endfor
```

It should be noted, that due to the way that the Octave assignment functions are written that the assignment will reallocate the memory used by the sparse matrix at each iteration of the above loop. Therefore the *spalloc* function ignores the *nz* argument and does not pre-assign the memory for the matrix. Therefore, it is vitally important that code using to above structure should be vectorized as much as possible to minimize the number of assignments and reduce the number of memory allocations.

FM = `full` (*SM*) [Built-in Function]

Return a full storage matrix from a sparse, diagonal, permutation matrix, or a range.

See also: [sparse], page 492, [issparse], page 493.

`s = spalloc (m, n, nz)` [Built-in Function]

Create an *m*-by-*n* sparse matrix with pre-allocated space for at most *nz* nonzero elements.

This is useful for building a matrix incrementally by a sequence of indexed assignments. Subsequent indexed assignments after `spalloc` will reuse the pre-allocated memory, provided they are of one of the simple forms

- `s(I:J) = x`
- `s(:,I:J) = x`
- `s(K:L,I:J) = x`

and that the following conditions are met:

- the assignment does not decrease nnz (*S*).
- after the assignment, nnz (*S*) does not exceed *nz*.
- no index is out of bounds.

Partial movement of data may still occur, but in general the assignment will be more memory and time efficient under these circumstances. In particular, it is possible to efficiently build a pre-allocated sparse matrix from a contiguous block of columns.

The amount of pre-allocated memory for a given matrix may be queried using the function `nzmax`.

See also: [nzmax], page 494, [sparse], page 492.

`s = sparse (a)` [Built-in Function]
`s = sparse (i, j, sv, m, n)` [Built-in Function]
`s = sparse (i, j, sv)` [Built-in Function]
`s = sparse (m, n)` [Built-in Function]
`s = sparse (i, j, s, m, n, "unique")` [Built-in Function]
`s = sparse (i, j, sv, m, n, nzmax)` [Built-in Function]

Create a sparse matrix from a full matrix or row, column, value triplets.

If *a* is a full matrix, convert it to a sparse matrix representation, removing all zero values in the process.

Given the integer index vectors *i* and *j*, and a 1-by-`nnz` vector of real or complex values *sv*, construct the sparse matrix `S(i(k),j(k)) = sv(k)` with overall dimensions *m* and *n*. If any of *sv*, *i* or *j* are scalars, they are expanded to have a common size.

If *m* or *n* are not specified their values are derived from the maximum index in the vectors *i* and *j* as given by `m = max (i)`, `n = max (j)`.

Note: if multiple values are specified with the same *i*, *j* indices, the corresponding value in *s* will be the sum of the values at the repeated location. See `accumarray` for an example of how to produce different behavior, such as taking the minimum instead.

If the option `"unique"` is given, and more than one value is specified at the same *i*, *j* indices, then the last specified value will be used.

`sparse (m, n)` will create an empty *m*x*n* sparse matrix and is equivalent to `sparse ([], [], [], m, n)`

The argument `nzmax` is ignored but accepted for compatibility with MATLAB.

Example 1 (sum at repeated indices):

```
i = [1 1 2]; j = [1 1 2]; sv = [3 4 5];
sparse (i, j, sv, 3, 4)
⇒
Compressed Column Sparse (rows = 3, cols = 4, nnz = 2 [17%])

  (1, 1) ->  7
  (2, 2) ->  5
```

Example 2 ("unique" option):

```
i = [1 1 2]; j = [1 1 2]; sv = [3 4 5];
sparse (i, j, sv, 3, 4, "unique")
⇒
Compressed Column Sparse (rows = 3, cols = 4, nnz = 2 [17%])

  (1, 1) ->  4
  (2, 2) ->  5
```

See also: [full], page 491, [accumarray], page 466, [spalloc], page 492, [spdiags], page 489, [speye], page 489, [spones], page 490, [sprand], page 490, [sprandn], page 490, [sprandsym], page 490, [spconvert], page 493, [spfun], page 464.

x = `spconvert` (*m*) [Function File]

Convert a simple sparse matrix format easily generated by other programs into Octave's internal sparse format.

The input *m* is either a 3 or 4 column real matrix, containing the row, column, real, and imaginary parts of the elements of the sparse matrix. An element with a zero real and imaginary part can be used to force a particular matrix size.

See also: [sparse], page 492.

The above problem of memory reallocation can be avoided in oct-files. However, the construction of a sparse matrix from an oct-file is more complex than can be discussed here. See Appendix A [External Code Interface], page 759, for a full description of the techniques involved.

22.1.3 Finding Information about Sparse Matrices

There are a number of functions that allow information concerning sparse matrices to be obtained. The most basic of these is *issparse* that identifies whether a particular Octave object is in fact a sparse matrix.

Another very basic function is *nnz* that returns the number of non-zero entries there are in a sparse matrix, while the function *nzmax* returns the amount of storage allocated to the sparse matrix. Note that Octave tends to crop unused memory at the first opportunity for sparse objects. There are some cases of user created sparse objects where the value returned by *nzmax* will not be the same as *nnz*, but in general they will give the same result. The function *spstats* returns some basic statistics on the columns of a sparse matrix including the number of elements, the mean and the variance of each column.

`issparse (x)` [Built-in Function]

Return true if *x* is a sparse matrix.

See also: [ismatrix], page 62.

`n = nnz (a)` [Built-in Function]

Return the number of non-zero elements in *a*.

See also: [nzmax], page 494, [nonzeros], page 494, [find], page 381.

`nonzeros (s)` [Function File]

Return a vector of the non-zero values of the sparse matrix *s*.

See also: [find], page 381, [nnz], page 494.

`n = nzmax (SM)` [Built-in Function]

Return the amount of storage allocated to the sparse matrix *SM*.

Note that Octave tends to crop unused memory at the first opportunity for sparse objects. Thus, in general the value of `nzmax` will be the the same as `nnz` except for some cases of user-created sparse objects.

See also: [nnz], page 494, [spalloc], page 492, [sparse], page 492.

`[count, mean, var] = spstats (S)` [Function File]
`[count, mean, var] = spstats (S, j)` [Function File]

Return the stats for the non-zero elements of the sparse matrix *S*. *count* is the number of non-zeros in each column, *mean* is the mean of the non-zeros in each column, and *var* is the variance of the non-zeros in each column.

Called with two input arguments, if *S* is the data and *j* is the bin number for the data, compute the stats for each bin. In this case, bins can contain data values of zero, whereas with `spstats (S)` the zeros may disappear.

When solving linear equations involving sparse matrices Octave determines the means to solve the equation based on the type of the matrix (see Section 22.2 [Sparse Linear Algebra], page 508). Octave probes the matrix type when the div (/) or ldiv (\) operator is first used with the matrix and then caches the type. However the *matrix_type* function can be used to determine the type of the sparse matrix prior to use of the div or ldiv operators. For example,

```
a = tril (sprandn (1024, 1024, 0.02), -1) ...
    + speye (1024);
matrix_type (a);
ans = Lower
```

shows that Octave correctly determines the matrix type for lower triangular matrices. *matrix_type* can also be used to force the type of a matrix to be a particular type. For example:

```
a = matrix_type (tril (sprandn (1024, ...
   1024, 0.02), -1) + speye (1024), "Lower");
```

This allows the cost of determining the matrix type to be avoided. However, incorrectly defining the matrix type will result in incorrect results from solutions of linear equations, and so it is entirely the responsibility of the user to correctly identify the matrix type

There are several graphical means of finding out information about sparse matrices. The first is the *spy* command, which displays the structure of the non-zero elements of the matrix. See Figure 22.1, for an example of the use of *spy*. More advanced graphical information can be obtained with the *treeplot*, *etreeplot* and *gplot* commands.

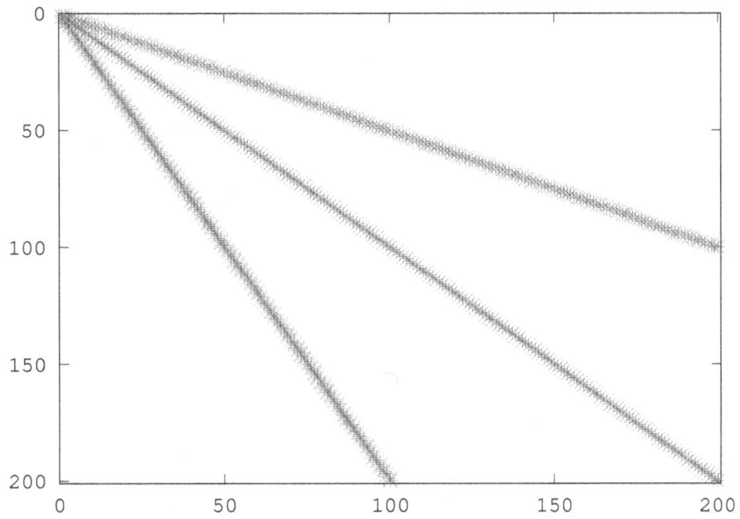

Figure 22.1: Structure of simple sparse matrix.

One use of sparse matrices is in graph theory, where the interconnections between nodes are represented as an adjacency matrix. That is, if the i-th node in a graph is connected to the j-th node. Then the ij-th node (and in the case of undirected graphs the ji-th node) of the sparse adjacency matrix is non-zero. If each node is then associated with a set of coordinates, then the *gplot* command can be used to graphically display the interconnections between nodes.

As a trivial example of the use of *gplot* consider the example,

```
A = sparse ([2,6,1,3,2,4,3,5,4,6,1,5],
    [1,1,2,2,3,3,4,4,5,5,6,6],1,6,6);
xy = [0,4,8,6,4,2;5,0,5,7,5,7]';
gplot (A,xy)
```

which creates an adjacency matrix `A` where node 1 is connected to nodes 2 and 6, node 2 with nodes 1 and 3, etc. The coordinates of the nodes are given in the n-by-2 matrix `xy`. See Figure 22.2.

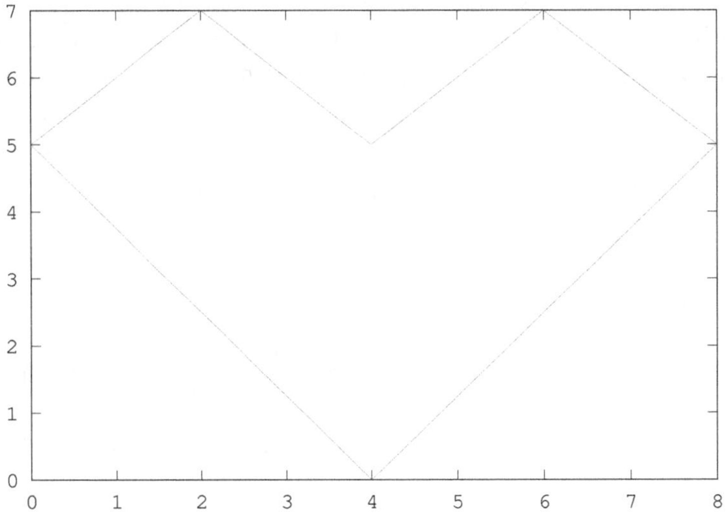

Figure 22.2: Simple use of the *gplot* command.

The dependencies between the nodes of a Cholesky factorization can be calculated in linear time without explicitly needing to calculate the Cholesky factorization by the `etree` command. This command returns the elimination tree of the matrix and can be displayed graphically by the command `treeplot (etree (A))` if `A` is symmetric or `treeplot (etree (A+A'))` otherwise.

`spy (`*x*`)` [Function File]
`spy (`*...*`, `*`markersize`*`)` [Function File]
`spy (`*...*`, `*`line_spec`*`)` [Function File]

Plot the sparsity pattern of the sparse matrix *x*.

If the argument *markersize* is given as a scalar value, it is used to determine the point size in the plot. If the string *line_spec* is given it is passed to `plot` and determines the appearance of the plot.

See also: [plot], page 260, [gplot], page 497.

p `= etree (`*S*`)` [Loadable Function]
p `= etree (`*S*`, `*typ*`)` [Loadable Function]
`[`*p*`, `*q*`] = etree (`*S*`, `*typ*`)` [Loadable Function]

Return the elimination tree for the matrix *S*. By default *S* is assumed to be symmetric and the symmetric elimination tree is returned. The argument *typ* controls whether a symmetric or column elimination tree is returned. Valid values of *typ* are `"sym"` or `"col"`, for symmetric or column elimination tree respectively.

Called with a second argument, `etree` also returns the postorder permutations on the tree.

`etreeplot (`*A*`)` [Function File]
`etreeplot (`*A*`, `*`node_style`*`, `*`edge_style`*`)` [Function File]

Plot the elimination tree of the matrix *A* or `A+A'` if *A* in not symmetric. The optional parameters *node_style* and *edge_style* define the output style.

See also: [treeplot], page 497, [gplot], page 497.

`gplot (A, xy)` [Function File]
`gplot (A, xy, line_style)` [Function File]
`[x, y] = gplot (A, xy)` [Function File]

Plot a graph defined by *A* and *xy* in the graph theory sense. *A* is the adjacency matrix of the array to be plotted and *xy* is an *n*-by-2 matrix containing the coordinates of the nodes of the graph.

The optional parameter *line_style* defines the output style for the plot. Called with no output arguments the graph is plotted directly. Otherwise, return the coordinates of the plot in *x* and *y*.

See also: [treeplot], page 497, [etreeplot], page 496, [spy], page 496.

`treeplot (tree)` [Function File]
`treeplot (tree, node_style, edge_style)` [Function File]

Produce a graph of tree or forest. The first argument is vector of predecessors, optional parameters *node_style* and *edge_style* define the output style. The complexity of the algorithm is O(n) in terms of is time and memory requirements.

See also: [etreeplot], page 496, [gplot], page 497.

`treelayout (tree)` [Function File]
`treelayout (tree, permutation)` [Function File]

treelayout lays out a tree or a forest. The first argument *tree* is a vector of predecessors, optional parameter *permutation* is an optional postorder permutation. The complexity of the algorithm is O(n) in terms of time and memory requirements.

See also: [etreeplot], page 496, [gplot], page 497, [treeplot], page 497.

22.1.4 Basic Operators and Functions on Sparse Matrices

22.1.4.1 Sparse Functions

Many Octave functions have been overloaded to work with either sparse or full matrices. There is no difference in calling convention when using an overloaded function with a sparse matrix, however, there is also no access to potentially sparse-specific features. At any time the sparse matrix specific version of a function can be used by explicitly calling its function name.

The table below lists all of the sparse functions of Octave. Note that the names of the specific sparse forms of the functions are typically the same as the general versions with a *sp* prefix. In the table below, and in the rest of this article, the specific sparse versions of functions are used.

Generate sparse matrices:
: *spalloc, spdiags, speye, sprand, sprandn, sprandsym*

Sparse matrix conversion:
: *full, sparse, spconvert*

Manipulate sparse matrices
: *issparse, nnz, nonzeros, nzmax, spfun, spones, spy*

Graph Theory:
: *etree*, *etreeplot*, *gplot*, *treeplot*

Sparse matrix reordering:
: *amd*, *ccolamd*, *colamd*, *colperm*, *csymamd*, *dmperm*, *symamd*, *randperm*, *symrcm*

Linear algebra:
: *condest*, *eigs*, *matrix_type*, *normest*, *sprank*, *spaugment*, *svds*

Iterative techniques:
: *luinc*, *pcg*, *pcr*

Miscellaneous:
: *spparms*, *symbfact*, *spstats*

In addition all of the standard Octave mapper functions (i.e., basic math functions that take a single argument) such as *abs*, etc. can accept sparse matrices. The reader is referred to the documentation supplied with these functions within Octave itself for further details.

22.1.4.2 Return Types of Operators and Functions

The two basic reasons to use sparse matrices are to reduce the memory usage and to not have to do calculations on zero elements. The two are closely related in that the computation time on a sparse matrix operator or function is roughly linear with the number of non-zero elements.

Therefore, there is a certain density of non-zero elements of a matrix where it no longer makes sense to store it as a sparse matrix, but rather as a full matrix. For this reason operators and functions that have a high probability of returning a full matrix will always return one. For example adding a scalar constant to a sparse matrix will almost always make it a full matrix, and so the example,

```
speye (3) + 0
⇒   1  0  0
  0  1  0
  0  0  1
```

returns a full matrix as can be seen.

Additionally, if `sparse_auto_mutate` is true, all sparse functions test the amount of memory occupied by the sparse matrix to see if the amount of storage used is larger than the amount used by the full equivalent. Therefore `speye (2) * 1` will return a full matrix as the memory used is smaller for the full version than the sparse version.

As all of the mixed operators and functions between full and sparse matrices exist, in general this does not cause any problems. However, one area where it does cause a problem is where a sparse matrix is promoted to a full matrix, where subsequent operations would resparsify the matrix. Such cases are rare, but can be artificially created, for example `(fliplr (speye (3)) + speye (3)) - speye (3)` gives a full matrix when it should give a sparse one. In general, where such cases occur, they impose only a small memory penalty.

There is however one known case where this behavior of Octave's sparse matrices will cause a problem. That is in the handling of the *diag* function. Whether *diag* returns a sparse or full matrix depending on the type of its input arguments. So

```
a = diag (sparse ([1,2,3]), -1);
```

should return a sparse matrix. To ensure this actually happens, the *sparse* function, and other functions based on it like *speye*, always returns a sparse matrix, even if the memory used will be larger than its full representation.

val = sparse_auto_mutate () [Built-in Function]
old_val = sparse_auto_mutate (*new_val*) [Built-in Function]
sparse_auto_mutate (*new_val*, "*local*") [Built-in Function]

Query or set the internal variable that controls whether Octave will automatically mutate sparse matrices to full matrices to save memory. For example:

```
s = speye (3);
sparse_auto_mutate (false);
s(:, 1) = 1;
typeinfo (s)
⇒ sparse matrix
sparse_auto_mutate (true);
s(1, :) = 1;
typeinfo (s)
⇒ matrix
```

When called from inside a function with the "local" option, the variable is changed locally for the function and any subroutines it calls. The original variable value is restored when exiting the function.

Note that the sparse_auto_mutate option is incompatible with MATLAB, and so it is off by default.

22.1.4.3 Mathematical Considerations

The attempt has been made to make sparse matrices behave in exactly the same manner as there full counterparts. However, there are certain differences and especially differences with other products sparse implementations.

First, the "./" and ".^" operators must be used with care. Consider what the examples

```
s = speye (4);
a1 = s .^ 2;
a2 = s .^ s;
a3 = s .^ -2;
a4 = s ./ 2;
a5 = 2 ./ s;
a6 = s ./ s;
```

will give. The first example of *s* raised to the power of 2 causes no problems. However *s* raised element-wise to itself involves a large number of terms 0 .^ 0 which is 1. There s .^ s is a full matrix.

Likewise s .^ -2 involves terms like 0 .^ -2 which is infinity, and so s .^ -2 is equally a full matrix.

For the "./" operator s ./ 2 has no problems, but 2 ./ s involves a large number of infinity terms as well and is equally a full matrix. The case of s ./ s involves terms like

`0 ./ 0` which is a `NaN` and so this is equally a full matrix with the zero elements of *s* filled with `NaN` values.

The above behavior is consistent with full matrices, but is not consistent with sparse implementations in other products.

A particular problem of sparse matrices comes about due to the fact that as the zeros are not stored, the sign-bit of these zeros is equally not stored. In certain cases the sign-bit of zero is important. For example:

```
a = 0 ./ [-1, 1; 1, -1];
b = 1 ./ a
⇒ -Inf            Inf
    Inf           -Inf
c = 1 ./ sparse (a)
⇒  Inf            Inf
    Inf            Inf
```

To correct this behavior would mean that zero elements with a negative sign-bit would need to be stored in the matrix to ensure that their sign-bit was respected. This is not done at this time, for reasons of efficiency, and so the user is warned that calculations where the sign-bit of zero is important must not be done using sparse matrices.

In general any function or operator used on a sparse matrix will result in a sparse matrix with the same or a larger number of non-zero elements than the original matrix. This is particularly true for the important case of sparse matrix factorizations. The usual way to address this is to reorder the matrix, such that its factorization is sparser than the factorization of the original matrix. That is the factorization of `L * U = P * S * Q` has sparser terms `L` and `U` than the equivalent factorization `L * U = S`.

Several functions are available to reorder depending on the type of the matrix to be factorized. If the matrix is symmetric positive-definite, then *symamd* or *csymamd* should be used. Otherwise *amd*, *colamd* or *ccolamd* should be used. For completeness the reordering functions *colperm* and *randperm* are also available.

See Figure 22.3, for an example of the structure of a simple positive definite matrix.

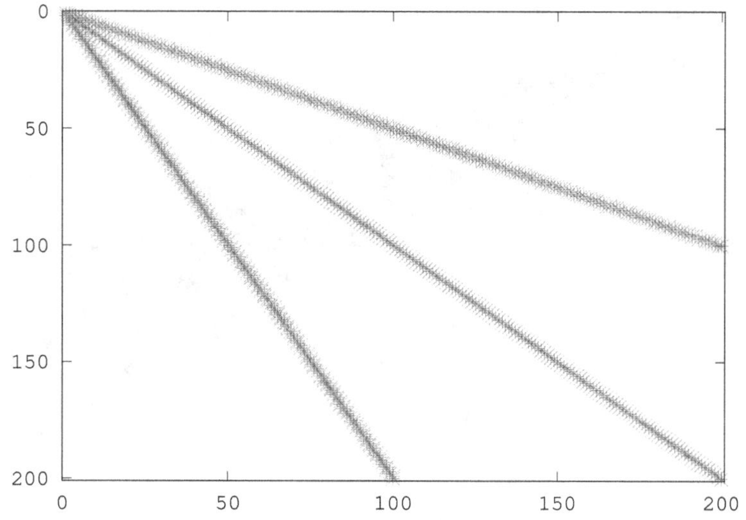

Figure 22.3: Structure of simple sparse matrix.

The standard Cholesky factorization of this matrix can be obtained by the same command that would be used for a full matrix. This can be visualized with the command `r = chol (A); spy (r);`. See Figure 22.4. The original matrix had 598 non-zero terms, while this Cholesky factorization has 10200, with only half of the symmetric matrix being stored. This is a significant level of fill in, and although not an issue for such a small test case, can represents a large overhead in working with other sparse matrices.

The appropriate sparsity preserving permutation of the original matrix is given by *symamd* and the factorization using this reordering can be visualized using the command `q = symamd (A); r = chol (A(q,q)); spy (r)`. This gives 399 non-zero terms which is a significant improvement.

The Cholesky factorization itself can be used to determine the appropriate sparsity preserving reordering of the matrix during the factorization, In that case this might be obtained with three return arguments as `[r, p, q] = chol (A); spy (r)`.

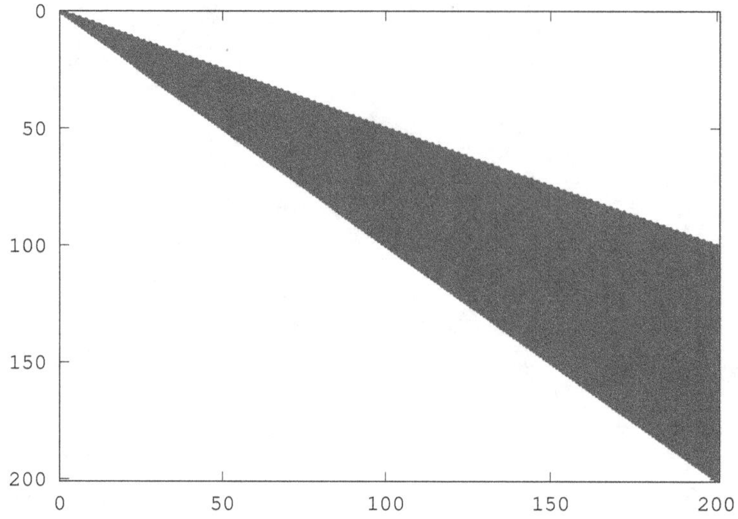

Figure 22.4: Structure of the unpermuted Cholesky factorization of the above matrix.

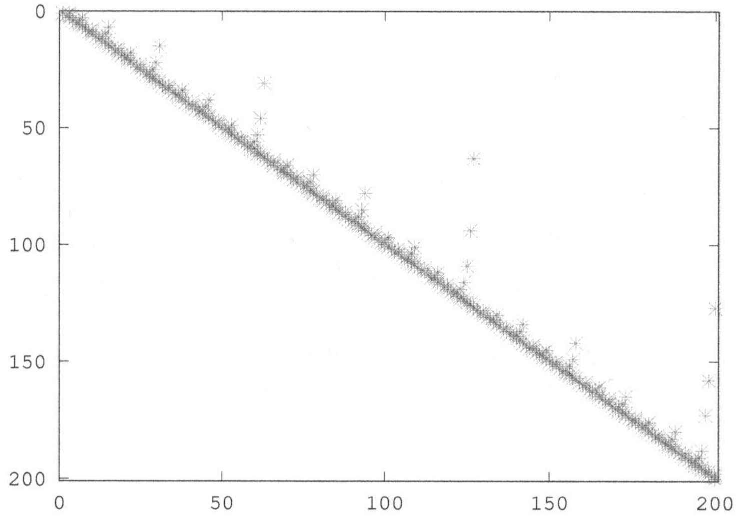

Figure 22.5: Structure of the permuted Cholesky factorization of the above matrix.

In the case of an asymmetric matrix, the appropriate sparsity preserving permutation is *colamd* and the factorization using this reordering can be visualized using the command `q = colamd (A); [l, u, p] = lu (A(:,q)); spy (l+u)`.

Finally, Octave implicitly reorders the matrix when using the div (/) and ldiv (\) operators, and so no the user does not need to explicitly reorder the matrix to maximize performance.

p = amd (*S*) [Loadable Function]
p = amd (*S*, *opts*) [Loadable Function]

Return the approximate minimum degree permutation of a matrix. This permutation such that the Cholesky factorization of `S (p, p)` tends to be sparser than the

Cholesky factorization of S itself. `amd` is typically faster than `symamd` but serves a similar purpose.

The optional parameter *opts* is a structure that controls the behavior of `amd`. The fields of the structure are

opts.dense Determines what `amd` considers to be a dense row or column of the input matrix. Rows or columns with more than `max(16, (dense * sqrt (n)` entries, where n is the order of the matrix S, are ignored by `amd` during the calculation of the permutation The value of dense must be a positive scalar and its default value is 10.0

opts.aggressive
If this value is a nonzero scalar, then `amd` performs aggressive absorption. The default is not to perform aggressive absorption.

The author of the code itself is Timothy A. Davis `davis@cise.ufl.edu`, University of Florida (see `http://www.cise.ufl.edu/research/sparse/amd`).

See also: [symamd], page 506, [colamd], page 504.

`p = ccolamd (S)` [Loadable Function]
`p = ccolamd (S, knobs)` [Loadable Function]
`p = ccolamd (S, knobs, cmember)` [Loadable Function]
`[p, stats] = ccolamd (...)` [Loadable Function]

Constrained column approximate minimum degree permutation. `p = ccolamd (S)` returns the column approximate minimum degree permutation vector for the sparse matrix S. For a non-symmetric matrix S, `S(:, p)` tends to have sparser LU factors than S. `chol (S(:, p)' * S(:, p))` also tends to be sparser than `chol (S' * S)`. `p = ccolamd (S, 1)` optimizes the ordering for `lu (S(:, p))`. The ordering is followed by a column elimination tree post-ordering.

knobs is an optional 1-element to 5-element input vector, with a default value of `[0 10 10 1 0]` if not present or empty. Entries not present are set to their defaults.

`knobs(1)` if nonzero, the ordering is optimized for `lu (S(:, p))`. It will be a poor ordering for `chol (S(:, p)' * S(:, p))`. This is the most important knob for ccolamd.

`knobs(2)` if S is m-by-n, rows with more than `max (16, knobs(2) * sqrt (n))` entries are ignored.

`knobs(3)` columns with more than `max (16, knobs(3) * sqrt (min (m, n)))` entries are ignored and ordered last in the output permutation (subject to the cmember constraints).

`knobs(4)` if nonzero, aggressive absorption is performed.

`knobs(5)` if nonzero, statistics and knobs are printed.

cmember is an optional vector of length n. It defines the constraints on the column ordering. If `cmember(j) = c`, then column j is in constraint set c (c must be in the range 1 to n). In the output permutation p, all columns in set 1 appear first, followed by all columns in set 2, and so on. `cmember = ones (1,n)` if not present or empty. `ccolamd (S, [], 1 : n)` returns `1 : n` `p = ccolamd (S)` is about the same as `p = colamd (S)`. *knobs* and its default values differ. `colamd` always does aggressive absorption, and it finds an ordering suitable for both `lu (S(:, p))` and `chol (S(:, p))' * S(:, p))`; it cannot optimize its ordering for `lu (S(:, p))` to the extent that `ccolamd (S, 1)` can.

stats is an optional 20-element output vector that provides data about the ordering and the validity of the input matrix *S*. Ordering statistics are in `stats(1 : 3)`. `stats(1)` and `stats(2)` are the number of dense or empty rows and columns ignored by CCOLAMD and `stats(3)` is the number of garbage collections performed on the internal data structure used by CCOLAMD (roughly of size `2.2 * nnz (S) + 4 * m + 7 * n` integers).

`stats(4 : 7)` provide information if CCOLAMD was able to continue. The matrix is OK if `stats(4)` is zero, or 1 if invalid. `stats(5)` is the rightmost column index that is unsorted or contains duplicate entries, or zero if no such column exists. `stats(6)` is the last seen duplicate or out-of-order row index in the column index given by `stats(5)`, or zero if no such row index exists. `stats(7)` is the number of duplicate or out-of-order row indices. `stats(8 : 20)` is always zero in the current version of CCOLAMD (reserved for future use).

The authors of the code itself are S. Larimore, T. Davis (Univ. of Florida) and S. Rajamanickam in collaboration with J. Bilbert and E. Ng. Supported by the National Science Foundation (DMS-9504974, DMS-9803599, CCR-0203270), and a grant from Sandia National Lab. See `http://www.cise.ufl.edu/research/sparse` for ccolamd, csymamd, amd, colamd, symamd, and other related orderings.

See also: [colamd], page 504, [csymamd], page 505.

`p = colamd (S)` [Loadable Function]
`p = colamd (S, knobs)` [Loadable Function]
`[p, stats] = colamd (S)` [Loadable Function]
`[p, stats] = colamd (S, knobs)` [Loadable Function]

Column approximate minimum degree permutation. `p = colamd (S)` returns the column approximate minimum degree permutation vector for the sparse matrix *S*. For a non-symmetric matrix *S*, `S(:,p)` tends to have sparser LU factors than *S*. The Cholesky factorization of `S(:,p)' * S(:,p)` also tends to be sparser than that of `S' * S`.

knobs is an optional one- to three-element input vector. If *S* is m-by-n, then rows with more than `max(16,knobs(1)*sqrt(n))` entries are ignored. Columns with more than `max (16,knobs(2)*sqrt(min(m,n)))` entries are removed prior to ordering, and ordered last in the output permutation *p*. Only completely dense rows or columns are removed if `knobs(1)` and `knobs(2)` are < 0, respectively. If `knobs(3)` is nonzero, *stats* and *knobs* are printed. The default is `knobs = [10 10 0]`. Note that *knobs* differs from earlier versions of colamd.

stats is an optional 20-element output vector that provides data about the ordering and the validity of the input matrix *S*. Ordering statistics are in `stats(1:3)`. `stats(1)` and `stats(2)` are the number of dense or empty rows and columns ignored by COLAMD and `stats(3)` is the number of garbage collections performed on the internal data structure used by COLAMD (roughly of size `2.2 * nnz(S) + 4 * m + 7 * n` integers).

Octave built-in functions are intended to generate valid sparse matrices, with no duplicate entries, with ascending row indices of the nonzeros in each column, with a non-negative number of entries in each column (!) and so on. If a matrix is invalid, then COLAMD may or may not be able to continue. If there are duplicate entries (a row index appears two or more times in the same column) or if the row indices in a column are out of order, then COLAMD can correct these errors by ignoring the duplicate entries and sorting each column of its internal copy of the matrix *S* (the input matrix *S* is not repaired, however). If a matrix is invalid in other ways then COLAMD cannot continue, an error message is printed, and no output arguments (*p* or *stats*) are returned. COLAMD is thus a simple way to check a sparse matrix to see if it's valid.

`stats(4:7)` provide information if COLAMD was able to continue. The matrix is OK if `stats(4)` is zero, or 1 if invalid. `stats(5)` is the rightmost column index that is unsorted or contains duplicate entries, or zero if no such column exists. `stats(6)` is the last seen duplicate or out-of-order row index in the column index given by `stats(5)`, or zero if no such row index exists. `stats(7)` is the number of duplicate or out-of-order row indices. `stats(8:20)` is always zero in the current version of COLAMD (reserved for future use).

The ordering is followed by a column elimination tree post-ordering.

The authors of the code itself are Stefan I. Larimore and Timothy A. Davis `davis@cise.ufl.edu`, University of Florida. The algorithm was developed in collaboration with John Gilbert, Xerox PARC, and Esmond Ng, Oak Ridge National Laboratory. (see `http://www.cise.ufl.edu/research/sparse/colamd`)

See also: [colperm], page 505, [symamd], page 506, [ccolamd], page 503.

p = `colperm` (*s*) [Function File]

Return the column permutations such that the columns of `s (:, p)` are ordered in terms of increase number of non-zero elements. If *s* is symmetric, then *p* is chosen such that `s (p, p)` orders the rows and columns with increasing number of non zeros elements.

p = `csymamd` (*S*) [Loadable Function]
p = `csymamd` (*S*, *knobs*) [Loadable Function]
p = `csymamd` (*S*, *knobs*, *cmember*) [Loadable Function]
[*p*, *stats*] = `csymamd` (...) [Loadable Function]

For a symmetric positive definite matrix *S*, returns the permutation vector *p* such that `S(p,p)` tends to have a sparser Cholesky factor than *S*. Sometimes `csymamd` works well for symmetric indefinite matrices too. The matrix *S* is assumed to be symmetric; only the strictly lower triangular part is referenced. *S* must be square. The ordering is followed by an elimination tree post-ordering.

knobs is an optional 1-element to 3-element input vector, with a default value of `[10 1 0]` if present or empty. Entries not present are set to their defaults.

`knobs(1)` If *S* is n-by-n, then rows and columns with more than `max(16,knobs(1)*sqrt(n))` entries are ignored, and ordered last in the output permutation (subject to the cmember constraints).

`knobs(2)` If nonzero, aggressive absorption is performed.

`knobs (3)` If nonzero, statistics and knobs are printed.

cmember is an optional vector of length n. It defines the constraints on the ordering. If `cmember (j) = S`, then row/column j is in constraint set *c* (*c* must be in the range 1 to n). In the output permutation *p*, rows/columns in set 1 appear first, followed by all rows/columns in set 2, and so on. `cmember = ones (1,n)` if not present or empty. `csymamd (S,[],1:n)` returns `1:n`.

`p = csymamd (S)` is about the same as `p = symamd (S)`. *knobs* and its default values differ.

`stats (4:7)` provide information if CCOLAMD was able to continue. The matrix is OK if `stats (4)` is zero, or 1 if invalid. `stats (5)` is the rightmost column index that is unsorted or contains duplicate entries, or zero if no such column exists. `stats (6)` is the last seen duplicate or out-of-order row index in the column index given by `stats (5)`, or zero if no such row index exists. `stats (7)` is the number of duplicate or out-of-order row indices. `stats (8:20)` is always zero in the current version of CCOLAMD (reserved for future use).

The authors of the code itself are S. Larimore, T. Davis (Uni of Florida) and S. Rajamanickam in collaboration with J. Bilbert and E. Ng. Supported by the National Science Foundation (DMS-9504974, DMS-9803599, CCR-0203270), and a grant from Sandia National Lab. See `http://www.cise.ufl.edu/research/sparse` for ccolamd, csymamd, amd, colamd, symamd, and other related orderings.

See also: [symamd], page 506, [ccolamd], page 503.

`p = dmperm (S)` [Loadable Function]
`[p, q, r, S] = dmperm (S)` [Loadable Function]

Perform a Dulmage-Mendelsohn permutation of the sparse matrix *S*. With a single output argument `dmperm` performs the row permutations *p* such that `S (p,:)` has no zero elements on the diagonal.

Called with two or more output arguments, returns the row and column permutations, such that `S (p, q)` is in block triangular form. The values of *r* and *S* define the boundaries of the blocks. If *S* is square then `r == S`.

The method used is described in: A. Pothen & C.-J. Fan. *Computing the Block Triangular Form of a Sparse Matrix*. ACM Trans. Math. Software, 16(4):303-324, 1990.

See also: [colamd], page 504, [ccolamd], page 503.

`p = symamd (S)` [Loadable Function]
`p = symamd (S, knobs)` [Loadable Function]
`[p, stats] = symamd (S)` [Loadable Function]
`[p, stats] = symamd (S, knobs)` [Loadable Function]

For a symmetric positive definite matrix *S*, returns the permutation vector p such that `S (p, p)` tends to have a sparser Cholesky factor than *S*. Sometimes `symamd` works well for symmetric indefinite matrices too. The matrix *S* is assumed to be symmetric; only the strictly lower triangular part is referenced. *S* must be square.

knobs is an optional one- to two-element input vector. If *S* is n-by-n, then rows and columns with more than `max (16,knobs (1)*sqrt(n))` entries are removed prior to ordering, and ordered last in the output permutation *p*. No rows/columns are removed if `knobs (1) < 0`. If `knobs (2)` is nonzero, `stats` and *knobs* are printed. The default is `knobs = [10 0]`. Note that *knobs* differs from earlier versions of symamd.

stats is an optional 20-element output vector that provides data about the ordering and the validity of the input matrix *S*. Ordering statistics are in `stats (1:3)`. `stats (1) = stats (2)` is the number of dense or empty rows and columns ignored by SYMAMD and `stats (3)` is the number of garbage collections performed on the internal data structure used by SYMAMD (roughly of size `8.4 * nnz (tril (S, -1)) + 9 * n` integers).

Octave built-in functions are intended to generate valid sparse matrices, with no duplicate entries, with ascending row indices of the nonzeros in each column, with a non-negative number of entries in each column (!) and so on. If a matrix is invalid, then SYMAMD may or may not be able to continue. If there are duplicate entries (a row index appears two or more times in the same column) or if the row indices in a column are out of order, then SYMAMD can correct these errors by ignoring the duplicate entries and sorting each column of its internal copy of the matrix S (the input matrix S is not repaired, however). If a matrix is invalid in other ways then SYMAMD cannot continue, an error message is printed, and no output arguments (*p* or *stats*) are returned. SYMAMD is thus a simple way to check a sparse matrix to see if it's valid.

`stats (4:7)` provide information if SYMAMD was able to continue. The matrix is OK if `stats (4)` is zero, or 1 if invalid. `stats (5)` is the rightmost column index that is unsorted or contains duplicate entries, or zero if no such column exists. `stats (6)` is the last seen duplicate or out-of-order row index in the column index given by `stats (5)`, or zero if no such row index exists. `stats (7)` is the number of duplicate or out-of-order row indices. `stats (8:20)` is always zero in the current version of SYMAMD (reserved for future use).

The ordering is followed by a column elimination tree post-ordering.

The authors of the code itself are Stefan I. Larimore and Timothy A. Davis `davis@cise.ufl.edu`, University of Florida. The algorithm was developed in collaboration with John Gilbert, Xerox PARC, and Esmond Ng, Oak Ridge National Laboratory. (see `http://www.cise.ufl.edu/research/sparse/colamd`)

See also: [colperm], page 505, [colamd], page 504.

p = `symrcm (S)` [Loadable Function]

Return the symmetric reverse Cuthill-McKee permutation of *S*. *p* is a permutation vector such that `S (p, p)` tends to have its diagonal elements closer to the diagonal than *S*. This is a good preordering for LU or Cholesky factorization of matrices that come from "long, skinny" problems. It works for both symmetric and asymmetric *S*.

The algorithm represents a heuristic approach to the NP-complete bandwidth minimization problem. The implementation is based in the descriptions found in

E. Cuthill, J. McKee. *Reducing the Bandwidth of Sparse Symmetric Matrices.* Proceedings of the 24th ACM National Conference, 157–172 1969, Brandon Press, New Jersey.

A. George, J.W.H. Liu. *Computer Solution of Large Sparse Positive Definite Systems*, Prentice Hall Series in Computational Mathematics, ISBN 0-13-165274-5, 1981.

See also: [colperm], page 505, [colamd], page 504, [symamd], page 506.

22.2 Linear Algebra on Sparse Matrices

Octave includes a polymorphic solver for sparse matrices, where the exact solver used to factorize the matrix, depends on the properties of the sparse matrix itself. Generally, the cost of determining the matrix type is small relative to the cost of factorizing the matrix itself, but in any case the matrix type is cached once it is calculated, so that it is not re-determined each time it is used in a linear equation.

The selection tree for how the linear equation is solve is

1. If the matrix is diagonal, solve directly and goto 8
2. If the matrix is a permuted diagonal, solve directly taking into account the permutations. Goto 8
3. If the matrix is square, banded and if the band density is less than that given by `spparms ("bandden")` continue, else goto 4.
   a. If the matrix is tridiagonal and the right-hand side is not sparse continue, else goto 3b.
      1. If the matrix is Hermitian, with a positive real diagonal, attempt Cholesky factorization using LAPACK xPTSV.
      2. If the above failed or the matrix is not Hermitian with a positive real diagonal use Gaussian elimination with pivoting using LAPACK xGTSV, and goto 8.
   b. If the matrix is Hermitian with a positive real diagonal, attempt Cholesky factorization using LAPACK xPBTRF.
   c. if the above failed or the matrix is not Hermitian with a positive real diagonal use Gaussian elimination with pivoting using LAPACK xGBTRF, and goto 8.
4. If the matrix is upper or lower triangular perform a sparse forward or backward substitution, and goto 8
5. If the matrix is an upper triangular matrix with column permutations or lower triangular matrix with row permutations, perform a sparse forward or backward substitution, and goto 8
6. If the matrix is square, Hermitian with a real positive diagonal, attempt sparse Cholesky factorization using CHOLMOD.
7. If the sparse Cholesky factorization failed or the matrix is not Hermitian with a real positive diagonal, and the matrix is square, factorize using UMFPACK.
8. If the matrix is not square, or any of the previous solvers flags a singular or near singular matrix, find a minimum norm solution using CXSPARSE[3].

The band density is defined as the number of non-zero values in the band divided by the total number of values in the full band. The banded matrix solvers can be entirely disabled by using *spparms* to set `bandden` to 1 (i.e., `spparms ("bandden", 1)`).

---

[3] The CHOLMOD, UMFPACK and CXSPARSE packages were written by Tim Davis and are available at `http://www.cise.ufl.edu/research/sparse/`

The QR solver factorizes the problem with a Dulmage-Mendelsohn decomposition, to separate the problem into blocks that can be treated as over-determined, multiple well determined blocks, and a final over-determined block. For matrices with blocks of strongly connected nodes this is a big win as LU decomposition can be used for many blocks. It also significantly improves the chance of finding a solution to over-determined problems rather than just returning a vector of *NaN*'s.

All of the solvers above, can calculate an estimate of the condition number. This can be used to detect numerical stability problems in the solution and force a minimum norm solution to be used. However, for narrow banded, triangular or diagonal matrices, the cost of calculating the condition number is significant, and can in fact exceed the cost of factoring the matrix. Therefore the condition number is not calculated in these cases, and Octave relies on simpler techniques to detect singular matrices or the underlying LAPACK code in the case of banded matrices.

The user can force the type of the matrix with the `matrix_type` function. This overcomes the cost of discovering the type of the matrix. However, it should be noted that identifying the type of the matrix incorrectly will lead to unpredictable results, and so `matrix_type` should be used with care.

n = `normest` (*A*) [Function File]
n = `normest` (*A*, *tol*) [Function File]
[*n*, *c*] = `normest` (...) [Function File]

Estimate the 2-norm of the matrix *A* using a power series analysis. This is typically used for large matrices, where the cost of calculating `norm (A)` is prohibitive and an approximation to the 2-norm is acceptable.

tol is the tolerance to which the 2-norm is calculated. By default *tol* is 1e-6. *c* returns the number of iterations needed for `normest` to converge.

[*est*, *v*, *w*, *iter*] = `onenormest` (*A*, *t*) [Function File]
[*est*, *v*, *w*, *iter*] = `onenormest` (*apply*, *apply_t*, *n*, *t*) [Function File]

Apply Higham and Tisseur's randomized block 1-norm estimator to matrix *A* using *t* test vectors. If *t* exceeds 5, then only 5 test vectors are used.

If the matrix is not explicit, e.g., when estimating the norm of `inv (A)` given an LU factorization, `onenormest` applies *A* and its conjugate transpose through a pair of functions *apply* and *apply_t*, respectively, to a dense matrix of size *n* by *t*. The implicit version requires an explicit dimension *n*.

Returns the norm estimate *est*, two vectors *v* and *w* related by norm `(w, 1) = est * norm (v, 1)`, and the number of iterations *iter*. The number of iterations is limited to 10 and is at least 2.

References:

- N.J. Higham and F. Tisseur, *A Block Algorithm for Matrix 1-Norm Estimation, with an Application to 1-Norm Pseudospectra.* SIMAX vol 21, no 4, pp 1185-1201. `http://dx.doi.org/10.1137/S0895479899356080`
- N.J. Higham and F. Tisseur, *A Block Algorithm for Matrix 1-Norm Estimation, with an Application to 1-Norm Pseudospectra.* `http://citeseer.ist.psu.edu/223007.html`

See also: [condest], page 510, [norm], page 439, [cond], page 436.

`condest (A)` [Function File]
`condest (A, t)` [Function File]
`[est, v] = condest (...)` [Function File]
`[est, v] = condest (A, solve, solve_t, t)` [Function File]
`[est, v] = condest (apply, apply_t, solve, solve_t, n, t)` [Function File]

Estimate the 1-norm condition number of a matrix A using t test vectors using a randomized 1-norm estimator. If t exceeds 5, then only 5 test vectors are used.

If the matrix is not explicit, e.g., when estimating the condition number of A given an LU factorization, `condest` uses the following functions:

apply `A*x` for a matrix `x` of size n by t.

apply_t `A'*x` for a matrix `x` of size n by t.

solve `A \ b` for a matrix `b` of size n by t.

solve_t `A' \ b` for a matrix `b` of size n by t.

The implicit version requires an explicit dimension n.

`condest` uses a randomized algorithm to approximate the 1-norms.

`condest` returns the 1-norm condition estimate *est* and a vector *v* satisfying `norm (A*v, 1) == norm (A, 1) * norm (v, 1) / est`. When *est* is large, *v* is an approximate null vector.

References:

- N.J. Higham and F. Tisseur, *A Block Algorithm for Matrix 1-Norm Estimation, with an Application to 1-Norm Pseudospectra.* SIMAX vol 21, no 4, pp 1185-1201. `http://dx.doi.org/10.1137/S0895479899356080`
- N.J. Higham and F. Tisseur, *A Block Algorithm for Matrix 1-Norm Estimation, with an Application to 1-Norm Pseudospectra.* `http://citeseer.ist.psu.edu/223007.html`

See also: [cond], page 436, [norm], page 439, [onenormest], page 509.

`spparms ()` [Built-in Function]
`vals = spparms ()` [Built-in Function]
`[keys, vals] = spparms ()` [Built-in Function]
`val = spparms (key)` [Built-in Function]
`spparms (vals)` [Built-in Function]
`spparms ("defaults")` [Built-in Function]
`spparms ("tight")` [Built-in Function]
`spparms (key, val)` [Built-in Function]

Query or set the parameters used by the sparse solvers and factorization functions. The first four calls above get information about the current settings, while the others change the current settings. The parameters are stored as pairs of keys and values, where the values are all floats and the keys are one of the following strings:

'`spumoni`' Printing level of debugging information of the solvers (default 0)

'`ths_rel`' Included for compatibility. Not used. (default 1)

'`ths_abs`' Included for compatibility. Not used. (default 1) 'exact_d' Included for compatibility. Not used. (default 0)

'supernd' Included for compatibility. Not used. (default 3)

'rreduce' Included for compatibility. Not used. (default 3)

'wh_frac' Included for compatibility. Not used. (default 0.5)

'autommd' Flag whether the LU/QR and the '\' and '/' operators will automatically use the sparsity preserving mmd functions (default 1)

'autoamd' Flag whether the LU and the '\' and '/' operators will automatically use the sparsity preserving amd functions (default 1)

'piv_tol' The pivot tolerance of the UMFPACK solvers (default 0.1)

'sym_tol' The pivot tolerance of the UMFPACK symmetric solvers (default 0.001)

'bandden' The density of non-zero elements in a banded matrix before it is treated by the LAPACK banded solvers (default 0.5)

'umfpack' Flag whether the UMFPACK or mmd solvers are used for the LU, '\' and '/' operations (default 1)

The value of individual keys can be set with `spparms (key, val)`. The default values can be restored with the special keyword `"defaults"`. The special keyword `"tight"` can be used to set the mmd solvers to attempt a sparser solution at the potential cost of longer running time.

See also: [chol], page 442, [colamd], page 504, [lu], page 444, [qr], page 446, [symamd], page 506.

`p = sprank (S)` [Loadable Function]

Calculate the structural rank of the sparse matrix *S*. Note that only the structure of the matrix is used in this calculation based on a Dulmage-Mendelsohn permutation to block triangular form. As such the numerical rank of the matrix *S* is bounded by `sprank (S) >= rank (S)`. Ignoring floating point errors `sprank (S) == rank (S)`.

See also: [dmperm], page 506.

`[count, h, parent, post, r] = symbfact (S)` [Loadable Function]
`[...] = symbfact (S, typ)` [Loadable Function]
`[...] = symbfact (S, typ, mode)` [Loadable Function]

Perform a symbolic factorization analysis on the sparse matrix *S*. Where

S *S* is a complex or real sparse matrix.

typ Is the type of the factorization and can be one of

'sym' Factorize *S*. This is the default.

'col' Factorize `S' * S`.

'row' Factorize `S * S'`.

'lo' Factorize `S'`

mode The default is to return the Cholesky factorization for *r*, and if *mode* is `'L'`, the conjugate transpose of the Cholesky factorization is returned. The conjugate transpose version is faster and uses less memory, but returns the same values for *count*, *h*, *parent* and *post* outputs.

The output variables are

count The row counts of the Cholesky factorization as determined by *typ*.

h The height of the elimination tree.

parent The elimination tree itself.

post A sparse boolean matrix whose structure is that of the Cholesky factorization as determined by *typ*.

For non square matrices, the user can also utilize the `spaugment` function to find a least squares solution to a linear equation.

s = `spaugment` (*A*, *c*) [Function File]

Create the augmented matrix of *A*.

This is given by

```
[c * eye(m, m), A;
            A', zeros(n, n)]
```

This is related to the least squares solution of `A \ b`, by

```
s * [ r / c; x] = [ b, zeros(n, columns(b)) ]
```

where *r* is the residual error

```
r = b - A * x
```

As the matrix *s* is symmetric indefinite it can be factorized with `lu`, and the minimum norm solution can therefore be found without the need for a `qr` factorization. As the residual error will be `zeros (m, m)` for underdetermined problems, and example can be

```
m = 11; n = 10; mn = max (m, n);
A = spdiags ([ones(mn,1), 10*ones(mn,1), -ones(mn,1)],
             [-1, 0, 1], m, n);
x0 = A \ ones (m,1);
s = spaugment (A);
[L, U, P, Q] = lu (s);
x1 = Q * (U \ (L \ (P  * [ones(m,1); zeros(n,1)])));
x1 = x1(end - n + 1 : end);
```

To find the solution of an overdetermined problem needs an estimate of the residual error *r* and so it is more complex to formulate a minimum norm solution using the `spaugment` function.

In general the left division operator is more stable and faster than using the `spaugment` function.

See also: [mldivide], page 139.

Finally, the function **eigs** can be used to calculate a limited number of eigenvalues and eigenvectors based on a selection criteria and likewise for **svds** which calculates a limited number of singular values and vectors.

d = eigs (*A*) [Function File]
d = eigs (*A*, *k*) [Function File]
d = eigs (*A*, *k*, *sigma*) [Function File]
d = eigs (*A*, *k*, *sigma*, *opts*) [Function File]
d = eigs (*A*, *B*) [Function File]
d = eigs (*A*, *B*, *k*) [Function File]
d = eigs (*A*, *B*, *k*, *sigma*) [Function File]
d = eigs (*A*, *B*, *k*, *sigma*, *opts*) [Function File]
d = eigs (*af*, *n*) [Function File]
d = eigs (*af*, *n*, *B*) [Function File]
d = eigs (*af*, *n*, *k*) [Function File]
d = eigs (*af*, *n*, *B*, *k*) [Function File]
d = eigs (*af*, *n*, *k*, *sigma*) [Function File]
d = eigs (*af*, *n*, *B*, *k*, *sigma*) [Function File]
d = eigs (*af*, *n*, *k*, *sigma*, *opts*) [Function File]
d = eigs (*af*, *n*, *B*, *k*, *sigma*, *opts*) [Function File]
[*V*, *d*] = eigs (*A*, ...) [Function File]
[*V*, *d*] = eigs (*af*, *n*, ...) [Function File]
[*V*, *d*, *flag*] = eigs (*A*, ...) [Function File]
[*V*, *d*, *flag*] = eigs (*af*, *n*, ...) [Function File]

Calculate a limited number of eigenvalues and eigenvectors of A, based on a selection criteria. The number of eigenvalues and eigenvectors to calculate is given by k and defaults to 6.

By default, **eigs** solve the equation $A\nu = \lambda\nu$, where λ is a scalar representing one of the eigenvalues, and ν is the corresponding eigenvector. If given the positive definite matrix B then **eigs** solves the general eigenvalue equation $A\nu = \lambda B\nu$.

The argument *sigma* determines which eigenvalues are returned. *sigma* can be either a scalar or a string. When *sigma* is a scalar, the k eigenvalues closest to *sigma* are returned. If *sigma* is a string, it must have one of the following values.

"lm" Largest Magnitude (default).

"sm" Smallest Magnitude.

"la" Largest Algebraic (valid only for real symmetric problems).

"sa" Smallest Algebraic (valid only for real symmetric problems).

"be" Both Ends, with one more from the high-end if k is odd (valid only for real symmetric problems).

"lr" Largest Real part (valid only for complex or unsymmetric problems).

"sr" Smallest Real part (valid only for complex or unsymmetric problems).

"li" Largest Imaginary part (valid only for complex or unsymmetric problems).

`"si"` Smallest Imaginary part (valid only for complex or unsymmetric problems).

If *opts* is given, it is a structure defining possible options that `eigs` should use. The fields of the *opts* structure are:

`issym` If *af* is given, then flags whether the function *af* defines a symmetric problem. It is ignored if *A* is given. The default is false.

`isreal` If *af* is given, then flags whether the function *af* defines a real problem. It is ignored if *A* is given. The default is true.

`tol` Defines the required convergence tolerance, calculated as `tol * norm (A)`. The default is `eps`.

`maxit` The maximum number of iterations. The default is 300.

`p` The number of Lanzcos basis vectors to use. More vectors will result in faster convergence, but a greater use of memory. The optimal value of `p` is problem dependent and should be in the range *k* to *n*. The default value is `2 * k`.

`v0` The starting vector for the algorithm. An initial vector close to the final vector will speed up convergence. The default is for ARPACK to randomly generate a starting vector. If specified, `v0` must be an *n*-by-1 vector where `n = rows (A)`

`disp` The level of diagnostic printout (0|1|2). If `disp` is 0 then diagnostics are disabled. The default value is 0.

`cholB` Flag if `chol (B)` is passed rather than *B*. The default is false.

`permB` The permutation vector of the Cholesky factorization of *B* if `cholB` is true. That is `chol (B(permB, permB))`. The default is `1:n`.

It is also possible to represent *A* by a function denoted *af*. *af* must be followed by a scalar argument *n* defining the length of the vector argument accepted by *af*. *af* can be a function handle, an inline function, or a string. When *af* is a string it holds the name of the function to use.

af is a function of the form `y = af (x)` where the required return value of *af* is determined by the value of *sigma*. The four possible forms are

`A * x` if *sigma* is not given or is a string other than "sm".

`A \ x` if *sigma* is 0 or "sm".

`(A - sigma * I) \ x`
for the standard eigenvalue problem, where `I` is the identity matrix of the same size as *A*.

`(A - sigma * B) \ x`
for the general eigenvalue problem.

The return arguments of `eigs` depend on the number of return arguments requested. With a single return argument, a vector *d* of length *k* is returned containing the *k* eigenvalues that have been found. With two return arguments, V is a n-by-k matrix whose columns are the k eigenvectors corresponding to the returned eigenvalues. The eigenvalues themselves are returned in d in the form of a n-by-k matrix, where the elements on the diagonal are the eigenvalues.

Given a third return argument *flag*, `eigs` returns the status of the convergence. If *flag* is 0 then all eigenvalues have converged. Any other value indicates a failure to converge.

This function is based on the ARPACK package, written by R. Lehoucq, K. Maschhoff, D. Sorensen, and C. Yang. For more information see `http://www.caam.rice.edu/software/ARPACK/`.

See also: [eig], page 437, [svds], page 515.

`s = svds (A)` [Function File]
`s = svds (A, k)` [Function File]
`s = svds (A, k, sigma)` [Function File]
`s = svds (A, k, sigma, opts)` [Function File]
`[u, s, v] = svds (...)` [Function File]
`[u, s, v, flag] = svds (...)` [Function File]

Find a few singular values of the matrix A. The singular values are calculated using

```
[m, n] = size (A);
s = eigs ([sparse(m, m), A;
                         A', sparse(n, n)])
```

The eigenvalues returned by `eigs` correspond to the singular values of A. The number of singular values to calculate is given by k and defaults to 6.

The argument *sigma* specifies which singular values to find. When *sigma* is the string `'L'`, the default, the largest singular values of A are found. Otherwise, *sigma* must be a real scalar and the singular values closest to *sigma* are found. As a corollary, `sigma = 0` finds the smallest singular values. Note that for relatively small values of *sigma*, there is a chance that the requested number of singular values will not be found. In that case *sigma* should be increased.

opts is a structure defining options that `svds` will pass to `eigs`. The possible fields of this structure are documented in `eigs`. By default, `svds` sets the following three fields:

`tol` The required convergence tolerance for the singular values. The default value is 1e-10. `eigs` is passed `tol / sqrt(2)`.

`maxit` The maximum number of iterations. The default is 300.

`disp` The level of diagnostic printout (0|1|2). If `disp` is 0 then diagnostics are disabled. The default value is 0.

If more than one output is requested then `svds` will return an approximation of the singular value decomposition of A

```
A_approx = u*s*v'
```

where *A_approx* is a matrix of size A but only rank k.

flag returns 0 if the algorithm has succesfully converged, and 1 otherwise. The test for convergence is

```
norm (A*v - u*s, 1) <= tol * norm (A, 1)
```

`svds` is best for finding only a few singular values from a large sparse matrix. Otherwise, `svd (full (A))` will likely be more efficient.

See also: [svd], page 450, [eigs], page 513.

22.3 Iterative Techniques Applied to Sparse Matrices

The left division \ and right division / operators, discussed in the previous section, use direct solvers to resolve a linear equation of the form `x = A \ b` or `x = b / A`. Octave equally includes a number of functions to solve sparse linear equations using iterative techniques.

`x = pcg (A, b, tol, maxit, m1, m2, x0, ...)` [Function File]
`[x, flag, relres, iter, resvec, eigest] = pcg (...)` [Function File]

Solve the linear system of equations `A * x = b` by means of the Preconditioned Conjugate Gradient iterative method. The input arguments are

- *A* can be either a square (preferably sparse) matrix or a function handle, inline function or string containing the name of a function which computes `A * x`. In principle, *A* should be symmetric and positive definite; if `pcg` finds *A* not to be positive definite, a warning is printed and the *flag* output will be set.
- *b* is the right-hand side vector.
- *tol* is the required relative tolerance for the residual error, `b - A * x`. The iteration stops if `norm (b - A * x)` $\leq$ *tol* * norm (*b*). If *tol* is omitted or empty then a tolerance of 1e-6 is used.
- *maxit* is the maximum allowable number of iterations; if *maxit* is omitted or empty then a value of 20 is used.
- *m* = *m1* * *m2* is the (left) preconditioning matrix, so that the iteration is (theoretically) equivalent to solving by `pcg P * x = m \ b`, with `P = m \ A`. Note that a proper choice of the preconditioner may dramatically improve the overall performance of the method. Instead of matrices *m1* and *m2*, the user may pass two functions which return the results of applying the inverse of *m1* and *m2* to a vector (usually this is the preferred way of using the preconditioner). If *m1* is omitted or empty `[]` then no preconditioning is applied. If *m2* is omitted, *m* = *m1* will be used as a preconditioner.
- *x0* is the initial guess. If *x0* is omitted or empty then the function sets *x0* to a zero vector by default.

The arguments which follow *x0* are treated as parameters, and passed in a proper way to any of the functions (*A* or *m*) which are passed to `pcg`. See the examples below for further details. The output arguments are

- *x* is the computed approximation to the solution of `A * x = b`.
- *flag* reports on the convergence. A value of 0 means the solution converged and the tolerance criterion given by *tol* is satisfied. A value of 1 means that the *maxit* limit for the iteration count was reached. A value of 3 indicates that the (preconditioned) matrix was found not to be positive definite.
- *relres* is the ratio of the final residual to its initial value, measured in the Euclidean norm.

- *iter* is the actual number of iterations performed.
- *resvec* describes the convergence history of the method. `resvec(i,1)` is the Euclidean norm of the residual, and `resvec(i,2)` is the preconditioned residual norm, after the (*i*-1)-th iteration, `i = 1, 2, ..., iter+1`. The preconditioned residual norm is defined as `norm (r) ^ 2 = r' * (m \ r)` where `r = b - A * x`, see also the description of *m*. If *eigest* is not required, only `resvec(:,1)` is returned.
- *eigest* returns the estimate for the smallest `eigest(1)` and largest `eigest(2)` eigenvalues of the preconditioned matrix `P = m \ A`. In particular, if no preconditioning is used, the estimates for the extreme eigenvalues of A are returned. `eigest(1)` is an overestimate and `eigest(2)` is an underestimate, so that `eigest(2) / eigest(1)` is a lower bound for `cond (P, 2)`, which nevertheless in the limit should theoretically be equal to the actual value of the condition number. The method which computes *eigest* works only for symmetric positive definite A and m, and the user is responsible for verifying this assumption.

Let us consider a trivial problem with a diagonal matrix (we exploit the sparsity of A)

```
n = 10;
A = diag (sparse (1:n));
b = rand (n, 1);
[l, u, p, q] = luinc (A, 1.e-3);
```

Example 1: Simplest use of pcg

```
x = pcg (A, b)
```

Example 2: pcg with a function which computes `A * x`

```
function y = apply_a (x)
  y = [1:N]' .* x;
endfunction

x = pcg ("apply_a", b)
```

Example 3: pcg with a preconditioner: *l* * *u*

```
x = pcg (A, b, 1.e-6, 500, l*u)
```

Example 4: pcg with a preconditioner: *l* * *u*. Faster than Example 3 since lower and upper triangular matrices are easier to invert

```
x = pcg (A, b, 1.e-6, 500, l, u)
```

Example 5: Preconditioned iteration, with full diagnostics. The preconditioner (quite strange, because even the original matrix A is trivial) is defined as a function

```
function y = apply_m (x)
  k = floor (length (x) - 2);
  y = x;
  y(1:k) = x(1:k) ./ [1:k]';
endfunction

[x, flag, relres, iter, resvec, eigest] = ...
                     pcg (A, b, [], [], "apply_m");
semilogy (1:iter+1, resvec);
```

EXAMPLE 6: Finally, a preconditioner which depends on a parameter *k*.

```
function y = apply_M (x, varargin)
  K = varargin{1};
  y = x;
  y(1:K) = x(1:K) ./ [1:K]';
endfunction

[x, flag, relres, iter, resvec, eigest] = ...
     pcg (A, b, [], [], "apply_m", [], [], 3)
```

References:

1. C.T. Kelley, *Iterative Methods for Linear and Nonlinear Equations*, SIAM, 1995. (the base PCG algorithm)
2. Y. Saad, *Iterative Methods for Sparse Linear Systems*, PWS 1996. (condition number estimate from PCG) Revised version of this book is available online at `http://www-users.cs.umn.edu/~saad/books.html`

See also: [sparse], page 492, [pcr], page 518.

`x = pcr (A, b, tol, maxit, m, x0, ...)` [Function File]
`[x, flag, relres, iter, resvec] = pcr (...)` [Function File]

Solve the linear system of equations `A * x = b` by means of the Preconditioned Conjugate Residuals iterative method. The input arguments are

- *A* can be either a square (preferably sparse) matrix or a function handle, inline function or string containing the name of a function which computes `A * x`. In principle *A* should be symmetric and non-singular; if `pcr` finds *A* to be numerically singular, you will get a warning message and the *flag* output parameter will be set.
- *b* is the right hand side vector.
- *tol* is the required relative tolerance for the residual error, `b - A * x`. The iteration stops if `norm (b - A * x) <= tol * norm (b - A * x0)`. If *tol* is empty or is omitted, the function sets `tol = 1e-6` by default.
- *maxit* is the maximum allowable number of iterations; if `[]` is supplied for `maxit`, or `pcr` has less arguments, a default value equal to 20 is used.
- *m* is the (left) preconditioning matrix, so that the iteration is (theoretically) equivalent to solving by `pcr P * x = m \ b`, with `P = m \ A`. Note that a proper choice of the preconditioner may dramatically improve the overall performance

of the method. Instead of matrix *m*, the user may pass a function which returns the results of applying the inverse of *m* to a vector (usually this is the preferred way of using the preconditioner). If `[]` is supplied for *m*, or *m* is omitted, no preconditioning is applied.

- *x0* is the initial guess. If *x0* is empty or omitted, the function sets *x0* to a zero vector by default.

The arguments which follow *x0* are treated as parameters, and passed in a proper way to any of the functions (*A* or *m*) which are passed to `pcr`. See the examples below for further details. The output arguments are

- *x* is the computed approximation to the solution of `A * x = b`.
- *flag* reports on the convergence. `flag = 0` means the solution converged and the tolerance criterion given by *tol* is satisfied. `flag = 1` means that the *maxit* limit for the iteration count was reached. `flag = 3` reports t `pcr` breakdown, see [1] for details.
- *relres* is the ratio of the final residual to its initial value, measured in the Euclidean norm.
- *iter* is the actual number of iterations performed.
- *resvec* describes the convergence history of the method, so that `resvec (i)` contains the Euclidean norms of the residual after the (*i*-1)-th iteration, `i = 1,2, ..., iter+1`.

Let us consider a trivial problem with a diagonal matrix (we exploit the sparsity of A)

```
n = 10;
A = sparse (diag (1:n));
b = rand (N, 1);
```

EXAMPLE 1: Simplest use of `pcr`

```
x = pcr (A, b)
```

EXAMPLE 2: `pcr` with a function which computes `A * x`.

```
function y = apply_a (x)
  y = [1:10]' .* x;
endfunction

x = pcr ("apply_a", b)
```

EXAMPLE 3: Preconditioned iteration, with full diagnostics. The preconditioner (quite strange, because even the original matrix *A* is trivial) is defined as a function

```
function y = apply_m (x)
  k = floor (length (x) - 2);
  y = x;
  y(1:k) = x(1:k) ./ [1:k]';
endfunction

[x, flag, relres, iter, resvec] = ...
                   pcr (A, b, [], [], "apply_m")
semilogy ([1:iter+1], resvec);
```

EXAMPLE 4: Finally, a preconditioner which depends on a parameter *k*.

```
function y = apply_m (x, varargin)
  k = varargin{1};
  y = x;
  y(1:k) = x(1:k) ./ [1:k]';
endfunction

[x, flag, relres, iter, resvec] = ...
                    pcr (A, b, [], [], "apply_m"', [], 3)
```

References:

[1] W. Hackbusch, *Iterative Solution of Large Sparse Systems of Equations*, section 9.5.4; Springer, 1994

See also: [sparse], page 492, [pcg], page 516.

The speed with which an iterative solver converges to a solution can be accelerated with the use of a pre-conditioning matrix *M*. In this case the linear equation `M^-1 * x = M^-1 * A \ b` is solved instead. Typical pre-conditioning matrices are partial factorizations of the original matrix.

`[L, U, P, Q] = luinc (A, '0')` [Built-in Function]
`[L, U, P, Q] = luinc (A, droptol)` [Built-in Function]
`[L, U, P, Q] = luinc (A, opts)` [Built-in Function]

Produce the incomplete LU factorization of the sparse matrix *A*. Two types of incomplete factorization are possible, and the type is determined by the second argument to `luinc`.

Called with a second argument of `'0'`, the zero-level incomplete LU factorization is produced. This creates a factorization of *A* where the position of the non-zero arguments correspond to the same positions as in the matrix *A*.

Alternatively, the fill-in of the incomplete LU factorization can be controlled through the variable *droptol* or the structure *opts*. The UMFPACK multifrontal factorization code by Tim A. Davis is used for the incomplete LU factorization, (availability `http://www.cise.ufl.edu/research/sparse/umfpack/`)

droptol determines the values below which the values in the LU factorization are dropped and replaced by zero. It must be a positive scalar, and any values in the factorization whose absolute value are less than this value are dropped, expect if leaving them increase the sparsity of the matrix. Setting *droptol* to zero results in a complete LU factorization which is the default.

opts is a structure containing one or more of the fields

`droptol` The drop tolerance as above. If *opts* only contains `droptol` then this is equivalent to using the variable *droptol*.

`milu` A logical variable flagging whether to use the modified incomplete LU factorization. In the case that `milu` is true, the dropped values are subtracted from the diagonal of the matrix *U* of the factorization. The default is `false`.

`udiag` A logical variable that flags whether zero elements on the diagonal of U should be replaced with *droptol* to attempt to avoid singular factors. The default is `false`.

`thresh` Defines the pivot threshold in the interval [0,1]. Values outside that range are ignored.

All other fields in *opts* are ignored. The outputs from `luinc` are the same as for `lu`.

Given the string argument `"vector"`, `luinc` returns the values of p q as vector values.

See also: [sparse], page 492, [lu], page 444.

22.4 Real Life Example using Sparse Matrices

A common application for sparse matrices is in the solution of Finite Element Models. Finite element models allow numerical solution of partial differential equations that do not have closed form solutions, typically because of the complex shape of the domain.

In order to motivate this application, we consider the boundary value Laplace equation. This system can model scalar potential fields, such as heat or electrical potential. Given a medium Ω with boundary $\partial\Omega$. At all points on the $\partial\Omega$ the boundary conditions are known, and we wish to calculate the potential in Ω. Boundary conditions may specify the potential (Dirichlet boundary condition), its normal derivative across the boundary (Neumann boundary condition), or a weighted sum of the potential and its derivative (Cauchy boundary condition).

In a thermal model, we want to calculate the temperature in Ω and know the boundary temperature (Dirichlet condition) or heat flux (from which we can calculate the Neumann condition by dividing by the thermal conductivity at the boundary). Similarly, in an electrical model, we want to calculate the voltage in Ω and know the boundary voltage (Dirichlet) or current (Neumann condition after diving by the electrical conductivity). In an electrical model, it is common for much of the boundary to be electrically isolated; this is a Neumann boundary condition with the current equal to zero.

The simplest finite element models will divide Ω into simplexes (triangles in 2D, pyramids in 3D). We take as a 3-D example a cylindrical liquid filled tank with a small non-conductive ball from the EIDORS project[4]. This is model is designed to reflect an application of electrical impedance tomography, where current patterns are applied to such a tank in order to image the internal conductivity distribution. In order to describe the FEM geometry, we have a matrix of vertices `nodes` and simplices `elems`.

The following example creates a simple rectangular 2-D electrically conductive medium with 10 V and 20 V imposed on opposite sides (Dirichlet boundary conditions). All other edges are electrically isolated.

---

[4] EIDORS - Electrical Impedance Tomography and Diffuse optical Tomography Reconstruction Software `http://eidors3d.sourceforge.net`

```
node_y = [1;1.2;1.5;1.8;2]*ones(1,11);
node_x = ones(5,1)*[1,1.05,1.1,1.2, ...
          1.3,1.5,1.7,1.8,1.9,1.95,2];
nodes = [node_x(:), node_y(:)];

[h,w] = size (node_x);
elems = [];
for idx = 1:w-1
  widx = (idx-1)*h;
  elems = [elems; ...
    widx+[(1:h-1);(2:h);h+(1:h-1)]'; ...
    widx+[(2:h);h+(2:h);h+(1:h-1)]' ];
endfor

E = size (elems,1); # No. of simplices
N = size (nodes,1); # No. of vertices
D = size (elems,2); # dimensions+1
```

This creates a N-by-2 matrix `nodes` and a E-by-3 matrix `elems` with values, which define finite element triangles:

```
nodes(1:7,:)'
  1.00 1.00 1.00 1.00 1.00 1.05 1.05 ...
  1.00 1.20 1.50 1.80 2.00 1.00 1.20 ...

elems(1:7,:)'
  1    2    3    4    2    3    4 ...
  2    3    4    5    7    8    9 ...
  6    7    8    9    6    7    8 ...
```

Using a first order FEM, we approximate the electrical conductivity distribution in Ω as constant on each simplex (represented by the vector `conductivity`). Based on the finite element geometry, we first calculate a system (or stiffness) matrix for each simplex (represented as 3-by-3 elements on the diagonal of the element-wise system matrix `SE`. Based on `SE` and a N-by-DE connectivity matrix `C`, representing the connections between simplices and vertices, the global connectivity matrix `S` is calculated.

```
## Element conductivity
conductivity = [1*ones(1,16), ...
       2*ones(1,48), 1*ones(1,16)];

## Connectivity matrix
C = sparse ((1:D*E), reshape (elems', ...
       D*E, 1), 1, D*E, N);

## Calculate system matrix
Siidx = floor ([0:D*E-1]'/D) * D * ...
       ones(1,D) + ones(D*E,1)*(1:D) ;
Sjidx = [1:D*E]'*ones (1,D);
Sdata = zeros (D*E,D);
```

```
dfact = factorial (D-1);
for j = 1:E
   a = inv ([ones(D,1), ...
       nodes(elems(j,:), :)]);
   const = conductivity(j) * 2 / ...
       dfact / abs (det (a));
   Sdata(D*(j-1)+(1:D),:) = const * ...
       a(2:D,:)' * a(2:D,:);
endfor
## Element-wise system matrix
SE = sparse(Siidx,Sjidx,Sdata);
## Global system matrix
S = C'* SE *C;
```

The system matrix acts like the conductivity S in Ohm's law $SV = I$. Based on the Dirichlet and Neumann boundary conditions, we are able to solve for the voltages at each vertex V.

```
## Dirichlet boundary conditions
D_nodes = [1:5, 51:55];
D_value = [10*ones(1,5), 20*ones(1,5)];

V = zeros (N,1);
V(D_nodes) = D_value;
idx = 1:N; # vertices without Dirichlet
           # boundary condns
idx(D_nodes) = [];

## Neumann boundary conditions.  Note that
## N_value must be normalized by the
## boundary length and element conductivity
N_nodes = [];
N_value = [];

Q = zeros (N,1);
Q(N_nodes) = N_value;

V(idx) = S(idx,idx) \ ( Q(idx) - ...
           S(idx,D_nodes) * V(D_nodes));
```

Finally, in order to display the solution, we show each solved voltage value in the z-axis for each simplex vertex. See Figure 22.6.

```
elemx = elems(:,[1,2,3,1])';
xelems = reshape (nodes(elemx, 1), 4, E);
yelems = reshape (nodes(elemx, 2), 4, E);
velems = reshape (V(elemx), 4, E);
plot3 (xelems,yelems,velems,"k");
print "grid.eps";
```

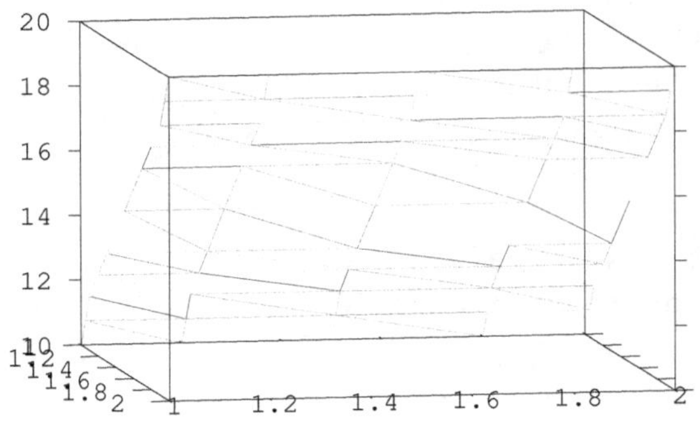

Figure 22.6: Example finite element model the showing triangular elements. The height of each vertex corresponds to the solution value.

23 Numerical Integration

Octave comes with several built-in functions for computing the integral of a function numerically (termed quadrature). These functions all solve 1-dimensional integration problems.

23.1 Functions of One Variable

Octave supports five different algorithms for computing the integral

$$\int_a^b f(x)dx$$

of a function f over the interval from a to b. These are

`quad` Numerical integration based on Gaussian quadrature.

`quadv` Numerical integration using an adaptive vectorized Simpson's rule.

`quadl` Numerical integration using an adaptive Lobatto rule.

`quadgk` Numerical integration using an adaptive Gauss-Konrod rule.

`quadcc` Numerical integration using adaptive Clenshaw-Curtis rules.

`trapz, cumtrapz`
Numerical integration of data using the trapezoidal method.

The best quadrature algorithm to use depends on the integrand. If you have empirical data, rather than a function, the choice is `trapz` or `cumtrapz`. If you are uncertain about the characteristics of the integrand, `quadcc` will be the most robust as it can handle discontinuities, singularities, oscillatory functions, and infinite intervals. When the integrand is smooth `quadgk` may be the fastest of the algorithms.

Function	**Characteristics**
quad	Low accuracy with nonsmooth integrands
quadv	Medium accuracy with smooth integrands
quadl	Medium accuracy with smooth integrands. Slower than quadgk.
quadgk	Medium accuracy ($1e^{-6}$–$1e^{-9}$) with smooth integrands.
	Handles oscillatory functions and infinite bounds
quadcc	Low to High accuracy with nonsmooth/smooth integrands
	Handles oscillatory functions, singularities, and infinite bounds

Here is an example of using `quad` to integrate the function

$$f(x) = x\sin(1/x)\sqrt{|1-x|}$$

from $x = 0$ to $x = 3$.

This is a fairly difficult integration (plot the function over the range of integration to see why).

The first step is to define the function:

```
function y = f (x)
  y = x .* sin (1./x) .* sqrt (abs (1 - x));
endfunction
```

Note the use of the 'dot' forms of the operators. This is not necessary for the `quad` integrator, but is required by the other integrators. In any case, it makes it much easier to generate a set of points for plotting because it is possible to call the function with a vector argument to produce a vector result.

The second step is to call quad with the limits of integration:

```
[q, ier, nfun, err] = quad ("f", 0, 3)
     ⇒ 1.9819
     ⇒ 1
     ⇒ 5061
     ⇒ 1.1522e-07
```

Although `quad` returns a nonzero value for *ier*, the result is reasonably accurate (to see why, examine what happens to the result if you move the lower bound to 0.1, then 0.01, then 0.001, etc.).

The function `"f"` can be the string name of a function, a function handle, or an inline function. These options make it quite easy to do integration without having to fully define a function in an m-file. For example:

```
# Verify integral (x^3) = x^4/4
f = inline ("x.^3");
quadgk (f, 0, 1)
     ⇒ 0.25000

# Verify gamma function = (n-1)! for n = 4
f = @(x) x.^3 .* exp (-x);
quadcc (f, 0, Inf)
     ⇒ 6.0000
```

q = quad (*f*, *a*, *b*) [Built-in Function]
q = quad (*f*, *a*, *b*, *tol*) [Built-in Function]
q = quad (*f*, *a*, *b*, *tol*, *sing*) [Built-in Function]
[*q*, *ier*, *nfun*, *err*] = quad (...) [Built-in Function]

Numerically evaluate the integral of *f* from *a* to *b* using Fortran routines from QUADPACK. *f* is a function handle, inline function, or a string containing the name of the function to evaluate. The function must have the form `y = f (x)` where *y* and *x* are scalars.

a and *b* are the lower and upper limits of integration. Either or both may be infinite.

The optional argument *tol* is a vector that specifies the desired accuracy of the result. The first element of the vector is the desired absolute tolerance, and the second element is the desired relative tolerance. To choose a relative test only, set the absolute tolerance to zero. To choose an absolute test only, set the relative tolerance to zero. Both tolerances default to `sqrt (eps)` or approximately $1.5e^{-8}$.

The optional argument *sing* is a vector of values at which the integrand is known to be singular.

The result of the integration is returned in *q*. *ier* contains an integer error code (0 indicates a successful integration). *nfun* indicates the number of function evaluations that were made, and *err* contains an estimate of the error in the solution.

The function `quad_options` can set other optional parameters for `quad`.

Note: because `quad` is written in Fortran it cannot be called recursively. This prevents its use in integrating over more than one variable by routines `dblquad` and `triplequad`.

See also: [quad_options], page 527, [quadv], page 527, [quadl], page 528, [quadgk], page 528, [quadcc], page 530, [trapz], page 531, [dblquad], page 533, [triplequad], page 534.

`quad_options ()` [Built-in Function]
`val = quad_options (opt)` [Built-in Function]
`quad_options (opt, val)` [Built-in Function]

Query or set options for the function `quad`. When called with no arguments, the names of all available options and their current values are displayed. Given one argument, return the value of the corresponding option. When called with two arguments, `quad_options` set the option *opt* to value *val*.

Options include

`"absolute tolerance"`
: Absolute tolerance; may be zero for pure relative error test.

`"relative tolerance"`
: Non-negative relative tolerance. If the absolute tolerance is zero, the relative tolerance must be greater than or equal to `max (50*eps, 0.5e-28)`.

`"single precision absolute tolerance"`
: Absolute tolerance for single precision; may be zero for pure relative error test.

`"single precision relative tolerance"`
: Non-negative relative tolerance for single precision. If the absolute tolerance is zero, the relative tolerance must be greater than or equal to `max (50*eps, 0.5e-28)`.

`q = quadv (f, a, b)` [Function File]
`q = quadv (f, a, b, tol)` [Function File]
`q = quadv (f, a, b, tol, trace)` [Function File]
`q = quadv (f, a, b, tol, trace, p1, p2, ...)` [Function File]
`[q, nfun] = quadv (...)` [Function File]

Numerically evaluate the integral of *f* from *a* to *b* using an adaptive Simpson's rule. *f* is a function handle, inline function, or string containing the name of the function to evaluate. `quadv` is a vectorized version of `quad` and the function defined by *f* must accept a scalar or vector as input and return a scalar, vector, or array as output.

a and *b* are the lower and upper limits of integration. Both limits must be finite.

The optional argument *tol* defines the tolerance used to stop the adaptation procedure. The default value is $1e^{-6}$.

The algorithm used by `quadv` involves recursively subdividing the integration interval and applying Simpson's rule on each subinterval. If *trace* is true then after computing each of these partial integrals display: (1) the total number of function evaluations, (2) the left end of the subinterval, (3) the length of the subinterval, (4) the approximation of the integral over the subinterval.

Additional arguments *p1*, etc., are passed directly to the function *f*. To use default values for *tol* and *trace*, one may pass empty matrices ([]).

The result of the integration is returned in *q*. *nfun* indicates the number of function evaluations that were made.

Note: `quadv` is written in Octave's scripting language and can be used recursively in `dblquad` and `triplequad`, unlike the similar `quad` function.

See also: [quad], page 526, [quadl], page 528, [quadgk], page 528, [quadcc], page 530, [trapz], page 531, [dblquad], page 533, [triplequad], page 534.

q = **quadl** (*f*, *a*, *b*) [Function File]
q = **quadl** (*f*, *a*, *b*, *tol*) [Function File]
q = **quadl** (*f*, *a*, *b*, *tol*, *trace*) [Function File]
q = **quadl** (*f*, *a*, *b*, *tol*, *trace*, *p1*, *p2*, ...) [Function File]

Numerically evaluate the integral of *f* from *a* to *b* using an adaptive Lobatto rule. *f* is a function handle, inline function, or string containing the name of the function to evaluate. The function *f* must be vectorized and return a vector of output values if given a vector of input values.

a and *b* are the lower and upper limits of integration. Both limits must be finite.

The optional argument *tol* defines the relative tolerance with which to perform the integration. The default value is `eps`.

The algorithm used by `quadl` involves recursively subdividing the integration interval. If *trace* is defined then for each subinterval display: (1) the left end of the subinterval, (2) the length of the subinterval, (3) the approximation of the integral over the subinterval.

Additional arguments *p1*, etc., are passed directly to the function *f*. To use default values for *tol* and *trace*, one may pass empty matrices ([]).

Reference: W. Gander and W. Gautschi, *Adaptive Quadrature - Revisited*, BIT Vol. 40, No. 1, March 2000, pp. 84–101. `http://www.inf.ethz.ch/personal/gander/`

See also: [quad], page 526, [quadv], page 527, [quadgk], page 528, [quadcc], page 530, [trapz], page 531, [dblquad], page 533, [triplequad], page 534.

q = **quadgk** (*f*, *a*, *b*) [Function File]
q = **quadgk** (*f*, *a*, *b*, *abstol*) [Function File]
q = **quadgk** (*f*, *a*, *b*, *abstol*, *trace*) [Function File]
q = **quadgk** (*f*, *a*, *b*, *prop*, *val*, ...) [Function File]
[*q*, *err*] = **quadgk** (...) [Function File]

Numerically evaluate the integral of *f* from *a* to *b* using adaptive Gauss-Konrod quadrature. *f* is a function handle, inline function, or string containing the name of the function to evaluate. The formulation is based on a proposal by L.F. Shampine, "*Vectorized adaptive quadrature in* MATLAB", *Journal of Computational and Applied*

Mathematics, pp131-140, Vol 211, Issue 2, Feb 2008 where all function evaluations at an iteration are calculated with a single call to *f*. Therefore, the function *f* must be vectorized and must accept a vector of input values x and return an output vector representing the function evaluations at the given values of x.

a and *b* are the lower and upper limits of integration. Either or both limits may be infinite or contain weak end singularities. Variable transformation will be used to treat any infinite intervals and weaken the singularities. For example:

```
quadgk (@(x) 1 ./ (sqrt (x) .* (x + 1)), 0, Inf)
```

Note that the formulation of the integrand uses the element-by-element operator `./` and all user functions to `quadgk` should do the same.

The optional argument *tol* defines the absolute tolerance used to stop the integration procedure. The default value is $1e^{-10}$.

The algorithm used by `quadgk` involves subdividing the integration interval and evaluating each subinterval. If *trace* is true then after computing each of these partial integrals display: (1) the number of subintervals at this step, (2) the current estimate of the error *err*, (3) the current estimate for the integral *q*.

Alternatively, properties of `quadgk` can be passed to the function as pairs `"prop", val`. Valid properties are

`AbsTol` Define the absolute error tolerance for the quadrature. The default absolute tolerance is 1e-10.

`RelTol` Define the relative error tolerance for the quadrature. The default relative tolerance is 1e-5.

`MaxIntervalCount`
: `quadgk` initially subdivides the interval on which to perform the quadrature into 10 intervals. Subintervals that have an unacceptable error are subdivided and re-evaluated. If the number of subintervals exceeds 650 subintervals at any point then a poor convergence is signaled and the current estimate of the integral is returned. The property `"MaxIntervalCount"` can be used to alter the number of subintervals that can exist before exiting.

`WayPoints`
: Discontinuities in the first derivative of the function to integrate can be flagged with the `"WayPoints"` property. This forces the ends of a subinterval to fall on the breakpoints of the function and can result in significantly improved estimation of the error in the integral, faster computation, or both. For example,

```
quadgk (@(x) abs (1 - x.^2), 0, 2, "Waypoints", 1)
```

signals the breakpoint in the integrand at `x = 1`.

`Trace` If logically true `quadgk` prints information on the convergence of the quadrature at each iteration.

If any of *a*, *b*, or *waypoints* is complex then the quadrature is treated as a contour integral along a piecewise continuous path defined by the above. In this case the integral is assumed to have no edge singularities. For example,

```
quadgk (@(z) log (z), 1+1i, 1+1i, "WayPoints",
        [1-1i, -1,-1i, -1+1i])
```

integrates `log (z)` along the square defined by `[1+1i, 1-1i, -1-1i, -1+1i]`

The result of the integration is returned in *q*. *err* is an approximate bound on the error in the integral `abs (q - I)`, where *I* is the exact value of the integral.

See also: [quad], page 526, [quadv], page 527, [quadl], page 528, [quadcc], page 530, [trapz], page 531, [dblquad], page 533, [triplequad], page 534.

`q = quadcc (f, a, b)` [Function File]
`q = quadcc (f, a, b, tol)` [Function File]
`q = quadcc (f, a, b, tol, sing)` [Function File]
`[q, err, nr_points] = quadcc (...)` [Function File]

Numerically evaluate the integral of *f* from *a* to *b* using the doubly-adaptive Clenshaw-Curtis quadrature described by P. Gonnet in *Increasing the Reliability of Adaptive Quadrature Using Explicit Interpolants*. *f* is a function handle, inline function, or string containing the name of the function to evaluate. The function *f* must be vectorized and must return a vector of output values if given a vector of input values. For example,

```
f = @(x) x .* sin (1./x) .* sqrt (abs (1 - x));
```

which uses the element-by-element "dot" form for all operators.

a and *b* are the lower and upper limits of integration. Either or both limits may be infinite. `quadcc` handles an inifinite limit by substituting the variable of integration with `x = tan (pi/2*u)`.

The optional argument *tol* defines the relative tolerance used to stop the integration procedure. The default value is $1e^{-6}$.

The optional argument *sing* contains a list of points where the integrand has known singularities, or discontinuities in any of its derivatives, inside the integration interval. For the example above, which has a discontinuity at x=1, the call to `quadcc` would be as follows

```
int = quadcc (f, a, b, 1.0e-6, [ 1 ]);
```

The result of the integration is returned in *q*. *err* is an estimate of the absolute integration error and *nr_points* is the number of points at which the integrand was evaluated. If the adaptive integration did not converge, the value of *err* will be larger than the requested tolerance. Therefore, it is recommended to verify this value for difficult integrands.

`quadcc` is capable of dealing with non-numeric values of the integrand such as `NaN` or `Inf`. If the integral diverges, and `quadcc` detects this, then a warning is issued and `Inf` or `-Inf` is returned.

Note: `quadcc` is a general purpose quadrature algorithm and, as such, may be less efficient for a smooth or otherwise well-behaved integrand than other methods such as `quadgk`.

The algorithm uses Clenshaw-Curtis quadrature rules of increasing degree in each interval and bisects the interval if either the function does not appear to be smooth or a rule of maximum degree has been reached. The error estimate is computed from the L2-norm of the difference between two successive interpolations of the integrand over the nodes of the respective quadrature rules.

Reference: P. Gonnet, *Increasing the Reliability of Adaptive Quadrature Using Explicit Interpolants*, ACM Transactions on Mathematical Software, Vol. 37, Issue 3, Article No. 3, 2010.

See also: [quad], page 526, [quadv], page 527, [quadl], page 528, [quadgk], page 528, [trapz], page 531, [dblquad], page 533, [triplequad], page 534.

Sometimes one does not have the function, but only the raw (x, y) points from which to perform an integration. This can occur when collecting data in an experiment. The `trapz` function can integrate these values as shown in the following example where "data" has been collected on the cosine function over the range [0, pi/2).

```
x = 0:0.1:pi/2;  # Uniformly spaced points
y = cos (x);
trapz (x, y)
     ⇒ 0.99666
```

The answer is reasonably close to the exact value of 1. Ordinary quadrature is sensitive to the characteristics of the integrand. Empirical integration depends not just on the integrand, but also on the particular points chosen to represent the function. Repeating the example above with the sine function over the range [0, pi/2) produces far inferior results.

```
x = 0:0.1:pi/2;  # Uniformly spaced points
y = sin (x);
trapz (x, y)
     ⇒ 0.92849
```

However, a slightly different choice of data points can change the result significantly. The same integration, with the same number of points, but spaced differently produces a more accurate answer.

```
x = linspace (0, pi/2, 16);  # Uniformly spaced, but including endpoint
y = sin (x);
trapz (x, y)
     ⇒ 0.99909
```

In general there may be no way of knowing the best distribution of points ahead of time. Or the points may come from an experiment where there is no freedom to select the best distribution. In any case, one must remain aware of this issue when using `trapz`.

q = **trapz** (*y*) [Function File]
q = **trapz** (*x*, *y*) [Function File]
q = **trapz** (..., *dim*) [Function File]

Numerically evaluate the integral of points *y* using the trapezoidal method. `trapz (y)` computes the integral of *y* along the first non-singleton dimension. When the argument *x* is omitted an equally spaced *x* vector with unit spacing (1) is assumed. `trapz (x, y)` evaluates the integral with respect to the spacing in *x* and the values in *y*. This is useful if the points in *y* have been sampled unevenly. If the optional *dim* argument is given, operate along this dimension.

If *x* is not specified then unit spacing will be used. To scale the integral to the correct value you must multiply by the actual spacing value (deltaX). As an example, the integral of x^3 over the range [0, 1] is $x^4/4$ or 0.25. The following code uses `trapz` to calculate the integral in three different ways.

```
x = 0:0.1:1;
y = x.^3;
q = trapz (y)
  ⇒ q = 2.525    # No scaling
q * 0.1
  ⇒ q = 0.2525   # Approximation to integral by scaling
trapz (x, y)
  ⇒ q = 0.2525   # Same result by specifying x
```

See also: [cumtrapz], page 532.

q = `cumtrapz` (*y*) [Function File]
q = `cumtrapz` (*x*, *y*) [Function File]
q = `cumtrapz` (..., *dim*) [Function File]

Cumulative numerical integration of points *y* using the trapezoidal method. `cumtrapz (y)` computes the cumulative integral of *y* along the first non-singleton dimension. Where `trapz` reports only the overall integral sum, `cumtrapz` reports the current partial sum value at each point of *y*. When the argument *x* is omitted an equally spaced *x* vector with unit spacing (1) is assumed. `cumtrapz (x, y)` evaluates the integral with respect to the spacing in *x* and the values in *y*. This is useful if the points in *y* have been sampled unevenly. If the optional *dim* argument is given, operate along this dimension.

If *x* is not specified then unit spacing will be used. To scale the integral to the correct value you must multiply by the actual spacing value (deltaX).

See also: [trapz], page 531, [cumsum], page 414.

23.2 Orthogonal Collocation

[*r*, *amat*, *bmat*, *q*] = `colloc` (*n*, "*left*", "*right*") [Built-in Function]

Compute derivative and integral weight matrices for orthogonal collocation using the subroutines given in J. Villadsen and M. L. Michelsen, *Solution of Differential Equation Models by Polynomial Approximation*.

Here is an example of using `colloc` to generate weight matrices for solving the second order differential equation $u' - \alpha u'' = 0$ with the boundary conditions $u(0) = 0$ and $u(1) = 1$.

First, we can generate the weight matrices for *n* points (including the endpoints of the interval), and incorporate the boundary conditions in the right hand side (for a specific value of α).

```
n = 7;
alpha = 0.1;
[r, a, b] = colloc (n-2, "left", "right");
at = a(2:n-1,2:n-1);
bt = b(2:n-1,2:n-1);
rhs = alpha * b(2:n-1,n) - a(2:n-1,n);
```

Then the solution at the roots *r* is

```
u = [ 0; (at - alpha * bt) \ rhs; 1]
     ⇒ [ 0.00; 0.004; 0.01 0.00; 0.12; 0.62; 1.00 ]
```

23.3 Functions of Multiple Variables

Octave does not have built-in functions for computing the integral of functions of multiple variables directly. It is possible, however, to compute the integral of a function of multiple variables using the existing functions for one-dimensional integrals.

To illustrate how the integration can be performed, we will integrate the function

$$f(x, y) = \sin(\pi xy)\sqrt{xy}$$

for x and y between 0 and 1.

The first approach creates a function that integrates f with respect to x, and then integrates that function with respect to y. Because `quad` is written in Fortran it cannot be called recursively. This means that `quad` cannot integrate a function that calls `quad`, and hence cannot be used to perform the double integration. Any of the other integrators, however, can be used which is what the following code demonstrates.

```
function q = g(y)
  q = ones (size (y));
  for i = 1:length (y)
    f = @(x) sin (pi*x.*y(i)) .* sqrt (x.*y(i));
    q(i) = quadgk (f, 0, 1);
  endfor
endfunction

I = quadgk ("g", 0, 1)
     ⇒ 0.30022
```

The above process can be simplified with the `dblquad` and `triplequad` functions for integrals over two and three variables. For example:

```
I = dblquad (@(x, y) sin (pi*x.*y) .* sqrt (x.*y), 0, 1, 0, 1)
     ⇒ 0.30022
```

`dblquad` (*f*, *xa*, *xb*, *ya*, *yb*) [Function File]
`dblquad` (*f*, *xa*, *xb*, *ya*, *yb*, *tol*) [Function File]
`dblquad` (*f*, *xa*, *xb*, *ya*, *yb*, *tol*, *quadf*) [Function File]
`dblquad` (*f*, *xa*, *xb*, *ya*, *yb*, *tol*, *quadf*, ...) [Function File]

Numerically evaluate the double integral of *f*. *f* is a function handle, inline function, or string containing the name of the function to evaluate. The function *f* must have the form $z = f(x, y)$ where x is a vector and y is a scalar. It should return a vector of the same length and orientation as x.

xa, *ya* and *xb*, *yb* are the lower and upper limits of integration for x and y respectively. The underlying integrator determines whether infinite bounds are accepted.

The optional argument *tol* defines the absolute tolerance used to integrate each sub-integral. The default value is $1e^{-6}$.

The optional argument *quadf* specifies which underlying integrator function to use. Any choice but `quad` is available and the default is `quadcc`.

Additional arguments, are passed directly to *f*. To use the default value for *tol* or *quadf* one may pass ':' or an empty matrix ([]).

See also: [triplequad], page 534, [quad], page 526, [quadv], page 527, [quadl], page 528, [quadgk], page 528, [quadcc], page 530, [trapz], page 531.

`triplequad (f, xa, xb, ya, yb, za, zb)` [Function File]
`triplequad (f, xa, xb, ya, yb, za, zb, tol)` [Function File]
`triplequad (f, xa, xb, ya, yb, za, zb, tol, quadf)` [Function File]
`triplequad (f, xa, xb, ya, yb, za, zb, tol, quadf, ...)` [Function File]

Numerically evaluate the triple integral of *f*. *f* is a function handle, inline function, or string containing the name of the function to evaluate. The function *f* must have the form $w = f(x, y, z)$ where either x or y is a vector and the remaining inputs are scalars. It should return a vector of the same length and orientation as x or y.

xa, *ya*, *za* and *xb*, *yb*, *zb* are the lower and upper limits of integration for x, y, and z respectively. The underlying integrator determines whether infinite bounds are accepted.

The optional argument *tol* defines the absolute tolerance used to integrate each sub-integral. The default value is $1e^{-6}$.

The optional argument *quadf* specifies which underlying integrator function to use. Any choice but `quad` is available and the default is `quadcc`.

Additional arguments, are passed directly to *f*. To use the default value for *tol* or *quadf* one may pass ':' or an empty matrix ([]).

See also: [dblquad], page 533, [quad], page 526, [quadv], page 527, [quadl], page 528, [quadgk], page 528, [quadcc], page 530, [trapz], page 531.

The above mentioned approach works, but is fairly slow, and that problem increases exponentially with the dimensionality of the integral. Another possible solution is to use Orthogonal Collocation as described in the previous section (see Section 23.2 [Orthogonal Collocation], page 532). The integral of a function $f(x, y)$ for x and y between 0 and 1 can be approximated using n points by

$$\int_0^1 \int_0^1 f(x,y)dxdy \approx \sum_{i=1}^{n}\sum_{j=1}^{n} q_i q_j f(r_i, r_j),$$

where q and r is as returned by `colloc (n)`. The generalization to more than two variables is straight forward. The following code computes the studied integral using $n = 8$ points.

```
f = @(x,y) sin (pi*x*y') .* sqrt (x*y');
n = 8;
[t, ~, ~, q] = colloc (n);
I = q'*f(t,t)*q;
     ⇒ 0.30022
```

It should be noted that the number of points determines the quality of the approximation. If the integration needs to be performed between a and b, instead of 0 and 1, then a change of variables is needed.

24 Differential Equations

Octave has built-in functions for solving ordinary differential equations, and differential-algebraic equations. All solvers are based on reliable ODE routines written in Fortran.

24.1 Ordinary Differential Equations

The function `lsode` can be used to solve ODEs of the form

$$\frac{dx}{dt} = f(x,t)$$

using Hindmarsh's ODE solver LSODE.

[*x*, *istate*, *msg*] = lsode (*fcn*, *x_0*, *t*) [Built-in Function]
[*x*, *istate*, *msg*] = lsode (*fcn*, *x_0*, *t*, *t_crit*) [Built-in Function]

Solve the set of differential equations

$$\frac{dx}{dt} = f(x,t)$$

with

$$x(t_0) = x_0$$

The solution is returned in the matrix *x*, with each row corresponding to an element of the vector *t*. The first element of *t* should be t_0 and should correspond to the initial state of the system *x_0*, so that the first row of the output is *x_0*.

The first argument, *fcn*, is a string, inline, or function handle that names the function f to call to compute the vector of right hand sides for the set of equations. The function must have the form

```
xdot = f (x, t)
```

in which *xdot* and *x* are vectors and *t* is a scalar.

If *fcn* is a two-element string array or a two-element cell array of strings, inline functions, or function handles, the first element names the function f described above, and the second element names a function to compute the Jacobian of f. The Jacobian function must have the form

```
jac = j (x, t)
```

in which *jac* is the matrix of partial derivatives

$$J = \frac{\partial f_i}{\partial x_j} = \begin{bmatrix} \frac{\partial f_1}{\partial x_1} & \frac{\partial f_1}{\partial x_2} & \cdots & \frac{\partial f_1}{\partial x_N} \\ \frac{\partial f_2}{\partial x_1} & \frac{\partial f_2}{\partial x_2} & \cdots & \frac{\partial f_2}{\partial x_N} \\ \vdots & \vdots & \ddots & \vdots \\ \frac{\partial f_3}{\partial x_1} & \frac{\partial f_3}{\partial x_2} & \cdots & \frac{\partial f_3}{\partial x_N} \end{bmatrix}$$

The second and third arguments specify the initial state of the system, x_0, and the initial value of the independent variable t_0.

The fourth argument is optional, and may be used to specify a set of times that the ODE solver should not integrate past. It is useful for avoiding difficulties with singularities and points where there is a discontinuity in the derivative.

After a successful computation, the value of *istate* will be 2 (consistent with the Fortran version of LSODE).

If the computation is not successful, *istate* will be something other than 2 and *msg* will contain additional information.

You can use the function `lsode_options` to set optional parameters for `lsode`.

See also: [daspk], page 538, [dassl], page 541, [dasrt], page 543.

`lsode_options ()` [Built-in Function]
`val = lsode_options (`*opt*`)` [Built-in Function]
`lsode_options (`*opt*`, `*val*`)` [Built-in Function]

Query or set options for the function `lsode`. When called with no arguments, the names of all available options and their current values are displayed. Given one argument, return the value of the corresponding option. When called with two arguments, `lsode_options` set the option *opt* to value *val*.

Options include

`"absolute tolerance"`
: Absolute tolerance. May be either vector or scalar. If a vector, it must match the dimension of the state vector.

`"relative tolerance"`
: Relative tolerance parameter. Unlike the absolute tolerance, this parameter may only be a scalar.

  The local error test applied at each integration step is

```
abs (local error in x(i)) <= ...
    rtol * abs (y(i)) + atol(i)
```

`"integration method"`
: A string specifying the method of integration to use to solve the ODE system. Valid values are

  `"adams"`
  `"non-stiff"`
  : No Jacobian used (even if it is available).

  `"bdf"`
  `"stiff"`
  : Use stiff backward differentiation formula (BDF) method. If a function to compute the Jacobian is not supplied, `lsode` will compute a finite difference approximation of the Jacobian matrix.

`"initial step size"`
: The step size to be attempted on the first step (default is determined automatically).

`"maximum order"`
: Restrict the maximum order of the solution method. If using the Adams method, this option must be between 1 and 12. Otherwise, it must be between 1 and 5, inclusive.

`"maximum step size"`
: Setting the maximum stepsize will avoid passing over very large regions (default is not specified).

`"minimum step size"`
: The minimum absolute step size allowed (default is 0).

`"step limit"`
: Maximum number of steps allowed (default is 100000).

Here is an example of solving a set of three differential equations using `lsode`. Given the function

```
## oregonator differential equation
function xdot = f (x, t)

  xdot = zeros (3,1);

  xdot(1) = 77.27 * (x(2) - x(1)*x(2) + x(1) \
            - 8.375e-06*x(1)^2);
  xdot(2) = (x(3) - x(1)*x(2) - x(2)) / 77.27;
  xdot(3) = 0.161*(x(1) - x(3));

endfunction
```

and the initial condition `x0 = [ 4; 1.1; 4 ]`, the set of equations can be integrated using the command

```
t = linspace (0, 500, 1000);

y = lsode ("f", x0, t);
```

If you try this, you will see that the value of the result changes dramatically between $t = 0$ and 5, and again around $t = 305$. A more efficient set of output points might be

```
t = [0, logspace(-1, log10(303), 150), \
        logspace(log10(304), log10(500), 150)];
```

See Alan C. Hindmarsh, *ODEPACK, A Systematized Collection of ODE Solvers*, in Scientific Computing, R. S. Stepleman, editor, (1983) for more information about the inner workings of `lsode`.

An m-file for the differential equation used above is included with the Octave distribution in the examples directory under the name '`oregonator.m`'.

24.2 Differential-Algebraic Equations

The function `daspk` can be used to solve DAEs of the form

$$0 = f(\dot{x}, x, t), \qquad x(t = 0) = x_0, \dot{x}(t = 0) = \dot{x}_0$$

where $\dot{x} = \frac{dx}{dt}$ is the derivative of x. The equation is solved using Petzold's DAE solver DASPK.

[x, xdot, istate, msg] = daspk (*fcn*, *x_0*, *xdot_0*, *t*, *t_crit*) [Built-in Function]

Solve the set of differential-algebraic equations

$$0 = f(x, \dot{x}, t)$$

with

$$x(t_0) = x_0, \dot{x}(t_0) = \dot{x}_0$$

The solution is returned in the matrices *x* and *xdot*, with each row in the result matrices corresponding to one of the elements in the vector *t*. The first element of *t* should be t_0 and correspond to the initial state of the system *x_0* and its derivative *xdot_0*, so that the first row of the output *x* is *x_0* and the first row of the output *xdot* is *xdot_0*.

The first argument, *fcn*, is a string, inline, or function handle that names the function f to call to compute the vector of residuals for the set of equations. It must have the form

```
res = f (x, xdot, t)
```

in which *x*, *xdot*, and *res* are vectors, and *t* is a scalar.

If *fcn* is a two-element string array or a two-element cell array of strings, inline functions, or function handles, the first element names the function f described above, and the second element names a function to compute the modified Jacobian

$$J = \frac{\partial f}{\partial x} + c\frac{\partial f}{\partial \dot{x}}$$

The modified Jacobian function must have the form

```
jac = j (x, xdot, t, c)
```

The second and third arguments to `daspk` specify the initial condition of the states and their derivatives, and the fourth argument specifies a vector of output times at which the solution is desired, including the time corresponding to the initial condition. The set of initial states and derivatives are not strictly required to be consistent. If they are not consistent, you must use the `daspk_options` function to provide additional information so that `daspk` can compute a consistent starting point.

The fifth argument is optional, and may be used to specify a set of times that the DAE solver should not integrate past. It is useful for avoiding difficulties with singularities and points where there is a discontinuity in the derivative.

After a successful computation, the value of *istate* will be greater than zero (consistent with the Fortran version of DASPK).

If the computation is not successful, the value of *istate* will be less than zero and *msg* will contain additional information.

You can use the function `daspk_options` to set optional parameters for `daspk`.

See also: [dassl], page 541.

daspk_options () [Built-in Function]
val = daspk_options (*opt*) [Built-in Function]
daspk_options (*opt*, *val*) [Built-in Function]

Query or set options for the function `daspk`. When called with no arguments, the names of all available options and their current values are displayed. Given one argument, return the value of the corresponding option. When called with two arguments, `daspk_options` set the option *opt* to value *val*.

Options include

`"absolute tolerance"`
: Absolute tolerance. May be either vector or scalar. If a vector, it must match the dimension of the state vector, and the relative tolerance must also be a vector of the same length.

`"relative tolerance"`
: Relative tolerance. May be either vector or scalar. If a vector, it must match the dimension of the state vector, and the absolute tolerance must also be a vector of the same length.

  The local error test applied at each integration step is

```
abs (local error in x(i))
     <= rtol(i) * abs (Y(i)) + atol(i)
```

`"compute consistent initial condition"`
: Denoting the differential variables in the state vector by '`Y_d`' and the algebraic variables by '`Y_a`', `ddaspk` can solve one of two initialization problems:

  1. Given Y_d, calculate Y_a and Y'_d
  2. Given Y', calculate Y.

  In either case, initial values for the given components are input, and initial guesses for the unknown components must also be provided as input. Set this option to 1 to solve the first problem, or 2 to solve the second (the default is 0, so you must provide a set of initial conditions that are consistent).

  If this option is set to a nonzero value, you must also set the `"algebraic variables"` option to declare which variables in the problem are algebraic.

`"use initial condition heuristics"`
: Set to a nonzero value to use the initial condition heuristics options described below.

`"initial condition heuristics"`
: A vector of the following parameters that can be used to control the initial condition calculation.

  `MXNIT` Maximum number of Newton iterations (default is 5).

  `MXNJ` Maximum number of Jacobian evaluations (default is 6).

MXNH
: Maximum number of values of the artificial stepsize parameter to be tried if the "compute consistent initial condition" option has been set to 1 (default is 5).

  Note that the maximum total number of Newton iterations allowed is MXNIT*MXNJ*MXNH if the "compute consistent initial condition" option has been set to 1 and MXNIT*MXNJ if it is set to 2.

LSOFF
: Set to a nonzero value to disable the linesearch algorithm (default is 0).

STPTOL
: Minimum scaled step in linesearch algorithm (default is eps^(2/3)).

EPINIT
: Swing factor in the Newton iteration convergence test. The test is applied to the residual vector, premultiplied by the approximate Jacobian. For convergence, the weighted RMS norm of this vector (scaled by the error weights) must be less than EPINIT*EPCON, where EPCON = 0.33 is the analogous test constant used in the time steps. The default is EPINIT = 0.01.

"print initial condition info"
: Set this option to a nonzero value to display detailed information about the initial condition calculation (default is 0).

"exclude algebraic variables from error test"
: Set to a nonzero value to exclude algebraic variables from the error test. You must also set the "algebraic variables" option to declare which variables in the problem are algebraic (default is 0).

"algebraic variables"
: A vector of the same length as the state vector. A nonzero element indicates that the corresponding element of the state vector is an algebraic variable (i.e., its derivative does not appear explicitly in the equation set.

  This option is required by the compute consistent initial condition" and "exclude algebraic variables from error test" options.

"enforce inequality constraints"
: Set to one of the following values to enforce the inequality constraints specified by the "inequality constraint types" option (default is 0).

  1. To have constraint checking only in the initial condition calculation.
  2. To enforce constraint checking during the integration.
  3. To enforce both options 1 and 2.

"inequality constraint types"
: A vector of the same length as the state specifying the type of inequality constraint. Each element of the vector corresponds to an element of the state and should be assigned one of the following codes

  -2 Less than zero.

-1	Less than or equal to zero.
0	Not constrained.
1	Greater than or equal to zero.
2	Greater than zero.

This option only has an effect if the `"enforce inequality constraints"` option is nonzero.

`"initial step size"`
: Differential-algebraic problems may occasionally suffer from severe scaling difficulties on the first step. If you know a great deal about the scaling of your problem, you can help to alleviate this problem by specifying an initial stepsize (default is computed automatically).

`"maximum order"`
: Restrict the maximum order of the solution method. This option must be between 1 and 5, inclusive (default is 5).

`"maximum step size"`
: Setting the maximum stepsize will avoid passing over very large regions (default is not specified).

Octave also includes DASSL, an earlier version of DASPK, and DASRT, which can be used to solve DAEs with constraints (stopping conditions).

[*x*, *xdot*, *istate*, *msg*] = dassl (*fcn*, *x_0*, *xdot_0*, *t*, *t_crit*) [Built-in Function]

Solve the set of differential-algebraic equations

$$0 = f(x, \dot{x}, t)$$

with

$$x(t_0) = x_0, \dot{x}(t_0) = \dot{x}_0$$

The solution is returned in the matrices *x* and *xdot*, with each row in the result matrices corresponding to one of the elements in the vector *t*. The first element of *t* should be t_0 and correspond to the initial state of the system *x_0* and its derivative *xdot_0*, so that the first row of the output *x* is *x_0* and the first row of the output *xdot* is *xdot_0*.

The first argument, *fcn*, is a string, inline, or function handle that names the function *f* to call to compute the vector of residuals for the set of equations. It must have the form

```
res = f (x, xdot, t)
```

in which *x*, *xdot*, and *res* are vectors, and *t* is a scalar.

If *fcn* is a two-element string array or a two-element cell array of strings, inline functions, or function handles, the first element names the function *f* described above, and the second element names a function to compute the modified Jacobian

$$J = \frac{\partial f}{\partial x} + c\frac{\partial f}{\partial \dot{x}}$$

The modified Jacobian function must have the form

```
jac = j (x, xdot, t, c)
```

The second and third arguments to `dassl` specify the initial condition of the states and their derivatives, and the fourth argument specifies a vector of output times at which the solution is desired, including the time corresponding to the initial condition.

The set of initial states and derivatives are not strictly required to be consistent. In practice, however, DASSL is not very good at determining a consistent set for you, so it is best if you ensure that the initial values result in the function evaluating to zero.

The fifth argument is optional, and may be used to specify a set of times that the DAE solver should not integrate past. It is useful for avoiding difficulties with singularities and points where there is a discontinuity in the derivative.

After a successful computation, the value of *istate* will be greater than zero (consistent with the Fortran version of DASSL).

If the computation is not successful, the value of *istate* will be less than zero and *msg* will contain additional information.

You can use the function `dassl_options` to set optional parameters for `dassl`.

See also: [daspk], page 538, [dasrt], page 543, [lsode], page 535.

`dassl_options ()` [Built-in Function]
`val = dassl_options (opt)` [Built-in Function]
`dassl_options (opt, val)` [Built-in Function]

Query or set options for the function `dassl`. When called with no arguments, the names of all available options and their current values are displayed. Given one argument, return the value of the corresponding option. When called with two arguments, `dassl_options` set the option *opt* to value *val*.

Options include

`"absolute tolerance"`
: Absolute tolerance. May be either vector or scalar. If a vector, it must match the dimension of the state vector, and the relative tolerance must also be a vector of the same length.

`"relative tolerance"`
: Relative tolerance. May be either vector or scalar. If a vector, it must match the dimension of the state vector, and the absolute tolerance must also be a vector of the same length.

The local error test applied at each integration step is

```
abs (local error in x(i))
     <= rtol(i) * abs (Y(i)) + atol(i)
```

`"compute consistent initial condition"`
: If nonzero, `dassl` will attempt to compute a consistent set of initial conditions. This is generally not reliable, so it is best to provide a consistent set and leave this option set to zero.

"enforce nonnegativity constraints"
: If you know that the solutions to your equations will always be non-negative, it may help to set this parameter to a nonzero value. However, it is probably best to try leaving this option set to zero first, and only setting it to a nonzero value if that doesn't work very well.

"initial step size"
: Differential-algebraic problems may occasionally suffer from severe scaling difficulties on the first step. If you know a great deal about the scaling of your problem, you can help to alleviate this problem by specifying an initial stepsize.

"maximum order"
: Restrict the maximum order of the solution method. This option must be between 1 and 5, inclusive.

"maximum step size"
: Setting the maximum stepsize will avoid passing over very large regions (default is not specified).

"step limit"
: Maximum number of integration steps to attempt on a single call to the underlying Fortran code.

[*x*, *xdot*, *t_out*, *istat*, *msg*] = dasrt (*fcn*, [], *x_0*, *xdot_0*, *t*) [Built-in Function]
... = dasrt (*fcn*, *g*, *x_0*, *xdot_0*, *t*) [Built-in Function]
... = dasrt (*fcn*, [], *x_0*, *xdot_0*, *t*, *t_crit*) [Built-in Function]
... = dasrt (*fcn*, *g*, *x_0*, *xdot_0*, *t*, *t_crit*) [Built-in Function]

Solve the set of differential-algebraic equations

$$0 = f(x, \dot{x}, t)$$

with

$$x(t_0) = x_0, \dot{x}(t_0) = \dot{x}_0$$

with functional stopping criteria (root solving).

The solution is returned in the matrices *x* and *xdot*, with each row in the result matrices corresponding to one of the elements in the vector *t_out*. The first element of *t* should be t_0 and correspond to the initial state of the system *x_0* and its derivative *xdot_0*, so that the first row of the output *x* is *x_0* and the first row of the output *xdot* is *xdot_0*.

The vector *t* provides an upper limit on the length of the integration. If the stopping condition is met, the vector *t_out* will be shorter than *t*, and the final element of *t_out* will be the point at which the stopping condition was met, and may not correspond to any element of the vector *t*.

The first argument, *fcn*, is a string, inline, or function handle that names the function *f* to call to compute the vector of residuals for the set of equations. It must have the form

```
res = f (x, xdot, t)
```

in which x, *xdot*, and *res* are vectors, and *t* is a scalar.

If *fcn* is a two-element string array or a two-element cell array of strings, inline functions, or function handles, the first element names the function *f* described above, and the second element names a function to compute the modified Jacobian

$$J = \frac{\partial f}{\partial x} + c\frac{\partial f}{\partial \dot{x}}$$

The modified Jacobian function must have the form

```
jac = j (x, xdot, t, c)
```

The optional second argument names a function that defines the constraint functions whose roots are desired during the integration. This function must have the form

```
g_out = g (x, t)
```

and return a vector of the constraint function values. If the value of any of the constraint functions changes sign, DASRT will attempt to stop the integration at the point of the sign change.

If the name of the constraint function is omitted, `dasrt` solves the same problem as `daspk` or `dassl`.

Note that because of numerical errors in the constraint functions due to round-off and integration error, DASRT may return false roots, or return the same root at two or more nearly equal values of *T*. If such false roots are suspected, the user should consider smaller error tolerances or higher precision in the evaluation of the constraint functions.

If a root of some constraint function defines the end of the problem, the input to DASRT should nevertheless allow integration to a point slightly past that root, so that DASRT can locate the root by interpolation.

The third and fourth arguments to `dasrt` specify the initial condition of the states and their derivatives, and the fourth argument specifies a vector of output times at which the solution is desired, including the time corresponding to the initial condition.

The set of initial states and derivatives are not strictly required to be consistent. In practice, however, DASSL is not very good at determining a consistent set for you, so it is best if you ensure that the initial values result in the function evaluating to zero.

The sixth argument is optional, and may be used to specify a set of times that the DAE solver should not integrate past. It is useful for avoiding difficulties with singularities and points where there is a discontinuity in the derivative.

After a successful computation, the value of *istate* will be greater than zero (consistent with the Fortran version of DASSL).

If the computation is not successful, the value of *istate* will be less than zero and *msg* will contain additional information.

You can use the function `dasrt_options` to set optional parameters for `dasrt`.

See also: [dasrt_options], page 545, [daspk], page 538, [dasrt], page 543, [lsode], page 535.

dasrt_options () [Built-in Function]
val = **dasrt_options** (*opt*) [Built-in Function]
dasrt_options (*opt*, *val*) [Built-in Function]

Query or set options for the function **dasrt**. When called with no arguments, the names of all available options and their current values are displayed. Given one argument, return the value of the corresponding option. When called with two arguments, `dasrt_options` set the option *opt* to value *val*.

Options include

`"absolute tolerance"`
: Absolute tolerance. May be either vector or scalar. If a vector, it must match the dimension of the state vector, and the relative tolerance must also be a vector of the same length.

`"relative tolerance"`
: Relative tolerance. May be either vector or scalar. If a vector, it must match the dimension of the state vector, and the absolute tolerance must also be a vector of the same length.

  The local error test applied at each integration step is

```
abs (local error in x(i)) <= ...
    rtol(i) * abs (Y(i)) + atol(i)
```

`"initial step size"`
: Differential-algebraic problems may occasionally suffer from severe scaling difficulties on the first step. If you know a great deal about the scaling of your problem, you can help to alleviate this problem by specifying an initial stepsize.

`"maximum order"`
: Restrict the maximum order of the solution method. This option must be between 1 and 5, inclusive.

`"maximum step size"`
: Setting the maximum stepsize will avoid passing over very large regions.

`"step limit"`
: Maximum number of integration steps to attempt on a single call to the underlying Fortran code.

See K. E. Brenan, et al., *Numerical Solution of Initial-Value Problems in Differential-Algebraic Equations*, North-Holland (1989) for more information about the implementation of DASSL.

25 Optimization

Octave comes with support for solving various kinds of optimization problems. Specifically Octave can solve problems in Linear Programming, Quadratic Programming, Nonlinear Programming, and Linear Least Squares Minimization.

25.1 Linear Programming

Octave can solve Linear Programming problems using the `glpk` function. That is, Octave can solve

$$\min_x c^T x$$

subject to the linear constraints $Ax = b$ where $x \geq 0$.

The `glpk` function also supports variations of this problem.

[*xopt*, *fmin*, *errnum*, *extra*] = glpk (*c*, *A*, *b*, *lb*, *ub*, *ctype*, *vartype*, *sense*, *param*) [Function File]

Solve a linear program using the GNU GLPK library. Given three arguments, `glpk` solves the following standard LP:

$$\min_x C^T x$$

subject to

$$Ax = b \qquad x \geq 0$$

but may also solve problems of the form

$$[\min_x \mid \max_x] C^T x$$

subject to

$$Ax [= \mid \leq \mid \geq] b \qquad LB \leq x \leq UB$$

Input arguments:

c A column array containing the objective function coefficients.

A A matrix containing the constraints coefficients.

b A column array containing the right-hand side value for each constraint in the constraint matrix.

lb An array containing the lower bound on each of the variables. If *lb* is not supplied, the default lower bound for the variables is zero.

ub An array containing the upper bound on each of the variables. If *ub* is not supplied, the default upper bound is assumed to be infinite.

ctype An array of characters containing the sense of each constraint in the constraint matrix. Each element of the array may be one of the following values

"F" A free (unbounded) constraint (the constraint is ignored).

"U" An inequality constraint with an upper bound (`A(i,:)*x <= b(i)`).

"S" An equality constraint (`A(i,:)*x = b(i)`).

"L" An inequality with a lower bound (`A(i,:)*x >= b(i)`).

"D" An inequality constraint with both upper and lower bounds (`A(i,:)*x >= -b(i)` *and* (`A(i,:)*x <= b(i)`).

vartype A column array containing the types of the variables.

"C" A continuous variable.

"I" An integer variable.

sense If *sense* is 1, the problem is a minimization. If *sense* is -1, the problem is a maximization. The default value is 1.

param A structure containing the following parameters used to define the behavior of solver. Missing elements in the structure take on default values, so you only need to set the elements that you wish to change from the default.

Integer parameters:

`msglev (default: 1)`
Level of messages output by solver routines:

0 (GLP_MSG_OFF)
No output.

1 (GLP_MSG_ERR)
Error and warning messages only.

2 (GLP_MSG_ON)
Normal output.

3 (GLP_MSG_ALL)
Full output (includes informational messages).

`scale (default: 16)`
Scaling option. The values can be combined with the bitwise OR operator and may be the following:

1 (GLP_SF_GM)
Geometric mean scaling.

16 (GLP_SF_EQ)
Equilibration scaling.

32 (GLP_SF_2N)
Round scale factors to power of two.

64 (GLP_SF_SKIP)
Skip if problem is well scaled.

Alternatively, a value of 128 (`GLP_SF_AUTO`) may be also specified, in which case the routine chooses the scaling options automatically.

`dual (default: 1)`
Simplex method option:

1 (`GLP_PRIMAL`)
Use two-phase primal simplex.

2 (`GLP_DUALP`)
Use two-phase dual simplex, and if it fails, switch to the primal simplex.

3 (`GLP_DUAL`)
Use two-phase dual simplex.

`price (default: 34)`
Pricing option (for both primal and dual simplex):

17 (`GLP_PT_STD`)
Textbook pricing.

34 (`GLP_PT_PSE`)
Steepest edge pricing.

`itlim (default: intmax)`
Simplex iterations limit. It is decreased by one each time when one simplex iteration has been performed, and reaching zero value signals the solver to stop the search.

`outfrq (default: 200)`
Output frequency, in iterations. This parameter specifies how frequently the solver sends information about the solution to the standard output.

`branch (default: 4)`
Branching technique option (for MIP only):

1 (`GLP_BR_FFV`)
First fractional variable.

2 (`GLP_BR_LFV`)
Last fractional variable.

3 (`GLP_BR_MFV`)
Most fractional variable.

4 (`GLP_BR_DTH`)
Heuristic by Driebeck and Tomlin.

5 (`GLP_BR_PCH`)
Hybrid pseudocost heuristic.

`btrack (default: 4)`
Backtracking technique option (for MIP only):

1 (GLP_BT_DFS)
: Depth first search.

2 (GLP_BT_BFS)
: Breadth first search.

3 (GLP_BT_BLB)
: Best local bound.

4 (GLP_BT_BPH)
: Best projection heuristic.

`presol (default: 1)`
: If this flag is set, the simplex solver uses the built-in LP presolver. Otherwise the LP presolver is not used.

`lpsolver (default: 1)`
: Select which solver to use. If the problem is a MIP problem this flag will be ignored.

  1 Revised simplex method.

  2 Interior point method.

`rtest (default: 34)`
: Ratio test technique:

  17 (GLP_RT_STD)
  : Standard ("textbook").

  34 (GLP_RT_HAR)
  : Harris' two-pass ratio test.

`tmlim (default: intmax)`
: Searching time limit, in milliseconds.

`outdly (default: 0)`
: Output delay, in seconds. This parameter specifies how long the solver should delay sending information about the solution to the standard output.

`save (default: 0)`
: If this parameter is nonzero, save a copy of the problem in CPLEX LP format to the file '`"outpb.lp"`'. There is currently no way to change the name of the output file.

Real parameters:

`tolbnd (default: 1e-7)`
: Relative tolerance used to check if the current basic solution is primal feasible. It is not recommended that you change this parameter unless you have a detailed understanding of its purpose.

`toldj (default: 1e-7)`
: Absolute tolerance used to check if the current basic solution is dual feasible. It is not recommended that you change this

parameter unless you have a detailed understanding of its purpose.

`tolpiv (default: 1e-10)`
: Relative tolerance used to choose eligible pivotal elements of the simplex table. It is not recommended that you change this parameter unless you have a detailed understanding of its purpose.

`objll (default: -DBL_MAX)`
: Lower limit of the objective function. If the objective function reaches this limit and continues decreasing, the solver stops the search. This parameter is used in the dual simplex method only.

`objul (default: +DBL_MAX)`
: Upper limit of the objective function. If the objective function reaches this limit and continues increasing, the solver stops the search. This parameter is used in the dual simplex only.

`tolint (default: 1e-5)`
: Relative tolerance used to check if the current basic solution is integer feasible. It is not recommended that you change this parameter unless you have a detailed understanding of its purpose.

`tolobj (default: 1e-7)`
: Relative tolerance used to check if the value of the objective function is not better than in the best known integer feasible solution. It is not recommended that you change this parameter unless you have a detailed understanding of its purpose.

Output values:

xopt The optimizer (the value of the decision variables at the optimum).

fopt The optimum value of the objective function.

errnum Error code.

0 No error.

1 (`GLP_EBADB`)
: Invalid basis.

2 (`GLP_ESING`)
: Singular matrix.

3 (`GLP_ECOND`)
: Ill-conditioned matrix.

4 (`GLP_EBOUND`)
: Invalid bounds.

5 (GLP_EFAIL)
: Solver failed.

6 (GLP_EOBJLL)
: Objective function lower limit reached.

7 (GLP_EOBJUL)
: Objective function upper limit reached.

8 (GLP_EITLIM)
: Iterations limit exhausted.

9 (GLP_ETMLIM)
: Time limit exhausted.

10 (GLP_ENOPFS)
: No primal feasible solution.

11 (GLP_ENODFS)
: No dual feasible solution.

12 (GLP_EROOT)
: Root LP optimum not provided.

13 (GLP_ESTOP)
: Search terminated by application.

14 (GLP_EMIPGAP)
: Relative MIP gap tolerance reached.

15 (GLP_ENOFEAS)
: No primal/dual feasible solution.

16 (GLP_ENOCVG)
: No convergence.

17 (GLP_EINSTAB)
: Numerical instability.

18 (GLP_EDATA)
: Invalid data.

19 (GLP_ERANGE)
: Result out of range.

extra A data structure containing the following fields:

`lambda` Dual variables.

`redcosts` Reduced Costs.

`time` Time (in seconds) used for solving LP/MIP problem.

`status` Status of the optimization.

1 (GLP_UNDEF)
: Solution status is undefined.

2 (GLP_FEAS)
Solution is feasible.

3 (GLP_INFEAS)
Solution is infeasible.

4 (GLP_NOFEAS)
Problem has no feasible solution.

5 (GLP_OPT)
Solution is optimal.

6 (GLP_UNBND)
Problem has no unbounded solution.

Example:

```
c = [10, 6, 4]';
A = [ 1, 1, 1;
     10, 4, 5;
      2, 2, 6];
b = [100, 600, 300]';
lb = [0, 0, 0]';
ub = [];
ctype = "UUU";
vartype = "CCC";
s = -1;

param.msglev = 1;
param.itlim = 100;

[xmin, fmin, status, extra] = ...
   glpk (c, A, b, lb, ub, ctype, vartype, s, param);
```

25.2 Quadratic Programming

Octave can also solve Quadratic Programming problems, this is

$$\min_x \frac{1}{2} x^T H x + x^T q$$

subject to

$$Ax = b \qquad lb \leq x \leq ub \qquad A_{lb} \leq A_{in} \leq A_{ub}$$

[x, obj, info, lambda] = qp (x0, H) [Function File]
[x, obj, info, lambda] = qp (x0, H, q) [Function File]
[x, obj, info, lambda] = qp (x0, H, q, A, b) [Function File]
[x, obj, info, lambda] = qp (x0, H, q, A, b, lb, ub) [Function File]
[x, obj, info, lambda] = qp (x0, H, q, A, b, lb, ub, A_lb, A_in, A_ub) [Function File]
[x, obj, info, lambda] = qp (..., options) [Function File]

Solve the quadratic program

$$\min_x \frac{1}{2} x^T H x + x^T q$$

subject to

$$Ax = b \qquad lb \leq x \leq ub \qquad A_{lb} \leq A_{in} \leq A_{ub}$$

using a null-space active-set method.

Any bound (*A*, *b*, *lb*, *ub*, *A_lb*, *A_ub*) may be set to the empty matrix (`[]`) if not present. If the initial guess is feasible the algorithm is faster.

options An optional structure containing the following parameter(s) used to define the behavior of the solver. Missing elements in the structure take on default values, so you only need to set the elements that you wish to change from the default.

`MaxIter (default: 200)`
Maximum number of iterations.

info Structure containing run-time information about the algorithm. The following fields are defined:

`solveiter`
The number of iterations required to find the solution.

`info` An integer indicating the status of the solution.

0 The problem is feasible and convex. Global solution found.

1 The problem is not convex. Local solution found.

2 The problem is not convex and unbounded.

3 Maximum number of iterations reached.

6 The problem is infeasible.

`x = pqpnonneg (c, d)` [Function File]
`x = pqpnonneg (c, d, x0)` [Function File]
`[x, minval] = pqpnonneg (...)` [Function File]
`[x, minval, exitflag] = pqpnonneg (...)` [Function File]
`[x, minval, exitflag, output] = pqpnonneg (...)` [Function File]
`[x, minval, exitflag, output, lambda] = pqpnonneg (...)` [Function File]

Minimize `1/2*x'*c*x + d'*x` subject to `x >= 0`. *c* and *d* must be real, and *c* must be symmetric and positive definite. *x0* is an optional initial guess for *x*.

Outputs:

- minval

  The minimum attained model value, 1/2*xmin'*c*xmin + d'*xmin

- exitflag

  An indicator of convergence. 0 indicates that the iteration count was exceeded, and therefore convergence was not reached; >0 indicates that the algorithm converged. (The algorithm is stable and will converge given enough iterations.)

- output

  A structure with two fields:

    - "algorithm": The algorithm used ("nnls")
    - "iterations": The number of iterations taken.
- lambda

  Not implemented.

See also: [optimset], page 558, [lsqnonneg], page 558, [qp], page 553.

25.3 Nonlinear Programming

Octave can also perform general nonlinear minimization using a successive quadratic programming solver.

`[x, obj, info, iter, nf, lambda] = sqp (x0, phi)` [Function File]
`[...] = sqp (x0, phi, g)` [Function File]
`[...] = sqp (x0, phi, g, h)` [Function File]
`[...] = sqp (x0, phi, g, h, lb, ub)` [Function File]
`[...] = sqp (x0, phi, g, h, lb, ub, maxiter)` [Function File]
`[...] = sqp (x0, phi, g, h, lb, ub, maxiter, tol)` [Function File]

Solve the nonlinear program

$$\min_x \phi(x)$$

subject to

$$g(x) = 0 \qquad h(x) \geq 0 \qquad lb \leq x \leq ub$$

using a sequential quadratic programming method.

The first argument is the initial guess for the vector *x0*.

The second argument is a function handle pointing to the objective function *phi*. The objective function must accept one vector argument and return a scalar.

The second argument may also be a 2- or 3-element cell array of function handles. The first element should point to the objective function, the second should point to a function that computes the gradient of the objective function, and the third should point to a function that computes the Hessian of the objective function. If the gradient function is not supplied, the gradient is computed by finite differences. If the Hessian function is not supplied, a BFGS update formula is used to approximate the Hessian.

When supplied, the gradient function `phi{2}` must accept one vector argument and return a vector. When supplied, the Hessian function `phi{3}` must accept one vector argument and return a matrix.

The third and fourth arguments *g* and *h* are function handles pointing to functions that compute the equality constraints and the inequality constraints, respectively. If the problem does not have equality (or inequality) constraints, then use an empty matrix ([]) for *g* (or *h*). When supplied, these equality and inequality constraint functions must accept one vector argument and return a vector.

The third and fourth arguments may also be 2-element cell arrays of function handles. The first element should point to the constraint function and the second should point to a function that computes the gradient of the constraint function:

$$\left(\frac{\partial f(x)}{\partial x_1}, \frac{\partial f(x)}{\partial x_2}, \ldots, \frac{\partial f(x)}{\partial x_N}\right)^T$$

The fifth and sixth arguments, *lb* and *ub*, contain lower and upper bounds on *x*. These must be consistent with the equality and inequality constraints *g* and *h*. If the arguments are vectors then *x*(i) is bound by *lb*(i) and *ub*(i). A bound can also be a scalar in which case all elements of *x* will share the same bound. If only one bound (lb, ub) is specified then the other will default to (*-realmax*, *+realmax*).

The seventh argument *maxiter* specifies the maximum number of iterations. The default value is 100.

The eighth argument *tol* specifies the tolerance for the stopping criteria. The default value is `sqrt (eps)`.

The value returned in *info* may be one of the following:

101 The algorithm terminated normally. Either all constraints meet the requested tolerance, or the stepsize, Δx, is less than `tol * norm (x)`.

102 The BFGS update failed.

103 The maximum number of iterations was reached.

An example of calling `sqp`:

```
function r = g (x)
  r = [ sumsq(x)-10;
        x(2)*x(3)-5*x(4)*x(5);
        x(1)^3+x(2)^3+1 ];
endfunction

function obj = phi (x)
  obj = exp (prod (x)) - 0.5*(x(1)^3+x(2)^3+1)^2;
endfunction

x0 = [-1.8; 1.7; 1.9; -0.8; -0.8];

[x, obj, info, iter, nf, lambda] = sqp (x0, @phi, @g, [])

x =

  -1.71714
   1.59571
   1.82725
  -0.76364
  -0.76364

obj = 0.053950
info = 101
```

```
iter = 8
nf = 10
lambda =

  -0.0401627
   0.0379578
  -0.0052227
```

See also: [qp], page 553.

25.4 Linear Least Squares

Octave also supports linear least squares minimization. That is, Octave can find the parameter b such that the model $y = xb$ fits data (x, y) as well as possible, assuming zero-mean Gaussian noise. If the noise is assumed to be isotropic the problem can be solved using the '\' or '/' operators, or the `ols` function. In the general case where the noise is assumed to be anisotropic the `gls` is needed.

`[beta, sigma, r] = ols (y, x)` [Function File]

Ordinary least squares estimation for the multivariate model $y = xb + e$ with $\bar{e} = 0$, and cov(vec(e)) = kron (s, I) where y is a $t \times p$ matrix, x is a $t \times k$ matrix, b is a $k \times p$ matrix, and e is a $t \times p$ matrix.

Each row of *y* and *x* is an observation and each column a variable.

The return values *beta*, *sigma*, and *r* are defined as follows.

beta The OLS estimator for b. *beta* is calculated directly via $(x^T x)^{-1} x^T y$ if the matrix $x^T x$ is of full rank. Otherwise, `beta = pinv (x) * y` where `pinv (x)` denotes the pseudoinverse of *x*.

sigma The OLS estimator for the matrix *s*,

```
sigma = (y-x*beta)'
  * (y-x*beta)
  / (t-rank(x))
```

r The matrix of OLS residuals, `r = y - x*beta`.

See also: [gls], page 557, [pinv], page 441.

`[beta, v, r] = gls (y, x, o)` [Function File]

Generalized least squares estimation for the multivariate model $y = xb + e$ with $\bar{e} = 0$ and cov(vec(e)) = $(s^2)o$, where y is a $t \times p$ matrix, x is a $t \times k$ matrix, b is a $k \times p$ matrix, e is a $t \times p$ matrix, and o is a $tp \times tp$ matrix.

Each row of *y* and *x* is an observation and each column a variable. The return values *beta*, *v*, and *r* are defined as follows.

beta The GLS estimator for b.

v The GLS estimator for s^2.

r The matrix of GLS residuals, $r = y - x * beta$.

See also: [ols], page 557.

`x = lsqnonneg (c, d)` [Function File]
`x = lsqnonneg (c, d, x0)` [Function File]
`x = lsqnonneg (c, d, x0, options)` [Function File]
`[x, resnorm] = lsqnonneg (...)` [Function File]
`[x, resnorm, residual] = lsqnonneg (...)` [Function File]
`[x, resnorm, residual, exitflag] = lsqnonneg (...)` [Function File]
`[x, resnorm, residual, exitflag, output] = lsqnonneg (...)` [Function File]
`[x, resnorm, residual, exitflag, output, lambda] = lsqnonneg (...)` [Function File]

Minimize `norm (c*x - d)` subject to `x >= 0`. *c* and *d* must be real. *x0* is an optional initial guess for *x*. Currently, `lsqnonneg` recognizes these options: `"MaxIter"`, `"TolX"`. For a description of these options, see [optimset], page 558.

Outputs:

- resnorm

  The squared 2-norm of the residual: norm (*c***x*-*d*)^2
- residual

  The residual: *d*-*c***x*
- exitflag

  An indicator of convergence. 0 indicates that the iteration count was exceeded, and therefore convergence was not reached; >0 indicates that the algorithm converged. (The algorithm is stable and will converge given enough iterations.)
- output

  A structure with two fields:
  - `"algorithm"`: The algorithm used (`"nnls"`)
  - `"iterations"`: The number of iterations taken.
- lambda

  Not implemented.

See also: [optimset], page 558, [pqpnonneg], page 554.

`optimset ()` [Function File]
`optimset (par, val, ...)` [Function File]
`optimset (old, par, val, ...)` [Function File]
`optimset (old, new)` [Function File]

Create options struct for optimization functions.

Valid parameters are:

AutoScaling
ComplexEqn
Display Request verbose display of results from optimizations. Values are:

`"off"` [default]
No display.

`"iter"` Display intermediate results for every loop iteration.

"final" Display the result of the final loop iteration.

"notify" Display the result of the final loop iteration if the function has failed to converge.

FinDiffType

FunValCheck
When enabled, display an error if the objective function returns an invalid value (a complex number, NaN, or Inf). Must be set to "on" or "off" [default]. Note: the functions **fzero** and **fminbnd** correctly handle Inf values and only complex values or NaN will cause an error in this case.

GradObj When set to "on", the function to be minimized must return a second argument which is the gradient, or first derivative, of the function at the point *x*. If set to "off" [default], the gradient is computed via finite differences.

Jacobian When set to "on", the function to be minimized must return a second argument which is the Jacobian, or first derivative, of the function at the point *x*. If set to "off" [default], the Jacobian is computed via finite differences.

MaxFunEvals
Maximum number of function evaluations before optimization stops. Must be a positive integer.

MaxIter Maximum number of algorithm iterations before optimization stops. Must be a positive integer.

OutputFcn
A user-defined function executed once per algorithm iteration.

TolFun Termination criterion for the function output. If the difference in the calculated objective function between one algorithm iteration and the next is less than `TolFun` the optimization stops. Must be a positive scalar.

TolX Termination criterion for the function input. If the difference in *x*, the current search point, between one algorithm iteration and the next is less than `TolX` the optimization stops. Must be a positive scalar.

TypicalX

Updating

`optimget (`*`options`*`, `*`parname`*`)` [Function File]
`optimget (`*`options`*`, `*`parname`*`, `*`default`*`)` [Function File]

Return a specific option from a structure created by `optimset`. If *parname* is not a field of the *options* structure, return *default* if supplied, otherwise return an empty matrix.

26 Statistics

Octave has support for various statistical methods. This includes basic descriptive statistics, probability distributions, statistical tests, random number generation, and much more.

The functions that analyze data all assume that multi-dimensional data is arranged in a matrix where each row is an observation, and each column is a variable. Thus, the matrix defined by

```
a = [ 0.9, 0.7;
      0.1, 0.1;
      0.5, 0.4 ];
```

contains three observations from a two-dimensional distribution. While this is the default data arrangement, most functions support different arrangements.

It should be noted that the statistics functions don't test for data containing NaN, NA, or Inf. These values need to be detected and dealt with explicitly. See [isnan], page 380, [isna], page 42, [isinf], page 380, [isfinite], page 380.

26.1 Descriptive Statistics

One principal goal of descriptive statistics is to represent the essence of a large data set concisely. Octave provides the mean, median, and mode functions which all summarize a data set with just a single number corresponding to the central tendency of the data.

mean (*x*) [Function File]
mean (*x*, *dim*) [Function File]
mean (*x*, *opt*) [Function File]
mean (*x*, *dim*, *opt*) [Function File]

Compute the mean of the elements of the vector *x*.

$$\text{mean}(x) = \bar{x} = \frac{1}{N}\sum_{i=1}^{N} x_i$$

If *x* is a matrix, compute the mean for each column and return them in a row vector.

The optional argument *opt* selects the type of mean to compute. The following options are recognized:

"a" Compute the (ordinary) arithmetic mean. [default]

"g" Compute the geometric mean.

"h" Compute the harmonic mean.

If the optional argument *dim* is given, operate along this dimension.

Both *dim* and *opt* are optional. If both are supplied, either may appear first.

See also: [median], page 561, [mode], page 562.

median (*x*) [Function File]
median (*x*, *dim*) [Function File]

Compute the median value of the elements of the vector *x*. If the elements of *x* are sorted, the median is defined as

$$\text{median}(x) = \begin{cases} x(\lceil N/2 \rceil), & N \text{ odd;} \\ (x(N/2) + x(N/2+1))/2, & N \text{ even.} \end{cases}$$

If *x* is a matrix, compute the median value for each column and return them in a row vector. If the optional *dim* argument is given, operate along this dimension.

See also: [mean], page 561, [mode], page 562.

`mode (x)` [Function File]
`mode (x, dim)` [Function File]
`[m, f, c] = mode (...)` [Function File]

Compute the most frequently occurring value in a dataset (mode). `mode` determines the frequency of values along the first non-singleton dimension and returns the value with the highest frequency. If two, or more, values have the same frequency `mode` returns the smallest.

If the optional argument *dim* is given, operate along this dimension.

The return variable *f* is the number of occurrences of the mode in in the dataset. The cell array *c* contains all of the elements with the maximum frequency.

See also: [mean], page 561, [median], page 561.

Using just one number, such as the mean, to represent an entire data set may not give an accurate picture of the data. One way to characterize the fit is to measure the dispersion of the data. Octave provides several functions for measuring dispersion.

`range (x)` [Function File]
`range (x, dim)` [Function File]

Return the range, i.e., the difference between the maximum and the minimum of the input data. If *x* is a vector, the range is calculated over the elements of *x*. If *x* is a matrix, the range is calculated over each column of *x*.

If the optional argument *dim* is given, operate along this dimension.

The range is a quickly computed measure of the dispersion of a data set, but is less accurate than `iqr` if there are outlying data points.

See also: [iqr], page 562, [std], page 563.

`iqr (x)` [Function File]
`iqr (x, dim)` [Function File]

Return the interquartile range, i.e., the difference between the upper and lower quartile of the input data. If *x* is a matrix, do the above for first non-singleton dimension of *x*.

If the optional argument *dim* is given, operate along this dimension.

As a measure of dispersion, the interquartile range is less affected by outliers than either `range` or `std`.

See also: [range], page 562, [std], page 563.

`meansq (x)` [Function File]
`meansq (x, dim)` [Function File]

Compute the mean square of the elements of the vector *x*.

$$\text{meansq}(x) = \frac{\sum_{i=1}^{N} {x_i}^2}{N}$$

where $\bar{x}$ is the mean value of x. For matrix arguments, return a row vector containing the mean square of each column.

If the optional argument *dim* is given, operate along this dimension.

See also: [var], page 563, [std], page 563, [moment], page 565.

`std (x)` [Function File]
`std (x, opt)` [Function File]
`std (x, opt, dim)` [Function File]

Compute the standard deviation of the elements of the vector x.

$$\text{std}(x) = \sigma = \sqrt{\frac{\sum_{i=1}^{N}(x_i - \bar{x})^2}{N-1}}$$

where $\bar{x}$ is the mean value of x and N is the number of elements.

If x is a matrix, compute the standard deviation for each column and return them in a row vector.

The argument *opt* determines the type of normalization to use. Valid values are

0: normalize with $N - 1$, provides the square root of the best unbiased estimator of the variance [default]

1: normalize with N, this provides the square root of the second moment around the mean

If the optional argument *dim* is given, operate along this dimension.

See also: [var], page 563, [range], page 562, [iqr], page 562, [mean], page 561, [median], page 561.

In addition to knowing the size of a dispersion it is useful to know the shape of the data set. For example, are data points massed to the left or right of the mean? Octave provides several common measures to describe the shape of the data set. Octave can also calculate moments allowing arbitrary shape measures to be developed.

`var (x)` [Function File]
`var (x, opt)` [Function File]
`var (x, opt, dim)` [Function File]

Compute the variance of the elements of the vector x.

$$\text{var}(x) = \sigma^2 = \frac{\sum_{i=1}^{N}(x_i - \bar{x})^2}{N-1}$$

where $\bar{x}$ is the mean value of x. If x is a matrix, compute the variance for each column and return them in a row vector.

The argument *opt* determines the type of normalization to use. Valid values are

0: normalize with $N-1$, provides the best unbiased estimator of the variance [default]

1: normalizes with N, this provides the second moment around the mean

If $N == 1$ the value of *opt* is ignored and normalization by N is used.

If the optional argument *dim* is given, operate along this dimension.

See also: [cov], page 571, [std], page 563, [skewness], page 564, [kurtosis], page 564, [moment], page 565.

`skewness (x)` [Function File]
`skewness (x, flag)` [Function File]
`skewness (x, flag, dim)` [Function File]

Compute the sample skewness of the elements of *x*:

$$\text{skewness}(x) = \frac{\frac{1}{N}\sum_{i=1}^{N}(x_i - \bar{x})^3}{\sigma^3},$$

where N is the length of x, $\bar{x}$ its mean and σ its (uncorrected) standard deviation.

The optional argument *flag* controls which normalization is used. If *flag* is equal to 1 (default value, used when *flag* is omitted or empty), return the sample skewness as defined above. If *flag* is equal to 0, return the adjusted skewness coefficient instead:

$$\text{skewness}(x) = \frac{\sqrt{N(N-1)}}{N-2} \times \frac{\frac{1}{N}\sum_{i=1}^{N}(x_i - \bar{x})^3}{\sigma^3}$$

The adjusted skewness coefficient is obtained by replacing the sample second and third central moments by their bias-corrected versions.

If x is a matrix, or more generally a multi-dimensional array, return the skewness along the first non-singleton dimension. If the optional *dim* argument is given, operate along this dimension.

See also: [var], page 563, [kurtosis], page 564, [moment], page 565.

`kurtosis (x)` [Function File]
`kurtosis (x, flag)` [Function File]
`kurtosis (x, flag, dim)` [Function File]

Compute the sample kurtosis of the elements of *x*:

$$\kappa_1 = \frac{\frac{1}{N}\sum_{i=1}^{N}(x_i - \bar{x})^4}{\sigma^4},$$

where N is the length of x, $\bar{x}$ its mean, and σ its (uncorrected) standard deviation.

The optional argument *flag* controls which normalization is used. If *flag* is equal to 1 (default value, used when *flag* is omitted or empty), return the sample kurtosis as defined above. If *flag* is equal to 0, return the "bias-corrected" kurtosis coefficient instead:

$$\kappa_0 = 3 + \frac{N-1}{(N-2)(N-3)}\left((N+1)\,\kappa_1 - 3(N-1)\right)$$

The bias-corrected kurtosis coefficient is obtained by replacing the sample second and fourth central moments by their unbiased versions. It is an unbiased estimate of the population kurtosis for normal populations.

If x is a matrix, or more generally a multi-dimensional array, return the kurtosis along the first non-singleton dimension. If the optional *dim* argument is given, operate along this dimension.

See also: [var], page 563, [skewness], page 564, [moment], page 565.

`moment` (*x*, *p*) [Function File]
`moment` (*x*, *p*, *type*) [Function File]
`moment` (*x*, *p*, *dim*) [Function File]
`moment` (*x*, *p*, *type*, *dim*) [Function File]
`moment` (*x*, *p*, *dim*, *type*) [Function File]

Compute the *p*-th central moment of the vector *x*.

$$\frac{\sum_{i=1}^{N}(x_i - \bar{x})^p}{N}$$

If *x* is a matrix, return the row vector containing the *p*-th central moment of each column.

The optional string *type* specifies the type of moment to be computed. Valid options are:

`"c"` Central Moment (default).

`"a"`
`"ac"` Absolute Central Moment. The moment about the mean ignoring sign defined as

$$\frac{\sum_{i=1}^{N}|x_i - \bar{x}|^p}{N}$$

`"r"` Raw Moment. The moment about zero defined as

$$\text{moment}(x) = \frac{\sum_{i=1}^{N} {x_i}^p}{N}$$

`"ar"` Absolute Raw Moment. The moment about zero ignoring sign defined as

$$\frac{\sum_{i=1}^{N}|x_i|^p}{N}$$

If the optional argument *dim* is given, operate along this dimension.

If both *type* and *dim* are given they may appear in any order.

See also: [var], page 563, [skewness], page 564, [kurtosis], page 564.

`q = quantile` (*x*) [Function File]
`q = quantile` (*x*, *p*) [Function File]
`q = quantile` (*x*, *p*, *dim*) [Function File]
`q = quantile` (*x*, *p*, *dim*, *method*) [Function File]

For a sample, *x*, calculate the quantiles, *q*, corresponding to the cumulative probability values in *p*. All non-numeric values (NaNs) of *x* are ignored.

If *x* is a matrix, compute the quantiles for each column and return them in a matrix, such that the i-th row of *q* contains the *p*(i)th quantiles of each column of *x*.

If *p* is unspecified, return the quantiles for `[0.00 0.25 0.50 0.75 1.00]`. The optional argument *dim* determines the dimension along which the quantiles are calculated. If *dim* is omitted, and *x* is a vector or matrix, it defaults to 1 (column-wise quantiles). If *x* is an N-D array, *dim* defaults to the first non-singleton dimension.

The methods available to calculate sample quantiles are the nine methods used by R (`http://www.r-project.org/`). The default value is METHOD = 5.

Discontinuous sample quantile methods 1, 2, and 3

1. Method 1: Inverse of empirical distribution function.
2. Method 2: Similar to method 1 but with averaging at discontinuities.
3. Method 3: SAS definition: nearest even order statistic.

Continuous sample quantile methods 4 through 9, where p(k) is the linear interpolation function respecting each methods' representative cdf.

4. Method 4: p(k) = k / n. That is, linear interpolation of the empirical cdf.
5. Method 5: p(k) = (k - 0.5) / n. That is a piecewise linear function where the knots are the values midway through the steps of the empirical cdf.
6. Method 6: p(k) = k / (n + 1).
7. Method 7: p(k) = (k - 1) / (n - 1).
8. Method 8: p(k) = (k - 1/3) / (n + 1/3). The resulting quantile estimates are approximately median-unbiased regardless of the distribution of *x*.
9. Method 9: p(k) = (k - 3/8) / (n + 1/4). The resulting quantile estimates are approximately unbiased for the expected order statistics if *x* is normally distributed.

Hyndman and Fan (1996) recommend method 8. Maxima, S, and R (versions prior to 2.0.0) use 7 as their default. Minitab and SPSS use method 6. MATLAB uses method 5.

References:

- Becker, R. A., Chambers, J. M. and Wilks, A. R. (1988) The New S Language. Wadsworth & Brooks/Cole.
- Hyndman, R. J. and Fan, Y. (1996) Sample quantiles in statistical packages, American Statistician, 50, 361–365.
- R: A Language and Environment for Statistical Computing; `http://cran.r-project.org/doc/manuals/fullrefman.pdf`.

Examples:

```
x = randi (1000, [10, 1]);  # Create empirical data in range 1-1000
q = quantile (x, [0, 1]);   # Return minimum, maximum of distribution
q = quantile (x, [0.25 0.5 0.75]); # Return quartiles of distribution
```

See also: [prctile], page 566.

q = **prctile** (*x*) [Function File]
q = **prctile** (*x*, *p*) [Function File]
q = **prctile** (*x*, *p*, *dim*) [Function File]

For a sample *x*, compute the quantiles, *q*, corresponding to the cumulative probability values, *p*, in percent. All non-numeric values (NaNs) of *x* are ignored.

If *x* is a matrix, compute the percentiles for each column and return them in a matrix, such that the i-th row of *y* contains the *p*(i)th percentiles of each column of *x*.

If *p* is unspecified, return the quantiles for `[0 25 50 75 100]`. The optional argument *dim* determines the dimension along which the percentiles are calculated. If *dim* is omitted, and *x* is a vector or matrix, it defaults to 1 (column-wise quantiles). When *x* is an N-D array, *dim* defaults to the first non-singleton dimension.

See also: [quantile], page 565.

A summary view of a data set can be generated quickly with the `statistics` function.

`statistics (x)` [Function File]
`statistics (x, dim)` [Function File]

Return a vector with the minimum, first quartile, median, third quartile, maximum, mean, standard deviation, skewness, and kurtosis of the elements of the vector *x*.

If *x* is a matrix, calculate statistics over the first non-singleton dimension. If the optional argument *dim* is given, operate along this dimension.

See also: [min], page 416, [max], page 416, [median], page 561, [mean], page 561, [std], page 563, [skewness], page 564, [kurtosis], page 564.

26.2 Basic Statistical Functions

Octave supports various helpful statistical functions. Many are useful as initial steps to prepare a data set for further analysis. Others provide different measures from those of the basic descriptive statistics.

`center (x)` [Function File]
`center (x, dim)` [Function File]

If *x* is a vector, subtract its mean. If *x* is a matrix, do the above for each column. If the optional argument *dim* is given, operate along this dimension.

See also: [zscore], page 567.

`[z, mu, sigma] = zscore (x)` [Function File]
`[z, mu, sigma] = zscore (x, opt)` [Function File]
`[z, mu, sigma] = zscore (x, opt, dim)` [Function File]

If x is a vector, subtract its mean and divide by its standard deviation. If the standard deviation is zero, divide by 1 instead. The optional parameter *opt* determines the normalization to use when computing the standard deviation and is the same as the corresponding parameter for `std`.

If *x* is a matrix, do the above along the first non-singleton dimension. If the third optional argument *dim* is given, operate along this dimension.

The mean and standard deviation along *dim* are given in *mu* and *sigma* respectively.

See also: [mean], page 561, [std], page 563, [center], page 567.

`n = histc (x, edges)` [Function File]
`n = histc (x, edges, dim)` [Function File]
`[n, idx] = histc (...)` [Function File]

Produce histogram counts.

When *x* is a vector, the function counts the number of elements of *x* that fall in the histogram bins defined by *edges*. This must be a vector of monotonically increasing values that define the edges of the histogram bins. `n(k)` contains the number of elements in *x* for which `edges(k) <= x < edges(k+1)`. The final element of *n* contains the number of elements of *x* exactly equal to the last element of *edges*.

When *x* is an *N*-dimensional array, the computation is carried out along dimension *dim*. If not specified *dim* defaults to the first non-singleton dimension.

When a second output argument is requested an index matrix is also returned. The *idx* matrix has the same size as *x*. Each element of *idx* contains the index of the histogram bin in which the corresponding element of *x* was counted.

See also: [hist], page 266.

`c = nchoosek (n, k)` [Function File]
`c = nchoosek (set, k)` [Function File]

Compute the binomial coefficient or all combinations of a set of items.

If *n* is a scalar then calculate the binomial coefficient of *n* and *k* which is defined as

$$\binom{n}{k} = \frac{n(n-1)(n-2)\cdots(n-k+1)}{k!} = \frac{n!}{k!(n-k)!}$$

This is the number of combinations of *n* items taken in groups of size *k*.

If the first argument is a vector, *set*, then generate all combinations of the elements of *set*, taken *k* at a time, with one row per combination. The result *c* has *k* columns and `nchoosek (length (set), k)` rows.

For example:

How many ways can three items be grouped into pairs?

```
nchoosek (3, 2)
   ⇒ 3
```

What are the possible pairs?

```
nchoosek (1:3, 2)
   ⇒  1   2
       1   3
       2   3
```

`nchoosek` works only for non-negative, integer arguments. Use `bincoeff` for non-integer and negative scalar arguments, or for computing many binomial coefficients at once with vector inputs for *n* or *k*.

See also: [bincoeff], page 424, [perms], page 568.

`perms (v)` [Function File]

Generate all permutations of *v*, one row per permutation. The result has size `factorial (n) * n`, where *n* is the length of *v*.

As an example, `perms ([1, 2, 3])` returns the matrix

```
1   2   3
2   1   3
1   3   2
2   3   1
3   1   2
3   2   1
```

`ranks` (x, *dim*) [Function File]

Return the ranks of x along the first non-singleton dimension adjusted for ties. If the optional argument *dim* is given, operate along this dimension.

See also: [spearman], page 571, [kendall], page 572.

`run_count` (x, n) [Function File]
`run_count` (x, n, *dim*) [Function File]

Count the upward runs along the first non-singleton dimension of x of length 1, 2, ..., n-1 and greater than or equal to n.

If the optional argument *dim* is given then operate along this dimension.

`[count, value] = runlength` (x) [Function File]

Find the lengths of all sequences of common values. Return the vector of lengths and the value that was repeated.

```
runlength ([2, 2, 0, 4, 4, 4, 0, 1, 1, 1, 1])
⇒  [2, 1, 3, 1, 4]
```

`probit` (p) [Function File]

For each component of p, return the probit (the quantile of the standard normal distribution) of p.

`logit` (p) [Function File]

For each component of p, return the logit of p defined as

$$\mathrm{logit}(p) = \log\left(\frac{p}{1-p}\right)$$

See also: [logistic_cdf], page 577.

`cloglog` (x) [Function File]

Return the complementary log-log function of x, defined as

$$\mathrm{cloglog}(x) = -\log(-\log(x))$$

`mahalanobis` (x, y) [Function File]

Return the Mahalanobis' D-square distance between the multivariate samples x and y, which must have the same number of components (columns), but may have a different number of observations (rows).

`[t, l_x] = table` (x) [Function File]
`[t, l_x, l_y] = table` (x, y) [Function File]

Create a contingency table t from data vectors. The *l_x* and *l_y* vectors are the corresponding levels.

Currently, only 1- and 2-dimensional tables are supported.

26.3 Statistical Plots

Octave can create Quantile Plots (QQ-Plots), and Probability Plots (PP-Plots). These are simple graphical tests for determining if a data set comes from a certain distribution.

Note that Octave can also show histograms of data using the `hist` function as described in Section 15.2.1 [Two-Dimensional Plots], page 259.

`[q, s] = qqplot (x)` [Function File]
`[q, s] = qqplot (x, y)` [Function File]
`[q, s] = qqplot (x, dist)` [Function File]
`[q, s] = qqplot (x, y, params)` [Function File]
`qqplot (...)` [Function File]

Perform a QQ-plot (quantile plot).

If F is the CDF of the distribution *dist* with parameters *params* and G its inverse, and x a sample vector of length n, the QQ-plot graphs ordinate $s(i)$ = i-th largest element of x versus abscissa $q(if)$ = G((i - 0.5)/n).

If the sample comes from F, except for a transformation of location and scale, the pairs will approximately follow a straight line.

If the second argument is a vector y the empirical CDF of y is used as *dist*.

The default for *dist* is the standard normal distribution. The optional argument *params* contains a list of parameters of *dist*. For example, for a quantile plot of the uniform distribution on [2,4] and x, use

```
qqplot (x, "unif", 2, 4)
```

dist can be any string for which a function *distinv* or *dist_inv* exists that calculates the inverse CDF of distribution *dist*.

If no output arguments are given, the data are plotted directly.

`[p, y] = ppplot (x, dist, params)` [Function File]

Perform a PP-plot (probability plot).

If F is the CDF of the distribution *dist* with parameters *params* and x a sample vector of length n, the PP-plot graphs ordinate $y(i)$ = F (i-th largest element of x) versus abscissa $p(i)$ = (i - 0.5)/n. If the sample comes from F, the pairs will approximately follow a straight line.

The default for *dist* is the standard normal distribution. The optional argument *params* contains a list of parameters of *dist*. For example, for a probability plot of the uniform distribution on [2,4] and x, use

```
ppplot (x, "uniform", 2, 4)
```

dist can be any string for which a function *dist_cdf* that calculates the CDF of distribution *dist* exists.

If no output arguments are given, the data are plotted directly.

26.4 Correlation and Regression Analysis

`cov (x)` [Function File]
`cov (x, opt)` [Function File]
`cov (x, y)` [Function File]
`cov (x, y, opt)` [Function File]

Compute the covariance matrix.

If each row of *x* and *y* is an observation, and each column is a variable, then the (*i*, *j*)-th entry of `cov (x, y)` is the covariance between the *i*-th variable in *x* and the *j*-th variable in *y*.

$$\sigma_{ij} = \frac{1}{N-1} \sum_{i=1}^{N} (x_i - \bar{x})(y_i - \bar{y})$$

where $\bar{x}$ and $\bar{y}$ are the mean values of x and y.

If called with one argument, compute `cov (x, x)`, the covariance between the columns of *x*.

The argument *opt* determines the type of normalization to use. Valid values are

0: normalize with $N-1$, provides the best unbiased estimator of the covariance [default]

1: normalize with N, this provides the second moment around the mean

MATLAB compatibility: Octave always computes the covariance matrix. For two inputs, however, MATLAB will calculate `cov (x(:), y(:))` whenever the number of elements in *x* and *y* are equal. This will result in a scalar rather than a matrix output. Code relying on this odd definition will need to be changed when running in Octave.

See also: [corr], page 571.

`corr (x)` [Function File]
`corr (x, y)` [Function File]

Compute matrix of correlation coefficients.

If each row of *x* and *y* is an observation and each column is a variable, then the (*i*, *j*)-th entry of `corr (x, y)` is the correlation between the *i*-th variable in *x* and the *j*-th variable in *y*.

$$\mathrm{corr}(x, y) = \frac{\mathrm{cov}(x, y)}{\mathrm{std}(x)\mathrm{std}(y)}$$

If called with one argument, compute `corr (x, x)`, the correlation between the columns of *x*.

See also: [cov], page 571.

`spearman (x)` [Function File]
`spearman (x, y)` [Function File]

Compute Spearman's rank correlation coefficient *rho*.

For two data vectors *x* and *y*, Spearman's *rho* is the correlation coefficient of the ranks of *x* and *y*.

If *x* and *y* are drawn from independent distributions, *rho* has zero mean and variance `1 / (n - 1)`, and is asymptotically normally distributed.

`spearman (x)` is equivalent to `spearman (x, x)`.

See also: [ranks], page 569, [kendall], page 572.

`kendall (x)` [Function File]
`kendall (x, y)` [Function File]

Compute Kendall's *tau*.

For two data vectors *x*, *y* of common length *n*, Kendall's *tau* is the correlation of the signs of all rank differences of *x* and *y*; i.e., if both *x* and *y* have distinct entries, then

$$\tau = \frac{1}{n(n-1)} \sum_{i,j} \mathrm{sign}(q_i - q_j)\mathrm{sign}(r_i - r_j)$$

in which the q_i and r_i are the ranks of *x* and *y*, respectively.

If *x* and *y* are drawn from independent distributions, Kendall's *tau* is asymptotically normal with mean 0 and variance $\frac{2(2n+5)}{9n(n-1)}$.

`kendall (x)` is equivalent to `kendall (x, x)`.

See also: [ranks], page 569, [spearman], page 571.

`[theta, beta, dev, dl, d2l, p] = logistic_regression (y, x, print, theta, beta)` [Function File]

Perform ordinal logistic regression.

Suppose *y* takes values in *k* ordered categories, and let `gamma_i (x)` be the cumulative probability that *y* falls in one of the first *i* categories given the covariate *x*. Then

```
[theta, beta] = logistic_regression (y, x)
```

fits the model

```
logit (gamma_i (x)) = theta_i - beta' * x,    i = 1 ... k-1
```

The number of ordinal categories, *k*, is taken to be the number of distinct values of `round (y)`. If *k* equals 2, *y* is binary and the model is ordinary logistic regression. The matrix *x* is assumed to have full column rank.

Given *y* only, `theta = logistic_regression (y)` fits the model with baseline logit odds only.

The full form is

```
[theta, beta, dev, dl, d2l, gamma]
   = logistic_regression (y, x, print, theta, beta)
```

in which all output arguments and all input arguments except *y* are optional.

Setting *print* to 1 requests summary information about the fitted model to be displayed. Setting *print* to 2 requests information about convergence at each iteration. Other values request no information to be displayed. The input arguments *theta* and *beta* give initial estimates for *theta* and *beta*.

The returned value *dev* holds minus twice the log-likelihood.

The returned values *dl* and *d2l* are the vector of first and the matrix of second derivatives of the log-likelihood with respect to *theta* and *beta*.

p holds estimates for the conditional distribution of *y* given *x*.

26.5 Distributions

Octave has functions for computing the Probability Density Function (PDF), the Cumulative Distribution function (CDF), and the quantile (the inverse of the CDF) for a large number of distributions.

The following table summarizes the supported distributions (in alphabetical order).

Distribution	**PDF**	**CDF**	**Quantile**
Beta	betapdf	betacdf	betainv
Binomial	binopdf	binocdf	binoinv
Cauchy	cauchy_pdf	cauchy_cdf	cauchy_inv
Chi-Square	chi2pdf	chi2cdf	chi2inv
Univariate Discrete	discrete_pdf	discrete_cdf	discrete_inv
Empirical	empirical_pdf	empirical_cdf	empirical_inv
Exponential	exppdf	expcdf	expinv
F	fpdf	fcdf	finv
Gamma	gampdf	gamcdf	gaminv
Geometric	geopdf	geocdf	geoinv
Hypergeometric	hygepdf	hygecdf	hygeinv
Kolmogorov Smirnov	*Not Available*	kolmogorov_smirnov_cdf	*Not Available*
Laplace	laplace_pdf	laplace_cdf	laplace_inv
Logistic	logistic_pdf	logistic_cdf	logistic_inv
Log-Normal	lognpdf	logncdf	logninv
Univariate Normal	normpdf	normcdf	norminv
Pascal	nbinpdf	nbincdf	nbininv
Poisson	poisspdf	poisscdf	poissinv
Standard Normal	stdnormal_pdf	stdnormal_cdf	stdnormal_inv
t (Student)	tpdf	tcdf	tinv
Uniform Discrete	unidpdf	unidcdf	unidinv
Uniform	unifpdf	unifcdf	unifinv
Weibull	wblpdf	wblcdf	wblinv

`betapdf` (*x*, *a*, *b*) [Function File]

For each element of *x*, compute the probability density function (PDF) at *x* of the Beta distribution with parameters *a* and *b*.

`betacdf` (*x*, *a*, *b*) [Function File]

For each element of *x*, compute the cumulative distribution function (CDF) at *x* of the Beta distribution with parameters *a* and *b*.

`betainv` (*x*, *a*, *b*) [Function File]

For each element of *x*, compute the quantile (the inverse of the CDF) at *x* of the Beta distribution with parameters *a* and *b*.

`binopdf` (*x*, *n*, *p*) [Function File]

For each element of *x*, compute the probability density function (PDF) at *x* of the binomial distribution with parameters *n* and *p*, where *n* is the number of trials and *p* is the probability of success.

`binocdf` (*x*, *n*, *p*) [Function File]

For each element of *x*, compute the cumulative distribution function (CDF) at *x* of the binomial distribution with parameters *n* and *p*, where *n* is the number of trials and *p* is the probability of success.

`binoinv` (*x*, *n*, *p*) [Function File]

For each element of *x*, compute the quantile (the inverse of the CDF) at *x* of the binomial distribution with parameters *n* and *p*, where *n* is the number of trials and *p* is the probability of success.

`cauchy_pdf` (*x*) [Function File]
`cauchy_pdf` (*x*, *location*, *scale*) [Function File]

For each element of *x*, compute the probability density function (PDF) at *x* of the Cauchy distribution with location parameter *location* and scale parameter *scale* > 0. Default values are *location* = 0, *scale* = 1.

`cauchy_cdf` (*x*) [Function File]
`cauchy_cdf` (*x*, *location*, *scale*) [Function File]

For each element of *x*, compute the cumulative distribution function (CDF) at *x* of the Cauchy distribution with location parameter *location* and scale parameter *scale*. Default values are *location* = 0, *scale* = 1.

`cauchy_inv` (*x*) [Function File]
`cauchy_inv` (*x*, *location*, *scale*) [Function File]

For each element of *x*, compute the quantile (the inverse of the CDF) at *x* of the Cauchy distribution with location parameter *location* and scale parameter *scale*. Default values are *location* = 0, *scale* = 1.

`chi2pdf` (*x*, *n*) [Function File]

For each element of *x*, compute the probability density function (PDF) at *x* of the chi-square distribution with *n* degrees of freedom.

`chi2cdf` (*x*, *n*) [Function File]

For each element of *x*, compute the cumulative distribution function (CDF) at *x* of the chi-square distribution with *n* degrees of freedom.

`chi2inv` (*x*, *n*) [Function File]

For each element of *x*, compute the quantile (the inverse of the CDF) at *x* of the chi-square distribution with *n* degrees of freedom.

`discrete_pdf` (*x*, *v*, *p*) [Function File]

For each element of *x*, compute the probability density function (PDF) at *x* of a univariate discrete distribution which assumes the values in *v* with probabilities *p*.

discrete_cdf (*x*, *v*, *p*) [Function File]
For each element of *x*, compute the cumulative distribution function (CDF) at *x* of a univariate discrete distribution which assumes the values in *v* with probabilities *p*.

discrete_inv (*x*, *v*, *p*) [Function File]
For each element of *x*, compute the quantile (the inverse of the CDF) at *x* of the univariate distribution which assumes the values in *v* with probabilities *p*.

empirical_pdf (*x*, *data*) [Function File]
For each element of *x*, compute the probability density function (PDF) at *x* of the empirical distribution obtained from the univariate sample *data*.

empirical_cdf (*x*, *data*) [Function File]
For each element of *x*, compute the cumulative distribution function (CDF) at *x* of the empirical distribution obtained from the univariate sample *data*.

empirical_inv (*x*, *data*) [Function File]
For each element of *x*, compute the quantile (the inverse of the CDF) at *x* of the empirical distribution obtained from the univariate sample *data*.

exppdf (*x*, *lambda*) [Function File]
For each element of *x*, compute the probability density function (PDF) at *x* of the exponential distribution with mean *lambda*.

expcdf (*x*, *lambda*) [Function File]
For each element of *x*, compute the cumulative distribution function (CDF) at *x* of the exponential distribution with mean *lambda*.

The arguments can be of common size or scalars.

expinv (*x*, *lambda*) [Function File]
For each element of *x*, compute the quantile (the inverse of the CDF) at *x* of the exponential distribution with mean *lambda*.

fpdf (*x*, *m*, *n*) [Function File]
For each element of *x*, compute the probability density function (PDF) at *x* of the F distribution with *m* and *n* degrees of freedom.

fcdf (*x*, *m*, *n*) [Function File]
For each element of *x*, compute the cumulative distribution function (CDF) at *x* of the F distribution with *m* and *n* degrees of freedom.

finv (*x*, *m*, *n*) [Function File]
For each element of *x*, compute the quantile (the inverse of the CDF) at *x* of the F distribution with *m* and *n* degrees of freedom.

gampdf (*x*, *a*, *b*) [Function File]
For each element of *x*, return the probability density function (PDF) at *x* of the Gamma distribution with shape parameter *a* and scale *b*.

gamcdf (*x*, *a*, *b*) [Function File]
For each element of *x*, compute the cumulative distribution function (CDF) at *x* of the Gamma distribution with shape parameter *a* and scale *b*.

`gaminv (x, a, b)` [Function File]

For each element of *x*, compute the quantile (the inverse of the CDF) at *x* of the Gamma distribution with shape parameter *a* and scale *b*.

`geopdf (x, p)` [Function File]

For each element of *x*, compute the probability density function (PDF) at *x* of the geometric distribution with parameter *p*.

The geometric distribution models the number of failures (*x*-1) of a Bernoulli trial with probability *p* before the first success (*x*).

`geocdf (x, p)` [Function File]

For each element of *x*, compute the cumulative distribution function (CDF) at *x* of the geometric distribution with parameter *p*.

The geometric distribution models the number of failures (*x*-1) of a Bernoulli trial with probability *p* before the first success (*x*).

`geoinv (x, p)` [Function File]

For each element of *x*, compute the quantile (the inverse of the CDF) at *x* of the geometric distribution with parameter *p*.

The geometric distribution models the number of failures (*x*-1) of a Bernoulli trial with probability *p* before the first success (*x*).

`hygepdf (x, t, m, n)` [Function File]

Compute the probability density function (PDF) at *x* of the hypergeometric distribution with parameters *t*, *m*, and *n*. This is the probability of obtaining *x* marked items when randomly drawing a sample of size *n* without replacement from a population of total size *t* containing *m* marked items.

The parameters *t*, *m*, and *n* must be positive integers with *m* and *n* not greater than *t*.

`hygecdf (x, t, m, n)` [Function File]

Compute the cumulative distribution function (CDF) at *x* of the hypergeometric distribution with parameters *t*, *m*, and *n*. This is the probability of obtaining not more than *x* marked items when randomly drawing a sample of size *n* without replacement from a population of total size *t* containing *m* marked items.

The parameters *t*, *m*, and *n* must be positive integers with *m* and *n* not greater than *t*.

`hygeinv (x, t, m, n)` [Function File]

For each element of *x*, compute the quantile (the inverse of the CDF) at *x* of the hypergeometric distribution with parameters *t*, *m*, and *n*. This is the probability of obtaining *x* marked items when randomly drawing a sample of size *n* without replacement from a population of total size *t* containing *m* marked items.

The parameters *t*, *m*, and *n* must be positive integers with *m* and *n* not greater than *t*.

kolmogorov_smirnov_cdf (*x*, *tol*) [Function File]

Return the cumulative distribution function (CDF) at *x* of the Kolmogorov-Smirnov distribution,

$$Q(x) = \sum_{k=-\infty}^{\infty} (-1)^k \exp(-2k^2 x^2)$$

for *x* > 0.

The optional parameter *tol* specifies the precision up to which the series should be evaluated; the default is *tol* = `eps`.

laplace_pdf (*x*) [Function File]

For each element of *x*, compute the probability density function (PDF) at *x* of the Laplace distribution.

laplace_cdf (*x*) [Function File]

For each element of *x*, compute the cumulative distribution function (CDF) at *x* of the Laplace distribution.

laplace_inv (*x*) [Function File]

For each element of *x*, compute the quantile (the inverse of the CDF) at *x* of the Laplace distribution.

logistic_pdf (*x*) [Function File]

For each element of *x*, compute the PDF at *x* of the logistic distribution.

logistic_cdf (*x*) [Function File]

For each element of *x*, compute the cumulative distribution function (CDF) at *x* of the logistic distribution.

logistic_inv (*x*) [Function File]

For each element of *x*, compute the quantile (the inverse of the CDF) at *x* of the logistic distribution.

lognpdf (*x*) [Function File]
lognpdf (*x*, *mu*, *sigma*) [Function File]

For each element of *x*, compute the probability density function (PDF) at *x* of the lognormal distribution with parameters *mu* and *sigma*. If a random variable follows this distribution, its logarithm is normally distributed with mean *mu* and standard deviation *sigma*.

Default values are *mu* = 0, *sigma* = 1.

logncdf (*x*) [Function File]
logncdf (*x*, *mu*, *sigma*) [Function File]

For each element of *x*, compute the cumulative distribution function (CDF) at *x* of the lognormal distribution with parameters *mu* and *sigma*. If a random variable follows this distribution, its logarithm is normally distributed with mean *mu* and standard deviation *sigma*.

Default values are *mu* = 0, *sigma* = 1.

`logninv` (*x*) [Function File]
`logninv` (*x*, *mu*, *sigma*) [Function File]

For each element of *x*, compute the quantile (the inverse of the CDF) at *x* of the lognormal distribution with parameters *mu* and *sigma*. If a random variable follows this distribution, its logarithm is normally distributed with mean *mu* and standard deviation *sigma*.

Default values are *mu* = 0, *sigma* = 1.

`nbinpdf` (*x*, *n*, *p*) [Function File]

For each element of *x*, compute the probability density function (PDF) at *x* of the negative binomial distribution with parameters *n* and *p*.

When *n* is integer this is the Pascal distribution. When *n* is extended to real numbers this is the Polya distribution.

The number of failures in a Bernoulli experiment with success probability *p* before the *n*-th success follows this distribution.

`nbincdf` (*x*, *n*, *p*) [Function File]

For each element of *x*, compute the cumulative distribution function (CDF) at *x* of the negative binomial distribution with parameters *n* and *p*.

When *n* is integer this is the Pascal distribution. When *n* is extended to real numbers this is the Polya distribution.

The number of failures in a Bernoulli experiment with success probability *p* before the *n*-th success follows this distribution.

`nbininv` (*x*, *n*, *p*) [Function File]

For each element of *x*, compute the quantile (the inverse of the CDF) at *x* of the negative binomial distribution with parameters *n* and *p*.

When *n* is integer this is the Pascal distribution. When *n* is extended to real numbers this is the Polya distribution.

The number of failures in a Bernoulli experiment with success probability *p* before the *n*-th success follows this distribution.

`normpdf` (*x*) [Function File]
`normpdf` (*x*, *mu*, *sigma*) [Function File]

For each element of *x*, compute the probability density function (PDF) at *x* of the normal distribution with mean *mu* and standard deviation *sigma*.

Default values are *mu* = 0, *sigma* = 1.

`normcdf` (*x*) [Function File]
`normcdf` (*x*, *mu*, *sigma*) [Function File]

For each element of *x*, compute the cumulative distribution function (CDF) at *x* of the normal distribution with mean *mu* and standard deviation *sigma*.

Default values are *mu* = 0, *sigma* = 1.

`norminv` (*x*) [Function File]
`norminv` (*x*, *mu*, *sigma*) [Function File]

For each element of *x*, compute the quantile (the inverse of the CDF) at *x* of the normal distribution with mean *mu* and standard deviation *sigma*.

Default values are *mu* = 0, *sigma* = 1.

`poisspdf (x, lambda)` [Function File]

For each element of *x*, compute the probability density function (PDF) at *x* of the Poisson distribution with parameter *lambda*.

`poisscdf (x, lambda)` [Function File]

For each element of *x*, compute the cumulative distribution function (CDF) at *x* of the Poisson distribution with parameter lambda.

`poissinv (x, lambda)` [Function File]

For each element of *x*, compute the quantile (the inverse of the CDF) at *x* of the Poisson distribution with parameter *lambda*.

`stdnormal_pdf (x)` [Function File]

For each element of *x*, compute the probability density function (PDF) at *x* of the standard normal distribution (mean = 0, standard deviation = 1).

`stdnormal_cdf (x)` [Function File]

For each element of *x*, compute the cumulative distribution function (CDF) at *x* of the standard normal distribution (mean = 0, standard deviation = 1).

`stdnormal_inv (x)` [Function File]

For each element of *x*, compute the quantile (the inverse of the CDF) at *x* of the standard normal distribution (mean = 0, standard deviation = 1).

`tpdf (x, n)` [Function File]

For each element of *x*, compute the probability density function (PDF) at *x* of the *t* (Student) distribution with *n* degrees of freedom.

`tcdf (x, n)` [Function File]

For each element of *x*, compute the cumulative distribution function (CDF) at *x* of the t (Student) distribution with *n* degrees of freedom, i.e., PROB $(t(n) \leq x)$.

`tinv (x, n)` [Function File]

For each element of *x*, compute the quantile (the inverse of the CDF) at *x* of the t (Student) distribution with *n* degrees of freedom. This function is analogous to looking in a table for the t-value of a single-tailed distribution.

`unidpdf (x, n)` [Function File]

For each element of *x*, compute the probability density function (PDF) at *x* of a discrete uniform distribution which assumes the integer values 1–*n* with equal probability.

Warning: The underlying implementation uses the double class and will only be accurate for $n \leq$ `bitmax` ($2^{53} - 1$ on IEEE-754 compatible systems).

`unidcdf (x, n)` [Function File]

For each element of *x*, compute the cumulative distribution function (CDF) at *x* of a discrete uniform distribution which assumes the integer values 1–*n* with equal probability.

`unidinv (x, n)` [Function File]

For each element of x, compute the quantile (the inverse of the CDF) at x of the discrete uniform distribution which assumes the integer values 1–n with equal probability.

`unifpdf (x)` [Function File]
`unifpdf (x, a, b)` [Function File]

For each element of x, compute the probability density function (PDF) at x of the uniform distribution on the interval [a, b].

Default values are a = 0, $b = 1$.

`unifcdf (x)` [Function File]
`unifcdf (x, a, b)` [Function File]

For each element of x, compute the cumulative distribution function (CDF) at x of the uniform distribution on the interval [a, b].

Default values are a = 0, $b = 1$.

`unifinv (x)` [Function File]
`unifinv (x, a, b)` [Function File]

For each element of x, compute the quantile (the inverse of the CDF) at x of the uniform distribution on the interval [a, b].

Default values are a = 0, $b = 1$.

`wblpdf (x)` [Function File]
`wblpdf (x, scale)` [Function File]
`wblpdf (x, scale, shape)` [Function File]

Compute the probability density function (PDF) at x of the Weibull distribution with scale parameter *scale* and shape parameter *shape* which is given by

$$\frac{shape}{scale^{shape}} \cdot x^{shape-1} \cdot e^{-(\frac{x}{scale})^{shape}}$$

for x ≥ 0.

Default values are *scale* = 1, *shape* = 1.

`wblcdf (x)` [Function File]
`wblcdf (x, scale)` [Function File]
`wblcdf (x, scale, shape)` [Function File]

Compute the cumulative distribution function (CDF) at x of the Weibull distribution with scale parameter *scale* and shape parameter *shape*, which is

$$1 - e^{-(\frac{x}{scale})^{shape}}$$

for $x \geq 0$.

`wblinv (x)` [Function File]
`wblinv (x, scale)` [Function File]
`wblinv (x, scale, shape)` [Function File]

Compute the quantile (the inverse of the CDF) at x of the Weibull distribution with scale parameter *scale* and shape parameter *shape*.

Default values are *scale* = 1, *shape* = 1.

26.6 Tests

Octave can perform many different statistical tests. The following table summarizes the available tests.

Hypothesis	**Test Functions**
Equal mean values	anova, hotelling_test2, t_test_2, welch_test, wilcoxon_test, z_test_2
Equal medians	kruskal_wallis_test, sign_test
Equal variances	bartlett_test, manova, var_test
Equal distributions	chisquare_test_homogeneity, kolmogorov_smirnov_test_2, u_test
Equal marginal frequencies	mcnemar_test
Equal success probabilities	prop_test_2
Independent observations	chisquare_test_independence, run_test
Uncorrelated observations	cor_test
Given mean value	hotelling_test, t_test, z_test
Observations from distribution	kolmogorov_smirnov_test
Regression	f_test_regression, t_test_regression

The tests return a p-value that describes the outcome of the test. Assuming that the test hypothesis is true, the p-value is the probability of obtaining a worse result than the observed one. So large p-values corresponds to a successful test. Usually a test hypothesis is accepted if the p-value exceeds 0.05.

`[pval, f, df_b, df_w] = anova (y, g)` [Function File]

Perform a one-way analysis of variance (ANOVA). The goal is to test whether the population means of data taken from *k* different groups are all equal.

Data may be given in a single vector *y* with groups specified by a corresponding vector of group labels *g* (e.g., numbers from 1 to *k*). This is the general form which does not impose any restriction on the number of data in each group or the group labels.

If *y* is a matrix and *g* is omitted, each column of *y* is treated as a group. This form is only appropriate for balanced ANOVA in which the numbers of samples from each group are all equal.

Under the null of constant means, the statistic *f* follows an F distribution with *df_b* and *df_w* degrees of freedom.

The p-value (1 minus the CDF of this distribution at *f*) is returned in *pval*.

If no output argument is given, the standard one-way ANOVA table is printed.

`[pval, chisq, df] = bartlett_test (x1, ...)` [Function File]

Perform a Bartlett test for the homogeneity of variances in the data vectors *x1*, *x2*, ..., *xk*, where $k > 1$.

Under the null of equal variances, the test statistic *chisq* approximately follows a chi-square distribution with *df* degrees of freedom.

The p-value (1 minus the CDF of this distribution at *chisq*) is returned in *pval*.

If no output argument is given, the p-value is displayed.

`[pval, chisq, df] = chisquare_test_homogeneity (x, y, c)` [Function File]

Given two samples *x* and *y*, perform a chisquare test for homogeneity of the null hypothesis that *x* and *y* come from the same distribution, based on the partition induced by the (strictly increasing) entries of *c*.

For large samples, the test statistic *chisq* approximately follows a chisquare distribution with *df* = `length (c)` degrees of freedom.

The p-value (1 minus the CDF of this distribution at *chisq*) is returned in *pval*.

If no output argument is given, the p-value is displayed.

`[pval, chisq, df] = chisquare_test_independence (x)` [Function File]

Perform a chi-square test for independence based on the contingency table *x*. Under the null hypothesis of independence, *chisq* approximately has a chi-square distribution with *df* degrees of freedom.

The p-value (1 minus the CDF of this distribution at chisq) of the test is returned in *pval*.

If no output argument is given, the p-value is displayed.

`cor_test (x, y, alt, method)` [Function File]

Test whether two samples *x* and *y* come from uncorrelated populations.

The optional argument string *alt* describes the alternative hypothesis, and can be `"!="` or `"<>"` (non-zero), `">"` (greater than 0), or `"<"` (less than 0). The default is the two-sided case.

The optional argument string *method* specifies which correlation coefficient to use for testing. If *method* is `"pearson"` (default), the (usual) Pearson's product moment correlation coefficient is used. In this case, the data should come from a bivariate normal distribution. Otherwise, the other two methods offer nonparametric alternatives. If *method* is `"kendall"`, then Kendall's rank correlation tau is used. If *method* is `"spearman"`, then Spearman's rank correlation rho is used. Only the first character is necessary.

The output is a structure with the following elements:

pval The p-value of the test.

stat The value of the test statistic.

dist The distribution of the test statistic.

params The parameters of the null distribution of the test statistic.

alternative
The alternative hypothesis.

method The method used for testing.

If no output argument is given, the p-value is displayed.

`[pval, f, df_num, df_den] = f_test_regression (y, x, rr, r)` [Function File]

Perform an F test for the null hypothesis rr * b = r in a classical normal regression model y = X * b + e.

Under the null, the test statistic *f* follows an F distribution with *df_num* and *df_den* degrees of freedom.

The p-value (1 minus the CDF of this distribution at *f*) is returned in *pval*.

If not given explicitly, $r = 0$.

If no output argument is given, the p-value is displayed.

`[pval, tsq] = hotelling_test (x, m)` [Function File]

For a sample *x* from a multivariate normal distribution with unknown mean and covariance matrix, test the null hypothesis that `mean (x) == m`.

Hotelling's T^2 is returned in *tsq*. Under the null, $(n - p)T^2/(p(n - 1))$ has an F distribution with p and $n - p$ degrees of freedom, where n and p are the numbers of samples and variables, respectively.

The p-value of the test is returned in *pval*.

If no output argument is given, the p-value of the test is displayed.

`[pval, tsq] = hotelling_test_2 (x, y)` [Function File]

For two samples *x* from multivariate normal distributions with the same number of variables (columns), unknown means and unknown equal covariance matrices, test the null hypothesis `mean (x) == mean (y)`.

Hotelling's two-sample T^2 is returned in *tsq*. Under the null,

$$\frac{n_x + n_y - p - 1)T^2}{p(n_x + n_y - 2)}$$

has an F distribution with p and $n_x + n_y - p - 1$ degrees of freedom, where n_x and n_y are the sample sizes and p is the number of variables.

The p-value of the test is returned in *pval*.

If no output argument is given, the p-value of the test is displayed.

`[pval, ks] = kolmogorov_smirnov_test (x, dist, params, alt)` [Function File]

Perform a Kolmogorov-Smirnov test of the null hypothesis that the sample *x* comes from the (continuous) distribution dist. I.e., if F and G are the CDFs corresponding to the sample and dist, respectively, then the null is that F == G.

The optional argument *params* contains a list of parameters of *dist*. For example, to test whether a sample *x* comes from a uniform distribution on [2,4], use

```
kolmogorov_smirnov_test (x, "unif", 2, 4)
```

dist can be any string for which a function *dist_cdf* that calculates the CDF of distribution *dist* exists.

With the optional argument string *alt*, the alternative of interest can be selected. If *alt* is `"!="` or `"<>"`, the null is tested against the two-sided alternative F != G. In this case, the test statistic *ks* follows a two-sided Kolmogorov-Smirnov distribution. If *alt* is ">", the one-sided alternative F > G is considered. Similarly for "<", the one-sided alternative F > G is considered. In this case, the test statistic *ks* has a one-sided Kolmogorov-Smirnov distribution. The default is the two-sided case.

The p-value of the test is returned in *pval*.

If no output argument is given, the p-value is displayed.

`[pval, ks, d] = kolmogorov_smirnov_test_2 (x, y, alt)` [Function File]

Perform a 2-sample Kolmogorov-Smirnov test of the null hypothesis that the samples *x* and *y* come from the same (continuous) distribution. I.e., if F and G are the CDFs corresponding to the *x* and *y* samples, respectively, then the null is that F == G.

With the optional argument string *alt*, the alternative of interest can be selected. If *alt* is "!=" or "<>", the null is tested against the two-sided alternative F != G. In this case, the test statistic *ks* follows a two-sided Kolmogorov-Smirnov distribution. If *alt* is ">", the one-sided alternative F > G is considered. Similarly for "<", the one-sided alternative F < G is considered. In this case, the test statistic *ks* has a one-sided Kolmogorov-Smirnov distribution. The default is the two-sided case.

The p-value of the test is returned in *pval*.

The third returned value, *d*, is the test statistic, the maximum vertical distance between the two cumulative distribution functions.

If no output argument is given, the p-value is displayed.

`[pval, k, df] = kruskal_wallis_test (x1, ...)` [Function File]

Perform a Kruskal-Wallis one-factor analysis of variance.

Suppose a variable is observed for *k* > 1 different groups, and let *x1*, ..., *xk* be the corresponding data vectors.

Under the null hypothesis that the ranks in the pooled sample are not affected by the group memberships, the test statistic *k* is approximately chi-square with *df* = *k* - 1 degrees of freedom.

If the data contains ties (some value appears more than once) *k* is divided by

1 - *sum_ties* / (*n*^3 - *n*)

where *sum_ties* is the sum of *t*^2 - *t* over each group of ties where *t* is the number of ties in the group and *n* is the total number of values in the input data. For more info on this adjustment see William H. Kruskal and W. Allen Wallis, *Use of Ranks in One-Criterion Variance Analysis*, Journal of the American Statistical Association, Vol. 47, No. 260 (Dec 1952).

The p-value (1 minus the CDF of this distribution at *k*) is returned in *pval*.

If no output argument is given, the p-value is displayed.

`manova (x, g)` [Function File]

Perform a one-way multivariate analysis of variance (MANOVA). The goal is to test whether the p-dimensional population means of data taken from *k* different groups are all equal. All data are assumed drawn independently from p-dimensional normal distributions with the same covariance matrix.

The data matrix is given by *x*. As usual, rows are observations and columns are variables. The vector *g* specifies the corresponding group labels (e.g., numbers from 1 to *k*).

The LR test statistic (Wilks' Lambda) and approximate p-values are computed and displayed.

`[pval, chisq, df] = mcnemar_test (x)` [Function File]

For a square contingency table *x* of data cross-classified on the row and column variables, McNemar's test can be used for testing the null hypothesis of symmetry of the classification probabilities.

Under the null, *chisq* is approximately distributed as chisquare with *df* degrees of freedom.

The p-value (1 minus the CDF of this distribution at *chisq*) is returned in *pval*.

If no output argument is given, the p-value of the test is displayed.

`[pval, z] = prop_test_2 (x1, n1, x2, n2, alt)` [Function File]

If *x1* and *n1* are the counts of successes and trials in one sample, and *x2* and *n2* those in a second one, test the null hypothesis that the success probabilities *p1* and *p2* are the same. Under the null, the test statistic *z* approximately follows a standard normal distribution.

With the optional argument string *alt*, the alternative of interest can be selected. If *alt* is `"!="` or `"<>"`, the null is tested against the two-sided alternative *p1* != *p2*. If *alt* is `">"`, the one-sided alternative *p1* > *p2* is used. Similarly for `"<"`, the one-sided alternative *p1* < *p2* is used. The default is the two-sided case.

The p-value of the test is returned in *pval*.

If no output argument is given, the p-value of the test is displayed.

`[pval, chisq] = run_test (x)` [Function File]

Perform a chi-square test with 6 degrees of freedom based on the upward runs in the columns of *x*. Can be used to test whether *x* contains independent data.

The p-value of the test is returned in *pval*.

If no output argument is given, the p-value is displayed.

`[pval, b, n] = sign_test (x, y, alt)` [Function File]

For two matched-pair samples *x* and *y*, perform a sign test of the null hypothesis PROB (*x* > *y*) == PROB (*x* < *y*) == 1/2. Under the null, the test statistic *b* roughly follows a binomial distribution with parameters `n = sum (x != y)` and *p* = 1/2.

With the optional argument `alt`, the alternative of interest can be selected. If *alt* is `"!="` or `"<>"`, the null hypothesis is tested against the two-sided alternative PROB (*x* < *y*) != 1/2. If *alt* is `">"`, the one-sided alternative PROB (*x* > *y*) > 1/2 ("x is stochastically greater than y") is considered. Similarly for `"<"`, the one-sided alternative PROB (*x* > *y*) < 1/2 ("x is stochastically less than y") is considered. The default is the two-sided case.

The p-value of the test is returned in *pval*.

If no output argument is given, the p-value of the test is displayed.

[*pval*, *t*, *df*] = t_test (*x*, *m*, *alt*) [Function File]

For a sample *x* from a normal distribution with unknown mean and variance, perform a t-test of the null hypothesis `mean (x) == m`. Under the null, the test statistic *t* follows a Student distribution with `df = length (x) - 1` degrees of freedom.

With the optional argument string *alt*, the alternative of interest can be selected. If *alt* is `"!="` or `"<>"`, the null is tested against the two-sided alternative `mean (x) != m`. If *alt* is `">"`, the one-sided alternative `mean (x) > m` is considered. Similarly for `"<"`, the one-sided alternative `mean (x) < m` is considered. The default is the two-sided case.

The p-value of the test is returned in *pval*.

If no output argument is given, the p-value of the test is displayed.

[*pval*, *t*, *df*] = t_test_2 (*x*, *y*, *alt*) [Function File]

For two samples x and y from normal distributions with unknown means and unknown equal variances, perform a two-sample t-test of the null hypothesis of equal means. Under the null, the test statistic *t* follows a Student distribution with *df* degrees of freedom.

With the optional argument string *alt*, the alternative of interest can be selected. If *alt* is `"!="` or `"<>"`, the null is tested against the two-sided alternative `mean (x) != mean (y)`. If *alt* is `">"`, the one-sided alternative `mean (x) > mean (y)` is used. Similarly for `"<"`, the one-sided alternative `mean (x) < mean (y)` is used. The default is the two-sided case.

The p-value of the test is returned in *pval*.

If no output argument is given, the p-value of the test is displayed.

[*pval*, *t*, *df*] = t_test_regression (*y*, *x*, *rr*, *r*, *alt*) [Function File]

Perform a t test for the null hypothesis `rr * b = r` in a classical normal regression model `y = x * b + e`. Under the null, the test statistic *t* follows a *t* distribution with *df* degrees of freedom.

If *r* is omitted, a value of 0 is assumed.

With the optional argument string *alt*, the alternative of interest can be selected. If *alt* is `"!="` or `"<>"`, the null is tested against the two-sided alternative `rr * b != r`. If *alt* is `">"`, the one-sided alternative `rr * b > r` is used. Similarly for `"<"`, the one-sided alternative `rr * b < r` is used. The default is the two-sided case.

The p-value of the test is returned in *pval*.

If no output argument is given, the p-value of the test is displayed.

[*pval*, *z*] = u_test (*x*, *y*, *alt*) [Function File]

For two samples *x* and *y*, perform a Mann-Whitney U-test of the null hypothesis PROB ($x > y$) == 1/2 == PROB ($x < y$). Under the null, the test statistic *z* approximately follows a standard normal distribution. Note that this test is equivalent to the Wilcoxon rank-sum test.

With the optional argument string *alt*, the alternative of interest can be selected. If *alt* is `"!="` or `"<>"`, the null is tested against the two-sided alternative PROB ($x > y$) != 1/2. If *alt* is `">"`, the one-sided alternative PROB ($x > y$) > 1/2 is considered.

Similarly for "<", the one-sided alternative PROB $(x > y) < 1/2$ is considered. The default is the two-sided case.

The p-value of the test is returned in *pval*.

If no output argument is given, the p-value of the test is displayed.

`[pval, f, df_num, df_den] = var_test (x, y, alt)` [Function File]

For two samples *x* and *y* from normal distributions with unknown means and unknown variances, perform an F-test of the null hypothesis of equal variances. Under the null, the test statistic *f* follows an F-distribution with *df_num* and *df_den* degrees of freedom.

With the optional argument string *alt*, the alternative of interest can be selected. If *alt* is "!=" or "<>", the null is tested against the two-sided alternative `var (x) != var (y)`. If *alt* is ">", the one-sided alternative `var (x) > var (y)` is used. Similarly for "<", the one-sided alternative `var (x) > var (y)` is used. The default is the two-sided case.

The p-value of the test is returned in *pval*.

If no output argument is given, the p-value of the test is displayed.

`[pval, t, df] = welch_test (x, y, alt)` [Function File]

For two samples *x* and *y* from normal distributions with unknown means and unknown and not necessarily equal variances, perform a Welch test of the null hypothesis of equal means. Under the null, the test statistic *t* approximately follows a Student distribution with *df* degrees of freedom.

With the optional argument string *alt*, the alternative of interest can be selected. If *alt* is "!=" or "<>", the null is tested against the two-sided alternative `mean (x) != m`. If *alt* is ">", the one-sided alternative mean(x) > *m* is considered. Similarly for "<", the one-sided alternative mean(x) < *m* is considered. The default is the two-sided case.

The p-value of the test is returned in *pval*.

If no output argument is given, the p-value of the test is displayed.

`[pval, z] = wilcoxon_test (x, y, alt)` [Function File]

For two matched-pair sample vectors *x* and *y*, perform a Wilcoxon signed-rank test of the null hypothesis PROB $(x > y) == 1/2$. Under the null, the test statistic *z* approximately follows a standard normal distribution when $n > 25$.

Caution: This function assumes a normal distribution for *z* and thus is invalid for $n \leq 25$.

With the optional argument string *alt*, the alternative of interest can be selected. If *alt* is "!=" or "<>", the null is tested against the two-sided alternative PROB $(x > y) != 1/2$. If alt is ">", the one-sided alternative PROB $(x > y) > 1/2$ is considered. Similarly for "<", the one-sided alternative PROB $(x > y) < 1/2$ is considered. The default is the two-sided case.

The p-value of the test is returned in *pval*.

If no output argument is given, the p-value of the test is displayed.

`[pval, z] = z_test (x, m, v, alt)` [Function File]

Perform a Z-test of the null hypothesis `mean (x) == m` for a sample *x* from a normal distribution with unknown mean and known variance *v*. Under the null, the test statistic *z* follows a standard normal distribution.

With the optional argument string *alt*, the alternative of interest can be selected. If *alt* is `"!="` or `"<>"`, the null is tested against the two-sided alternative `mean (x) != m`. If *alt* is `">"`, the one-sided alternative `mean (x) > m` is considered. Similarly for `"<"`, the one-sided alternative `mean (x) < m` is considered. The default is the two-sided case.

The p-value of the test is returned in *pval*.

If no output argument is given, the p-value of the test is displayed along with some information.

`[pval, z] = z_test_2 (x, y, v_x, v_y, alt)` [Function File]

For two samples *x* and *y* from normal distributions with unknown means and known variances *v_x* and *v_y*, perform a Z-test of the hypothesis of equal means. Under the null, the test statistic *z* follows a standard normal distribution.

With the optional argument string *alt*, the alternative of interest can be selected. If *alt* is `"!="` or `"<>"`, the null is tested against the two-sided alternative `mean (x) != mean (y)`. If alt is `">"`, the one-sided alternative `mean (x) > mean (y)` is used. Similarly for `"<"`, the one-sided alternative `mean (x) < mean (y)` is used. The default is the two-sided case.

The p-value of the test is returned in *pval*.

If no output argument is given, the p-value of the test is displayed along with some information.

26.7 Random Number Generation

Octave can generate random numbers from a large number of distributions. The random number generators are based on the random number generators described in Section 16.3 [Special Utility Matrices], page 391.

The following table summarizes the available random number generators (in alphabetical order).

Distribution	Function
Beta Distribution	betarnd
Binomial Distribution	binornd
Cauchy Distribution	cauchy_rnd
Chi-Square Distribution	chi2rnd
Univariate Discrete Distribution	discrete_rnd
Empirical Distribution	empirical_rnd
Exponential Distribution	exprnd
F Distribution	frnd
Gamma Distribution	gamrnd
Geometric Distribution	geornd
Hypergeometric Distribution	hygernd
Laplace Distribution	laplace_rnd
Logistic Distribution	logistic_rnd
Log-Normal Distribution	lognrnd
Pascal Distribution	nbinrnd
Univariate Normal Distribution	normrnd
Poisson Distribution	poissrnd
Standard Normal Distribution	stdnormal_rnd
t (Student) Distribution	trnd
Univariate Discrete Distribution	unidrnd
Uniform Distribution	unifrnd
Weibull Distribution	wblrnd
Wiener Process	wienrnd

betarnd (*a*, *b*) [Function File]
betarnd (*a*, *b*, *r*) [Function File]
betarnd (*a*, *b*, *r*, *c*, ...) [Function File]
betarnd (*a*, *b*, [*sz*]) [Function File]

Return a matrix of random samples from the Beta distribution with parameters *a* and *b*.

When called with a single size argument, return a square matrix with the dimension specified. When called with more than one scalar argument the first two arguments are taken as the number of rows and columns and any further arguments specify additional matrix dimensions. The size may also be specified with a vector of dimensions *sz*.

If no size arguments are given then the result matrix is the common size of *a* and *b*.

binornd (*n*, *p*) [Function File]
binornd (*n*, *p*, *r*) [Function File]
binornd (*n*, *p*, *r*, *c*, ...) [Function File]
binornd (*n*, *p*, [*sz*]) [Function File]

Return a matrix of random samples from the binomial distribution with parameters *n* and *p*, where *n* is the number of trials and *p* is the probability of success.

When called with a single size argument, return a square matrix with the dimension specified. When called with more than one scalar argument the first two arguments are taken as the number of rows and columns and any further arguments specify additional matrix dimensions. The size may also be specified with a vector of dimensions *sz*.

If no size arguments are given then the result matrix is the common size of *n* and *p*.

`cauchy_rnd (location, scale)` [Function File]
`cauchy_rnd (location, scale, r)` [Function File]
`cauchy_rnd (location, scale, r, c, ...)` [Function File]
`cauchy_rnd (location, scale, [sz])` [Function File]

Return a matrix of random samples from the Cauchy distribution with parameters *location* and *scale*.

When called with a single size argument, return a square matrix with the dimension specified. When called with more than one scalar argument the first two arguments are taken as the number of rows and columns and any further arguments specify additional matrix dimensions. The size may also be specified with a vector of dimensions *sz*.

If no size arguments are given then the result matrix is the common size of *location* and *scale*.

`chi2rnd (n)` [Function File]
`chi2rnd (n, r)` [Function File]
`chi2rnd (n, r, c, ...)` [Function File]
`chi2rnd (n, [sz])` [Function File]

Return a matrix of random samples from the chi-square distribution with *n* degrees of freedom.

When called with a single size argument, return a square matrix with the dimension specified. When called with more than one scalar argument the first two arguments are taken as the number of rows and columns and any further arguments specify additional matrix dimensions. The size may also be specified with a vector of dimensions *sz*.

If no size arguments are given then the result matrix is the size of *n*.

`discrete_rnd (v, p)` [Function File]
`discrete_rnd (v, p, r)` [Function File]
`discrete_rnd (v, p, r, c, ...)` [Function File]
`discrete_rnd (v, p, [sz])` [Function File]

Return a matrix of random samples from the univariate distribution which assumes the values in *v* with probabilities *p*.

When called with a single size argument, return a square matrix with the dimension specified. When called with more than one scalar argument the first two arguments are taken as the number of rows and columns and any further arguments specify additional matrix dimensions. The size may also be specified with a vector of dimensions *sz*.

If no size arguments are given then the result matrix is the common size of *v* and *p*.

`empirical_rnd (data)` [Function File]
`empirical_rnd (data, r)` [Function File]
`empirical_rnd (data, r, c, ...)` [Function File]
`empirical_rnd (data, [sz])` [Function File]

Return a matrix of random samples from the empirical distribution obtained from the univariate sample *data*.

When called with a single size argument, return a square matrix with the dimension specified. When called with more than one scalar argument the first two arguments are taken as the number of rows and columns and any further arguments specify additional matrix dimensions. The size may also be specified with a vector of dimensions *sz*.

If no size arguments are given then the result matrix is a random ordering of the sample *data*.

`exprnd (`*lambda*`)` [Function File]
`exprnd (`*lambda*, *r*`)` [Function File]
`exprnd (`*lambda*, *r*, *c*, ...`)` [Function File]
`exprnd (`*lambda*, [*sz*]`)` [Function File]

Return a matrix of random samples from the exponential distribution with mean *lambda*.

When called with a single size argument, return a square matrix with the dimension specified. When called with more than one scalar argument the first two arguments are taken as the number of rows and columns and any further arguments specify additional matrix dimensions. The size may also be specified with a vector of dimensions *sz*.

If no size arguments are given then the result matrix is the size of *lambda*.

`frnd (`*m*, *n*`)` [Function File]
`frnd (`*m*, *n*, *r*`)` [Function File]
`frnd (`*m*, *n*, *r*, *c*, ...`)` [Function File]
`frnd (`*m*, *n*, [*sz*]`)` [Function File]

Return a matrix of random samples from the F distribution with *m* and *n* degrees of freedom.

When called with a single size argument, return a square matrix with the dimension specified. When called with more than one scalar argument the first two arguments are taken as the number of rows and columns and any further arguments specify additional matrix dimensions. The size may also be specified with a vector of dimensions *sz*.

If no size arguments are given then the result matrix is the common size of *m* and *n*.

`gamrnd (`*a*, *b*`)` [Function File]
`gamrnd (`*a*, *b*, *r*`)` [Function File]
`gamrnd (`*a*, *b*, *r*, *c*, ...`)` [Function File]
`gamrnd (`*a*, *b*, [*sz*]`)` [Function File]

Return a matrix of random samples from the Gamma distribution with shape parameter *a* and scale *b*.

When called with a single size argument, return a square matrix with the dimension specified. When called with more than one scalar argument the first two arguments are taken as the number of rows and columns and any further arguments specify additional matrix dimensions. The size may also be specified with a vector of dimensions *sz*.

If no size arguments are given then the result matrix is the common size of *a* and *b*.

`geornd (`*p*`)` [Function File]
`geornd (`*p*, *r*`)` [Function File]
`geornd (`*p*, *r*, *c*, ...`)` [Function File]
`geornd (`*p*, [*sz*]`)` [Function File]

Return a matrix of random samples from the geometric distribution with parameter *p*.

When called with a single size argument, return a square matrix with the dimension specified. When called with more than one scalar argument the first two arguments are taken as the number of rows and columns and any further arguments specify additional matrix dimensions. The size may also be specified with a vector of dimensions *sz*.

If no size arguments are given then the result matrix is the size of *p*.

The geometric distribution models the number of failures (*x*-1) of a Bernoulli trial with probability *p* before the first success (*x*).

hygernd (*t*, *m*, *n*) [Function File]
hygernd (*t*, *m*, *n*, *r*) [Function File]
hygernd (*t*, *m*, *n*, *r*, *c*, ...) [Function File]
hygernd (*t*, *m*, *n*, [*sz*]) [Function File]

Return a matrix of random samples from the hypergeometric distribution with parameters *t*, *m*, and *n*.

The parameters *t*, *m*, and *n* must be positive integers with *m* and *n* not greater than *t*.

When called with a single size argument, return a square matrix with the dimension specified. When called with more than one scalar argument the first two arguments are taken as the number of rows and columns and any further arguments specify additional matrix dimensions. The size may also be specified with a vector of dimensions *sz*.

If no size arguments are given then the result matrix is the common size of *t*, *m*, and *n*.

laplace_rnd (*r*) [Function File]
laplace_rnd (*r*, *c*, ...) [Function File]
laplace_rnd ([*sz*]) [Function File]

Return a matrix of random samples from the Laplace distribution.

When called with a single size argument, return a square matrix with the dimension specified. When called with more than one scalar argument the first two arguments are taken as the number of rows and columns and any further arguments specify additional matrix dimensions. The size may also be specified with a vector of dimensions *sz*.

logistic_rnd (*r*) [Function File]
logistic_rnd (*r*, *c*, ...) [Function File]
logistic_rnd ([*sz*]) [Function File]

Return a matrix of random samples from the logistic distribution.

When called with a single size argument, return a square matrix with the dimension specified. When called with more than one scalar argument the first two arguments are taken as the number of rows and columns and any further arguments specify additional matrix dimensions. The size may also be specified with a vector of dimensions *sz*.

lognrnd (*mu*, *sigma*) [Function File]
lognrnd (*mu*, *sigma*, *r*) [Function File]
lognrnd (*mu*, *sigma*, *r*, *c*, ...) [Function File]
lognrnd (*mu*, *sigma*, [*sz*]) [Function File]

Return a matrix of random samples from the lognormal distribution with parameters *mu* and *sigma*.

When called with a single size argument, return a square matrix with the dimension specified. When called with more than one scalar argument the first two arguments are taken as the number of rows and columns and any further arguments specify additional matrix dimensions. The size may also be specified with a vector of dimensions *sz*.

If no size arguments are given then the result matrix is the common size of *mu* and *sigma*.

`nbinrnd` (*n*, *p*) [Function File]
`nbinrnd` (*n*, *p*, *r*) [Function File]
`nbinrnd` (*n*, *p*, *r*, *c*, ...) [Function File]
`nbinrnd` (*n*, *p*, [*sz*]) [Function File]

Return a matrix of random samples from the negative binomial distribution with parameters *n* and *p*.

When called with a single size argument, return a square matrix with the dimension specified. When called with more than one scalar argument the first two arguments are taken as the number of rows and columns and any further arguments specify additional matrix dimensions. The size may also be specified with a vector of dimensions *sz*.

If no size arguments are given then the result matrix is the common size of *n* and *p*.

`normrnd` (*mu*, *sigma*) [Function File]
`normrnd` (*mu*, *sigma*, *r*) [Function File]
`normrnd` (*mu*, *sigma*, *r*, *c*, ...) [Function File]
`normrnd` (*mu*, *sigma*, [*sz*]) [Function File]

Return a matrix of random samples from the normal distribution with parameters mean *mu* and standard deviation *sigma*.

When called with a single size argument, return a square matrix with the dimension specified. When called with more than one scalar argument the first two arguments are taken as the number of rows and columns and any further arguments specify additional matrix dimensions. The size may also be specified with a vector of dimensions *sz*.

If no size arguments are given then the result matrix is the common size of *mu* and *sigma*.

`poissrnd` (*lambda*) [Function File]
`poissrnd` (*lambda*, *r*) [Function File]
`poissrnd` (*lambda*, *r*, *c*, ...) [Function File]
`poissrnd` (*lambda*, [*sz*]) [Function File]

Return a matrix of random samples from the Poisson distribution with parameter *lambda*.

When called with a single size argument, return a square matrix with the dimension specified. When called with more than one scalar argument the first two arguments are taken as the number of rows and columns and any further arguments specify additional matrix dimensions. The size may also be specified with a vector of dimensions *sz*.

If no size arguments are given then the result matrix is the size of *lambda*.

`stdnormal_rnd` (*r*) [Function File]
`stdnormal_rnd` (*r*, *c*, ...) [Function File]

`stdnormal_rnd ([sz])` [Function File]

Return a matrix of random samples from the standard normal distribution (mean = 0, standard deviation = 1).

When called with a single size argument, return a square matrix with the dimension specified. When called with more than one scalar argument the first two arguments are taken as the number of rows and columns and any further arguments specify additional matrix dimensions. The size may also be specified with a vector of dimensions *sz*.

`trnd (n)` [Function File]
`trnd (n, r)` [Function File]
`trnd (n, r, c, ...)` [Function File]
`trnd (n, [sz])` [Function File]

Return a matrix of random samples from the t (Student) distribution with *n* degrees of freedom.

When called with a single size argument, return a square matrix with the dimension specified. When called with more than one scalar argument the first two arguments are taken as the number of rows and columns and any further arguments specify additional matrix dimensions. The size may also be specified with a vector of dimensions *sz*.

If no size arguments are given then the result matrix is the size of *n*.

`unidrnd (n)` [Function File]
`unidrnd (n, r)` [Function File]
`unidrnd (n, r, c, ...)` [Function File]
`unidrnd (n, [sz])` [Function File]

Return a matrix of random samples from the discrete uniform distribution which assumes the integer values 1–*n* with equal probability. *n* may be a scalar or a multi-dimensional array.

When called with a single size argument, return a square matrix with the dimension specified. When called with more than one scalar argument the first two arguments are taken as the number of rows and columns and any further arguments specify additional matrix dimensions. The size may also be specified with a vector of dimensions *sz*.

If no size arguments are given then the result matrix is the size of *n*.

`unifrnd (a, b)` [Function File]
`unifrnd (a, b, r)` [Function File]
`unifrnd (a, b, r, c, ...)` [Function File]
`unifrnd (a, b, [sz])` [Function File]

Return a matrix of random samples from the uniform distribution on [*a*, *b*].

When called with a single size argument, return a square matrix with the dimension specified. When called with more than one scalar argument the first two arguments are taken as the number of rows and columns and any further arguments specify additional matrix dimensions. The size may also be specified with a vector of dimensions *sz*.

If no size arguments are given then the result matrix is the common size of *a* and *b*.

`wblrnd (scale, shape)` [Function File]
`wblrnd (scale, shape, r)` [Function File]
`wblrnd (scale, shape, r, c, ...)` [Function File]

wblrnd (*scale*, *shape*, [*sz*]) [Function File]

Return a matrix of random samples from the Weibull distribution with parameters *scale* and *shape*.

When called with a single size argument, return a square matrix with the dimension specified. When called with more than one scalar argument the first two arguments are taken as the number of rows and columns and any further arguments specify additional matrix dimensions. The size may also be specified with a vector of dimensions *sz*.

If no size arguments are given then the result matrix is the common size of *scale* and *shape*.

wienrnd (*t*, *d*, *n*) [Function File]

Return a simulated realization of the d-dimensional Wiener Process on the interval $[0, t]$. If d is omitted, $d = 1$ is used. The first column of the return matrix contains time, the remaining columns contain the Wiener process.

The optional parameter n gives the number of summands used for simulating the process over an interval of length 1. If n is omitted, $n = 1000$ is used.

27 Sets

Octave has a limited number of functions for managing sets of data, where a set is defined as a collection of unique elements. In Octave a set is represented as a vector of numbers.

`unique` (*x*) [Function File]
`unique` (*x*, "*rows*") [Function File]
`unique` (..., "*first*") [Function File]
`unique` (..., "*last*") [Function File]
[*y*, *i*, *j*] = `unique` (...) [Function File]

Return the unique elements of *x*, sorted in ascending order. If the input *x* is a vector then the output is also a vector with the same orientation (row or column) as the input. For a matrix input the output is always a column vector. *x* may also be a cell array of strings.

If the optional argument `"rows"` is supplied, return the unique rows of *x*, sorted in ascending order.

If requested, return index vectors *i* and *j* such that `x(i)==y` and `y(j)==x`.

Additionally, if *i* is a requested output then one of `"first"` or `"last"` may be given as an input. If `"last"` is specified, return the highest possible indices in *i*, otherwise, if `"first"` is specified, return the lowest. The default is `"last"`.

See also: [union], page 598, [intersect], page 598, [setdiff], page 598, [setxor], page 599, [ismember], page 597.

27.1 Set Operations

Octave supports the basic set operations. That is, Octave can compute the union, intersection, and difference of two sets. Octave also supports the *Exclusive Or* set operation, and membership determination. The functions for set operations all work in pretty much the same way. As an example, assume that `x` and `y` contains two sets, then

```
union (x, y)
```

computes the union of the two sets.

tf = `ismember` (*A*, *s*) [Function File]
[*tf*, *S_idx*] = `ismember` (*A*, *s*) [Function File]
[*tf*, *S_idx*] = `ismember` (*A*, *s*, "*rows*") [Function File]

Return a logical matrix *tf* with the same shape as *A* which is true (1) if `A(i,j)` is in *s* and false (0) if it is not. If a second output argument is requested, the index into *s* of each of the matching elements is also returned.

```
a = [3, 10, 1];
s = [0:9];
[tf, s_idx] = ismember (a, s)
     ⇒ tf = [1, 0, 1]
     ⇒ s_idx = [4, 0, 2]
```

The inputs, *A* and *s*, may also be cell arrays.

```
a = {"abc"};
s = {"abc", "def"};
[tf, s_idx] = ismember (a, s)
     ⇒ tf = [1, 0]
     ⇒ s_idx = [1, 0]
```

With the optional third argument "rows", and matrices *A* and *s* with the same number of columns, compare rows in *A* with the rows in *s*.

```
a = [1:3; 5:7; 4:6];
s = [0:2; 1:3; 2:4; 3:5; 4:6];
[tf, s_idx] = ismember (a, s, "rows")
     ⇒ tf = logical ([1; 0; 1])
     ⇒ s_idx = [2; 0; 5];
```

See also: [unique], page 597, [union], page 598, [intersect], page 598, [setxor], page 599, [setdiff], page 598.

`union (a, b)` [Function File]
`union (a, b, "rows")` [Function File]
`[c, ia, ib] = union (a, b)` [Function File]

Return the set of elements that are in either of the sets *a* and *b*. *a*, *b* may be cell arrays of strings. For example:

```
union ([1, 2, 4], [2, 3, 5])
    ⇒ [1, 2, 3, 4, 5]
```

If the optional third input argument is the string "rows" then each row of the matrices *a* and *b* will be considered as a single set element. For example:

```
union ([1, 2; 2, 3], [1, 2; 3, 4], "rows")
   ⇒  1   2
      2   3
      3   4
```

The optional outputs *ia* and *ib* are index vectors such that `a(ia)` and `b(ib)` are disjoint sets whose union is *c*.

See also: [intersect], page 598, [setdiff], page 598, [unique], page 597.

`intersect (a, b)` [Function File]
`[c, ia, ib] = intersect (a, b)` [Function File]

Return the elements in both *a* and *b*, sorted in ascending order. If *a* and *b* are both column vectors return a column vector, otherwise return a row vector. *a*, *b* may be cell arrays of string(s).

Return index vectors *ia* and *ib* such that `a(ia)==c` and `b(ib)==c`.

See also: [unique], page 597, [union], page 598, [setxor], page 599, [setdiff], page 598, [ismember], page 597.

`setdiff (a, b)` [Function File]
`setdiff (a, b, "rows")` [Function File]

`[c, i] = setdiff (a, b)` [Function File]

Return the elements in *a* that are not in *b*, sorted in ascending order. If *a* and *b* are both column vectors return a column vector, otherwise return a row vector. *a*, *b* may be cell arrays of string(s).

Given the optional third argument `"rows"`, return the rows in *a* that are not in *b*, sorted in ascending order by rows.

If requested, return *i* such that `c = a(i)`.

See also: [unique], page 597, [union], page 598, [intersect], page 598, [setxor], page 599, [ismember], page 597.

`setxor (a, b)` [Function File]
`setxor (a, b, "rows")` [Function File]
`[c, ia, ib] = setxor (a, b)` [Function File]

Return the elements exclusive to *a* or *b*, sorted in ascending order. If *a* and *b* are both column vectors return a column vector, otherwise return a row vector. *a*, *b* may be cell arrays of string(s).

With three output arguments, return index vectors *ia* and *ib* such that `a(ia)` and `b(ib)` are disjoint sets whose union is *c*.

See also: [unique], page 597, [union], page 598, [intersect], page 598, [setdiff], page 598, [ismember], page 597.

`powerset (a)` [Function File]
`powerset (a, "rows")` [Function File]

Compute the powerset (all subsets) of the set *a*.

The set *a* must be a numerical matrix or a cell array of strings. The output will always be a cell array of either vectors or strings.

With the optional second argument `"rows"`, each row of the set *a* is considered one element of the set. As a result, *a* must then be a numerical 2-D matrix.

See also: [unique], page 597, [union], page 598, [setxor], page 599, [setdiff], page 598, [ismember], page 597.

28 Polynomial Manipulations

In Octave, a polynomial is represented by its coefficients (arranged in descending order). For example, a vector *c* of length $N+1$ corresponds to the following polynomial of order N

$$p(x) = c_1 x^N + \ldots + c_N x + c_{N+1}.$$

28.1 Evaluating Polynomials

The value of a polynomial represented by the vector *c* can be evaluated at the point *x* very easily, as the following example shows:

```
N = length (c) - 1;
val = dot (x.^(N:-1:0), c);
```

While the above example shows how easy it is to compute the value of a polynomial, it isn't the most stable algorithm. With larger polynomials you should use more elegant algorithms, such as Horner's Method, which is exactly what the Octave function `polyval` does.

In the case where *x* is a square matrix, the polynomial given by *c* is still well-defined. As when *x* is a scalar the obvious implementation is easily expressed in Octave, but also in this case more elegant algorithms perform better. The `polyvalm` function provides such an algorithm.

y = `polyval` (*p*, *x*) [Function File]
y = `polyval` (*p*, *x*, [], *mu*) [Function File]
[*y*, *dy*] = `polyval` (*p*, *x*, *s*) [Function File]
[*y*, *dy*] = `polyval` (*p*, *x*, *s*, *mu*) [Function File]

Evaluate the polynomial *p* at the specified values of *x*. When *mu* is present, evaluate the polynomial for (*x*-*mu*(1))/*mu*(2). If *x* is a vector or matrix, the polynomial is evaluated for each of the elements of *x*.

In addition to evaluating the polynomial, the second output represents the prediction interval, *y* +/- *dy*, which contains at least 50% of the future predictions. To calculate the prediction interval, the structured variable *s*, originating from `polyfit`, must be supplied.

See also: [polyvalm], page 601, [polyaffine], page 606, [polyfit], page 607, [roots], page 602, [poly], page 615.

`polyvalm` (*c*, *x*) [Function File]

Evaluate a polynomial in the matrix sense.

`polyvalm (c, x)` will evaluate the polynomial in the matrix sense, i.e., matrix multiplication is used instead of element by element multiplication as used in `polyval`.

The argument *x* must be a square matrix.

See also: [polyval], page 601, [roots], page 602, [poly], page 615.

28.2 Finding Roots

Octave can find the roots of a given polynomial. This is done by computing the companion matrix of the polynomial (see the `compan` function for a definition), and then finding its eigenvalues.

`roots (v)` [Function File]

For a vector *v* with N components, return the roots of the polynomial

$$v_1 z^{N-1} + \cdots + v_{N-1} z + v_N.$$

As an example, the following code finds the roots of the quadratic polynomial

$$p(x) = x^2 - 5.$$

```
c = [1, 0, -5];
roots (c)
⇒  2.2361
⇒ -2.2361
```

Note that the true result is $\pm\sqrt{5}$ which is roughly ± 2.2361.

See also: [poly], page 615, [compan], page 602, [fzero], page 475.

`z = polyeig (C0, C1, ..., Cl)` [Function File]
`[v, z] = polyeig (C0, C1, ..., Cl)` [Function File]

Solve the polynomial eigenvalue problem of degree *l*.

Given an *n*n* matrix polynomial `C(s) = C0 + C1 s + ... + Cl s^l` polyeig solves the eigenvalue problem `(C0 + C1 + ... + Cl)v = 0`. Note that the eigenvalues *z* are the zeros of the matrix polynomial. *z* is an *lxn* vector and *v* is an (*n* x *n*)l matrix with columns that correspond to the eigenvectors.

See also: [eig], page 437, [eigs], page 513, [compan], page 602.

`compan (c)` [Function File]

Compute the companion matrix corresponding to polynomial coefficient vector *c*.

The companion matrix is

$$A = \begin{bmatrix} -c_2/c_1 & -c_3/c_1 & \cdots & -c_N/c_1 & -c_{N+1}/c_1 \\ 1 & 0 & \cdots & 0 & 0 \\ 0 & 1 & \cdots & 0 & 0 \\ \vdots & \vdots & \ddots & \vdots & \vdots \\ 0 & 0 & \cdots & 1 & 0 \end{bmatrix}.$$

The eigenvalues of the companion matrix are equal to the roots of the polynomial.

See also: [roots], page 602, [poly], page 615, [eig], page 437.

`[multp, idxp] = mpoles (p)` [Function File]
`[multp, idxp] = mpoles (p, tol)` [Function File]
`[multp, idxp] = mpoles (p, tol, reorder)` [Function File]

Identify unique poles in *p* and their associated multiplicity. The output is ordered from largest pole to smallest pole.

If the relative difference of two poles is less than *tol* then they are considered to be multiples. The default value for *tol* is 0.001.

If the optional parameter *reorder* is zero, poles are not sorted.

The output *multp* is a vector specifying the multiplicity of the poles. `multp(n)` refers to the multiplicity of the Nth pole `p(idxp(n))`.

For example:

```
p = [2 3 1 1 2];
[m, n] = mpoles (p)
   ⇒ m = [1; 1; 2; 1; 2]
   ⇒ n = [2; 5; 1; 4; 3]
   ⇒ p(n) = [3, 2, 2, 1, 1]
```

See also: [residue], page 604, [poly], page 615, [roots], page 602, [conv], page 603, [deconv], page 604.

28.3 Products of Polynomials

`conv` (*a*, *b*) [Function File]
`conv` (*a*, *b*, *shape*) [Function File]

Convolve two vectors *a* and *b*.

The output convolution is a vector with length equal to `length (a) + length (b) - 1`. When *a* and *b* are the coefficient vectors of two polynomials, the convolution represents the coefficient vector of the product polynomial.

The optional *shape* argument may be

shape = `"full"`
  Return the full convolution. (default)

shape = `"same"`
  Return the central part of the convolution with the same size as *a*.

See also: [deconv], page 604, [conv2], page 604, [convn], page 603, [fftconv], page 642.

C = `convn` (*A*, *B*) [Built-in Function]
C = `convn` (*A*, *B*, *shape*) [Built-in Function]

Return the n-D convolution of *A* and *B*. The size of the result is determined by the optional *shape* argument which takes the following values

shape = `"full"`
  Return the full convolution. (default)

shape = `"same"`
  Return central part of the convolution with the same size as *A*. The central part of the convolution begins at the indices `floor ([size(B)/2] + 1)`.

shape = `"valid"`
  Return only the parts which do not include zero-padded edges. The size of the result is `max (size (A) - size (B) + 1, 0)`.

See also: [conv2], page 604, [conv], page 603.

`deconv (y, a)` [Function File]

Deconvolve two vectors.

`[b, r] = deconv (y, a)` solves for *b* and *r* such that `y = conv (a, b) + r`.

If *y* and *a* are polynomial coefficient vectors, *b* will contain the coefficients of the polynomial quotient and *r* will be a remainder polynomial of lowest order.

See also: [conv], page 603, [residue], page 604.

`conv2 (A, B)` [Built-in Function]
`conv2 (v1, v2, m)` [Built-in Function]
`conv2 (..., shape)` [Built-in Function]

Return the 2-D convolution of *A* and *B*. The size of the result is determined by the optional *shape* argument which takes the following values

shape = `"full"`
: Return the full convolution. (default)

shape = `"same"`
: Return the central part of the convolution with the same size as *A*. The central part of the convolution begins at the indices `floor ([size(B)/2] + 1)`.

shape = `"valid"`
: Return only the parts which do not include zero-padded edges. The size of the result is `max (size (A) - size (B) + 1, 0)`.

When the third argument is a matrix, return the convolution of the matrix *m* by the vector *v1* in the column direction and by the vector *v2* in the row direction.

See also: [conv], page 603, [convn], page 603.

`q = polygcd (b, a)` [Function File]
`q = polygcd (b, a, tol)` [Function File]

Find the greatest common divisor of two polynomials. This is equivalent to the polynomial found by multiplying together all the common roots. Together with deconv, you can reduce a ratio of two polynomials. The tolerance *tol* defaults to `sqrt (eps)`.

Caution: This is a numerically unstable algorithm and should not be used on large polynomials.

Example code:

```
polygcd (poly (1:8), poly (3:12)) - poly (3:8)
⇒ [ 0, 0, 0, 0, 0, 0, 0 ]
deconv (poly (1:8), polygcd (poly (1:8), poly (3:12))) - poly (1:2)
⇒ [ 0, 0, 0 ]
```

See also: [poly], page 615, [roots], page 602, [conv], page 603, [deconv], page 604, [residue], page 604.

`[r, p, k, e] = residue (b, a)` [Function File]
`[b, a] = residue (r, p, k)` [Function File]

`[b, a] = residue (r, p, k, e)` [Function File]

The first calling form computes the partial fraction expansion for the quotient of the polynomials, b and a.

$$\frac{B(s)}{A(s)} = \sum_{m=1}^{M} \frac{r_m}{(s - p_m)_m^e} + \sum_{i=1}^{N} k_i s^{N-i}.$$

where M is the number of poles (the length of the r, p, and e), the k vector is a polynomial of order $N - 1$ representing the direct contribution, and the e vector specifies the multiplicity of the m-th residue's pole.

For example,

```
b = [1, 1, 1];
a = [1, -5, 8, -4];
[r, p, k, e] = residue (b, a)
   ⇒ r = [-2; 7; 3]
   ⇒ p = [2; 2; 1]
   ⇒ k = [](0x0)
   ⇒ e = [1; 2; 1]
```

which represents the following partial fraction expansion

$$\frac{s^2 + s + 1}{s^3 - 5s^2 + 8s - 4} = \frac{-2}{s - 2} + \frac{7}{(s - 2)^2} + \frac{3}{s - 1}$$

The second calling form performs the inverse operation and computes the reconstituted quotient of polynomials, $b(s)/a(s)$, from the partial fraction expansion; represented by the residues, poles, and a direct polynomial specified by r, p and k, and the pole multiplicity e.

If the multiplicity, e, is not explicitly specified the multiplicity is determined by the function `mpoles`.

For example:

```
r = [-2; 7; 3];
p = [2; 2; 1];
k = [1, 0];
[b, a] = residue (r, p, k)
   ⇒ b = [1, -5, 9, -3, 1]
   ⇒ a = [1, -5, 8, -4]

where mpoles is used to determine e = [1; 2; 1]
```

Alternatively the multiplicity may be defined explicitly, for example,

```
r = [7; 3; -2];
p = [2; 1; 2];
k = [1, 0];
e = [2; 1; 1];
[b, a] = residue (r, p, k, e)
   ⇒ b = [1, -5, 9, -3, 1]
   ⇒ a = [1, -5, 8, -4]
```

which represents the following partial fraction expansion

$$\frac{-2}{s-2} + \frac{7}{(s-2)^2} + \frac{3}{s-1} + s = \frac{s^4 - 5s^3 + 9s^2 - 3s + 1}{s^3 - 5s^2 + 8s - 4}$$

See also: [mpoles], page 602, [poly], page 615, [roots], page 602, [conv], page 603, [deconv], page 604.

28.4 Derivatives / Integrals / Transforms

Octave comes with functions for computing the derivative and the integral of a polynomial. The functions `polyder` and `polyint` both return new polynomials describing the result. As an example we'll compute the definite integral of $p(x) = x^2 + 1$ from 0 to 3.

```
c = [1, 0, 1];
integral = polyint (c);
area = polyval (integral, 3) - polyval (integral, 0)
⇒ 12
```

`polyder (p)` [Function File]
`[k] = polyder (a, b)` [Function File]
`[q, d] = polyder (b, a)` [Function File]

Return the coefficients of the derivative of the polynomial whose coefficients are given by the vector *p*. If a pair of polynomials is given, return the derivative of the product $a * b$. If two inputs and two outputs are given, return the derivative of the polynomial quotient b/a. The quotient numerator is in *q* and the denominator in *d*.

See also: [polyint], page 606, [polyval], page 601, [polyreduce], page 616.

`polyint (p)` [Function File]
`polyint (p, k)` [Function File]

Return the coefficients of the integral of the polynomial whose coefficients are represented by the vector *p*. The variable *k* is the constant of integration, which by default is set to zero.

See also: [polyder], page 606, [polyval], page 601.

`polyaffine (f, mu)` [Function File]

Return the coefficients of the polynomial vector *f* after an affine transformation. If *f* is the vector representing the polynomial f(x), then `g = polyaffine (f, mu)` is the vector representing:

```
g(x) = f( (x - mu(1)) / mu(2) )
```

See also: [polyval], page 601, [polyfit], page 607.

28.5 Polynomial Interpolation

Octave comes with good support for various kinds of interpolation, most of which are described in Chapter 29 [Interpolation], page 617. One simple alternative to the functions described in the aforementioned chapter, is to fit a single polynomial, or a piecewise polynomial (spline) to some given data points. To avoid a highly fluctuating polynomial, one most often wants to fit a low-order polynomial to data. This usually means that it is necessary to fit the polynomial in a least-squares sense, which just is what the `polyfit` function does.

p = **polyfit** (*x*, *y*, *n*) [Function File]
[*p*, *s*] = **polyfit** (*x*, *y*, *n*) [Function File]
[*p*, *s*, *mu*] = **polyfit** (*x*, *y*, *n*) [Function File]

Return the coefficients of a polynomial $p(x)$ of degree *n* that minimizes the least-squares-error of the fit to the points `[x, y]`. If *n* is a logical vector, it is used as a mask to selectively force the corresponding polynomial coefficients to be used or ignored.

The polynomial coefficients are returned in a row vector.

The optional output *s* is a structure containing the following fields:

'R' Triangular factor R from the QR decomposition.

'X' The Vandermonde matrix used to compute the polynomial coefficients.

'C' The unscaled covariance matrix, formally equal to the inverse of *x'***x*, but computed in a way minimizing roundoff error propagation.

'df' The degrees of freedom.

'normr' The norm of the residuals.

'yf' The values of the polynomial for each value of *x*.

The second output may be used by `polyval` to calculate the statistical error limits of the predicted values. In particular, the standard deviation of *p* coefficients is given by

`sqrt (diag (s.C)/s.df)*s.normr.`

When the third output, *mu*, is present the coefficients, *p*, are associated with a polynomial in *xhat* = (*x*-*mu*(1))/*mu*(2). Where *mu*(1) = mean (*x*), and *mu*(2) = std (*x*). This linear transformation of *x* improves the numerical stability of the fit.

See also: [polyval], page 601, [polyaffine], page 606, [roots], page 602, [vander], page 406, [zscore], page 567.

In situations where a single polynomial isn't good enough, a solution is to use several polynomials pieced together. The function `splinefit` fits a peicewise polynomial (spline) to a set of data.

pp = **splinefit** (*x*, *y*, *breaks*) [Function File]
pp = **splinefit** (*x*, *y*, *p*) [Function File]
pp = **splinefit** (..., "*periodic*", *periodic*) [Function File]
pp = **splinefit** (..., "*robust*", *robust*) [Function File]
pp = **splinefit** (..., "*beta*", *beta*) [Function File]
pp = **splinefit** (..., "*order*", *order*) [Function File]
pp = **splinefit** (..., "*constraints*", *constraints*) [Function File]

Fit a piecewise cubic spline with breaks (knots) *breaks* to the noisy data, *x* and *y*. *x* is a vector, and *y* is a vector or N-D array. If *y* is an N-D array, then *x*(j) is matched to *y*(:,...,:,j).

The fitted spline is returned as a piecewise polynomial, *pp*, and may be evaluated using `ppval`.

p is a positive integer defining the number of intervals along *x*, and *p*+1 is the number of breaks. The number of points in each interval differ by no more than 1.

The optional property *periodic* is a logical value which specifies whether a periodic boundary condition is applied to the spline. The length of the period is `max (breaks) - min (breaks)`. The default value is `false`.

The optional property *robust* is a logical value which specifies if robust fitting is to be applied to reduce the influence of outlying data points. Three iterations of weighted least squares are performed. Weights are computed from previous residuals. The sensitivity of outlier identification is controlled by the property *beta*. The value of *beta* is stricted to the range, 0 < *beta* < 1. The default value is *beta* = 1/2. Values close to 0 give all data equal weighting. Increasing values of *beta* reduce the influence of outlying data. Values close to unity may cause instability or rank deficiency.

The splines are constructed of polynomials with degree *order*. The default is a cubic, *order*=3. A spline with P pieces has P+*order* degrees of freedom. With periodic boundary conditions the degrees of freedom are reduced to P.

The optional property, *constaints*, is a structure specifying linear constraints on the fit. The structure has three fields, `"xc"`, `"yc"`, and `"cc"`.

`"xc"` Vector of the x-locations of the constraints.

`"yc"` Constraining values at the locations *xc*. The default is an array of zeros.

`"cc"` Coefficients (matrix). The default is an array of ones. The number of rows is limited to the order of the piecewise polynomials, *order*.

Constraints are linear combinations of derivatives of order 0 to *order*-1 according to

$$cc(1,j) \cdot y(xc(j)) + cc(2,j) \cdot y\prime(xc(j)) + ... = yc(:,\ldots,:,j).$$

See also: [interp1], page 617, [unmkpp], page 614, [ppval], page 614, [spline], page 620, [pchip], page 647, [ppder], page 615, [ppint], page 615, [ppjumps], page 615.

The number of *breaks* (or knots) used to construct the piecewise polynomial is a significant factor in suppressing the noise present in the input data, *x* and *y*. This is demonstrated by the example below.

```
x = 2 * pi * rand (1, 200);
y = sin (x) + sin (2 * x) + 0.2 * randn (size (x));
## Uniform breaks
breaks = linspace (0, 2 * pi, 41); % 41 breaks, 40 pieces
pp1 = splinefit (x, y, breaks);
## Breaks interpolated from data
pp2 = splinefit (x, y, 10);  % 11 breaks, 10 pieces
## Plot
xx = linspace (0, 2 * pi, 400);
y1 = ppval (pp1, xx);
y2 = ppval (pp2, xx);
plot (x, y, ".", xx, [y1; y2])
axis tight
ylim auto
legend ({"data", "41 breaks, 40 pieces", "11 breaks, 10 pieces"})
```

The result of which can be seen in Figure 28.1.

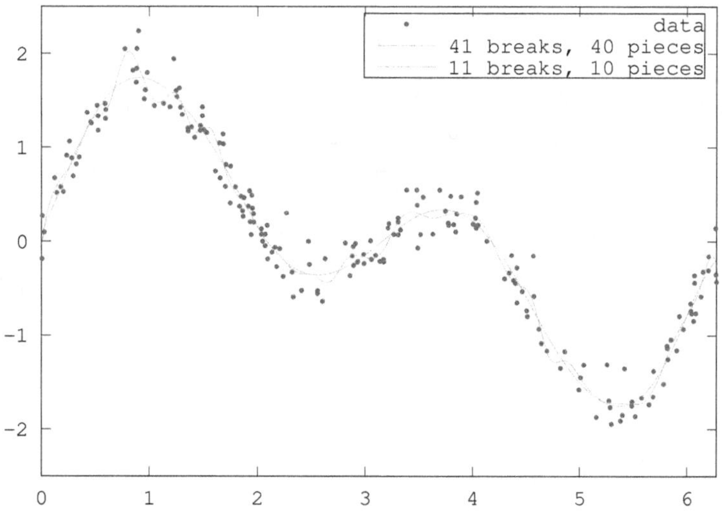

Figure 28.1: Comparison of a fitting a piecewise polynomial with 41 breaks to one with 11 breaks. The fit with the large number of breaks exhibits a fast ripple that is not present in the underlying function.

The piecewise polynomial fit, provided by **splinefit**, has continuous derivatives up to the *order*-1. For example, a cubic fit has continuous first and second derivatives. This is demonstrated by the code

```
## Data (200 points)
x = 2 * pi * rand (1, 200);
y = sin (x) + sin (2 * x) + 0.1 * randn (size (x));
## Piecewise constant
pp1 = splinefit (x, y, 8, "order", 0);
## Piecewise linear
pp2 = splinefit (x, y, 8, "order", 1);
```

```
## Piecewise quadratic
pp3 = splinefit (x, y, 8, "order", 2);
## Piecewise cubic
pp4 = splinefit (x, y, 8, "order", 3);
## Piecewise quartic
pp5 = splinefit (x, y, 8, "order", 4);
## Plot
xx = linspace (0, 2 * pi, 400);
y1 = ppval (pp1, xx);
y2 = ppval (pp2, xx);
y3 = ppval (pp3, xx);
y4 = ppval (pp4, xx);
y5 = ppval (pp5, xx);
plot (x, y, ".", xx, [y1; y2; y3; y4; y5])
axis tight
ylim auto
legend ({"data", "order 0", "order 1", "order 2", "order 3", "order 4"})
```

The result of which can be seen in Figure 28.2.

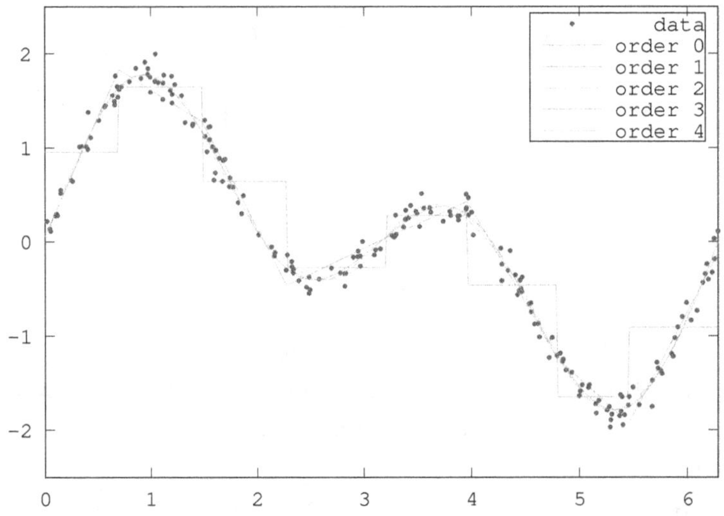

Figure 28.2: Comparison of a piecewise constant, linear, quadratic, cubic, and quartic polynomials with 8 breaks to noisy data. The higher order solutions more accurately represent the underlying function, but come with the expense of computational complexity.

When the underlying function to provide a fit to is periodic, `splinefit` is able to apply the boundary conditions needed to manifest a periodic fit. This is demonstrated by the code below.

```
## Data (100 points)
x = 2 * pi * [0, (rand (1, 98)), 1];
y = sin (x) - cos (2 * x) + 0.2 * randn (size (x));
## No constraints
pp1 = splinefit (x, y, 10, "order", 5);
## Periodic boundaries
pp2 = splinefit (x, y, 10, "order", 5, "periodic", true);
## Plot
xx = linspace (0, 2 * pi, 400);
y1 = ppval (pp1, xx);
y2 = ppval (pp2, xx);
plot (x, y, ".", xx, [y1; y2])
axis tight
ylim auto
legend ({"data", "no constraints", "periodic"})
```

The result of which can be seen in Figure 28.3.

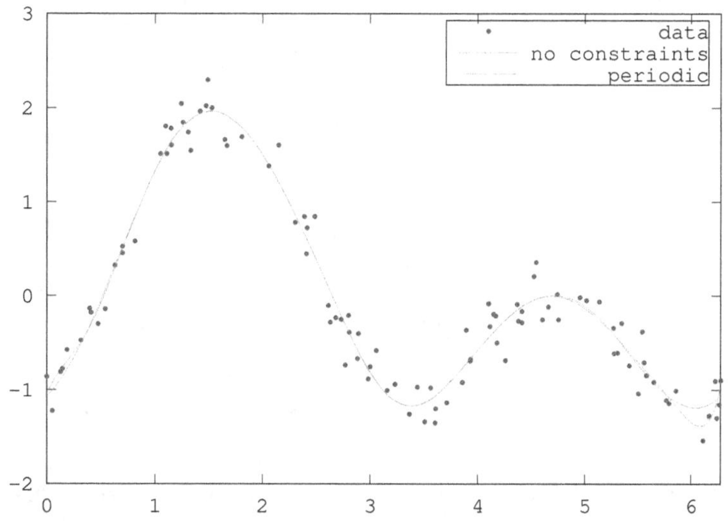

Figure 28.3: Comparison of piecewise polynomial fits to a noisy periodic function with, and without, periodic boundary conditions.

More complex constraints may be added as well. For example, the code below illustrates a periodic fit with values that have been clamped at the endpoints, and a second periodic fit which is hinged at the endpoints.

```
## Data (200 points)
x = 2 * pi * rand (1, 200);
y = sin (2 * x) + 0.1 * randn (size (x));
## Breaks
breaks = linspace (0, 2 * pi, 10);
## Clamped endpoints, y = y' = 0
xc = [0, 0, 2*pi, 2*pi];
cc = [(eye (2)), (eye (2))];
```

```
con = struct ("xc", xc, "cc", cc);
pp1 = splinefit (x, y, breaks, "constraints", con);
## Hinged periodic endpoints, y = 0
con = struct ("xc", 0);
pp2 = splinefit (x, y, breaks, "constraints", con, "periodic", true);
## Plot
xx = linspace (0, 2 * pi, 400);
y1 = ppval (pp1, xx);
y2 = ppval (pp2, xx);
plot (x, y, ".", xx, [y1; y2])
axis tight
ylim auto
legend ({"data", "clamped", "hinged periodic"})
```

The result of which can be seen in Figure 28.4.

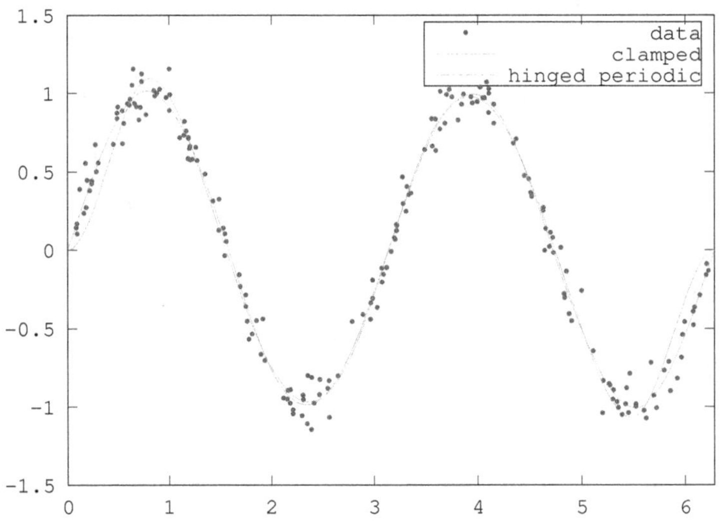

Figure 28.4: Comparison of two periodic piecewise cubic fits to a noisy periodic signal. One fit has its endpoints clamped and the second has its endpoints hinged.

The `splinefit` function also provides the convenience of a *robust* fitting, where the effect of outlying data is reduced. In the example below, three different fits are provided. Two with differing levels of outlier suppression and a third illustrating the non-robust solution.

```
## Data
x = linspace (0, 2*pi, 200);
y = sin (x) + sin (2 * x) + 0.05 * randn (size (x));
## Add outliers
x = [x, linspace(0,2*pi,60)];
y = [y, -ones(1,60)];
## Fit splines with hinged conditions
con = struct ("xc", [0, 2*pi]);
## Robust fitting, beta = 0.25
pp1 = splinefit (x, y, 8, "constraints", con, "beta", 0.25);
```

```
## Robust fitting, beta = 0.75
pp2 = splinefit (x, y, 8, "constraints", con, "beta", 0.75);
## No robust fitting
pp3 = splinefit (x, y, 8, "constraints", con);
## Plot
xx = linspace (0, 2*pi, 400);
y1 = ppval (pp1, xx);
y2 = ppval (pp2, xx);
y3 = ppval (pp3, xx);
plot (x, y, ".", xx, [y1; y2; y3])
legend ({"data with outliers","robust, beta = 0.25", ...
         "robust, beta = 0.75", "no robust fitting"})
axis tight
ylim auto
```

The result of which can be seen in Figure 28.5.

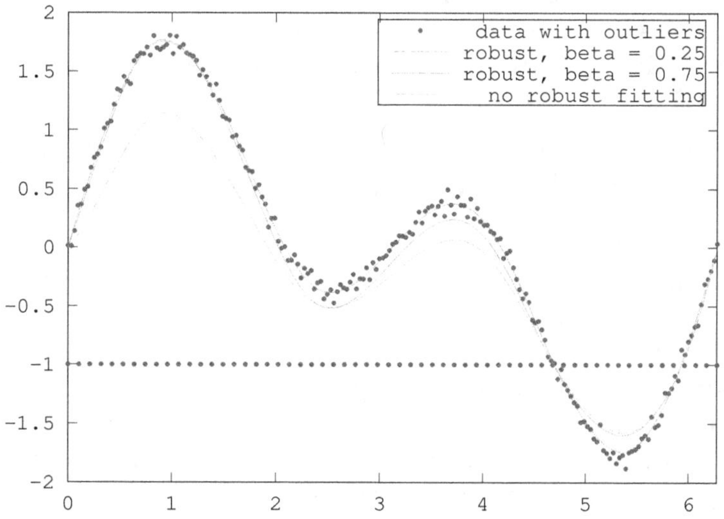

Figure 28.5: Comparison of two different levels of robust fitting (*beta* = 0.25 and 0.75) to noisy data combined with outlying data. A conventional fit, without robust fitting (*beta* = 0) is also included.

The function, `ppval`, evaluates the piecewise polynomials, created by `mkpp` or other means, and `unmkpp` returns detailed information about the piecewise polynomial.

The following example shows how to combine two linear functions and a quadratic into one function. Each of these functions is expressed on adjoined intervals.

```
x = [-2, -1, 1, 2];
p = [ 0,  1, 0;
      1, -2, 1;
      0, -1, 1 ];
pp = mkpp (x, p);
xi = linspace (-2, 2, 50);
yi = ppval (pp, xi);
plot (xi, yi);
```

pp = mkpp (*breaks*, *coefs*) [Function File]
pp = mkpp (*breaks*, *coefs*, *d*) [Function File]

Construct a piecewise polynomial (pp) structure from sample points *breaks* and coefficients *coefs*. *breaks* must be a vector of strictly increasing values. The number of intervals is given by `ni = length (breaks) - 1`. When *m* is the polynomial order *coefs* must be of size: *ni* x *m* + 1.

The i-th row of *coefs*, `coefs (i,:)`, contains the coefficients for the polynomial over the *i*-th interval, ordered from highest (*m*) to lowest (*0*).

coefs may also be a multi-dimensional array, specifying a vector-valued or array-valued polynomial. In that case the polynomial order is defined by the length of the last dimension of *coefs*. The size of first dimension(s) are given by the scalar or vector *d*. If *d* is not given it is set to `1`. In any case *coefs* is reshaped to a 2-D matrix of size `[ni*prod(d m)]`

See also: [unmkpp], page 614, [ppval], page 614, [spline], page 620, [pchip], page 647, [ppder], page 615, [ppint], page 615, [ppjumps], page 615.

[*x*, *p*, *n*, *k*, *d*] = unmkpp (*pp*) [Function File]

Extract the components of a piecewise polynomial structure *pp*. The components are:

x Sample points.

p Polynomial coefficients for points in sample interval. `p (i, :)` contains the coefficients for the polynomial over interval *i* ordered from highest to lowest. If `d > 1`, `p (r, i, :)` contains the coefficients for the r-th polynomial defined on interval *i*.

n Number of polynomial pieces.

k Order of the polynomial plus 1.

d Number of polynomials defined for each interval.

See also: [mkpp], page 614, [ppval], page 614, [spline], page 620, [pchip], page 647.

yi = ppval (*pp*, *xi*) [Function File]

Evaluate the piecewise polynomial structure *pp* at the points *xi*. If *pp* describes a scalar polynomial function, the result is an array of the same shape as *xi*. Otherwise, the size of the result is `[pp.dim, length(xi)]` if *xi* is a vector, or `[pp.dim, size(xi)]` if it is a multi-dimensional array.

See also: [mkpp], page 614, [unmkpp], page 614, [spline], page 620, [pchip], page 647.

`ppd = ppder (pp)` [Function File]
`ppd = ppder (pp, m)` [Function File]

Compute the piecewise *m*-th derivative of a piecewise polynomial struct *pp*. If *m* is omitted the first derivative is calculated.

See also: [mkpp], page 614, [ppval], page 614, [ppint], page 615.

`ppi = ppint (pp)` [Function File]
`ppi = ppint (pp, c)` [Function File]

Compute the integral of the piecewise polynomial struct *pp*. *c*, if given, is the constant of integration.

See also: [mkpp], page 614, [ppval], page 614, [ppder], page 615.

`jumps = ppjumps (pp)` [Function File]

Evaluate the boundary jumps of a piecewise polynomial. If there are *n* intervals, and the dimensionality of *pp* is *d*, the resulting array has dimensions `[d, n-1]`.

See also: [mkpp], page 614.

28.6 Miscellaneous Functions

`poly (A)` [Function File]
`poly (x)` [Function File]

If *A* is a square *N*-by-*N* matrix, `poly (A)` is the row vector of the coefficients of `det (z * eye (N) - A)`, the characteristic polynomial of *A*. For example, the following code finds the eigenvalues of *A* which are the roots of `poly (A)`.

```
roots (poly (eye (3)))
     ⇒ 1.00001 + 0.00001i
       1.00001 - 0.00001i
       0.99999 + 0.00000i
```

In fact, all three eigenvalues are exactly 1 which emphasizes that for numerical performance the `eig` function should be used to compute eigenvalues.

If *x* is a vector, `poly (x)` is a vector of the coefficients of the polynomial whose roots are the elements of *x*. That is, if *c* is a polynomial, then the elements of `d = roots (poly (c))` are contained in *c*. The vectors *c* and *d* are not identical, however, due to sorting and numerical errors.

See also: [roots], page 602, [eig], page 437.

`polyout (c)` [Function File]
`polyout (c, x)` [Function File]
`str = polyout (...)` [Function File]

Write formatted polynomial

$$c(x) = c_1 x^n + \ldots + c_n x + c_{n+1}$$

and return it as a string or write it to the screen (if *nargout* is zero). *x* defaults to the string `"s"`.

See also: [polyreduce], page 616.

`polyreduce (c)` [Function File]

Reduce a polynomial coefficient vector to a minimum number of terms by stripping off any leading zeros.

See also: [polyout], page 615.

29 Interpolation

29.1 One-dimensional Interpolation

Octave supports several methods for one-dimensional interpolation, most of which are described in this section. Section 28.5 [Polynomial Interpolation], page 606 and Section 30.4 [Interpolation on Scattered Data], page 637 describe additional methods.

yi = **interp1** (*x*, *y*, *xi*) [Function File]
yi = **interp1** (*y*, *xi*) [Function File]
yi = **interp1** (..., *method*) [Function File]
yi = **interp1** (..., *extrap*) [Function File]
yi = **interp1** (..., "*left*") [Function File]
yi = **interp1** (..., "*right*") [Function File]
pp = **interp1** (..., "*pp*") [Function File]

One-dimensional interpolation.

Interpolate input data to determine the value of *yi* at the points *xi*. If not specified, *x* is taken to be the indices of *y*. If *y* is a matrix or an N-dimensional array, the interpolation is performed on each column of *y*.

Method is one of:

`"nearest"`
: Return the nearest neighbor

`"linear"`
: Linear interpolation from nearest neighbors

`"pchip"`
: Piecewise cubic Hermite interpolating polynomial

`"cubic"`
: Cubic interpolation (same as `pchip`)

`"spline"`
: Cubic spline interpolation—smooth first and second derivatives throughout the curve

Adding '*' to the start of any method above forces `interp1` to assume that *x* is uniformly spaced, and only `x(1)` and `x(2)` are referenced. This is usually faster, and is never slower. The default method is `"linear"`.

If *extrap* is the string `"extrap"`, then extrapolate values beyond the endpoints using the current *method*. If *extrap* is a number, then replace values beyond the endpoints with that number. When unspecified, *extrap* defaults to NA.

If the string argument `"pp"` is specified, then *xi* should not be supplied and `interp1` returns a piecewise polynomial object. This object can later be used with `ppval` to evaluate the interpolation. There is an equivalence, such that `ppval (interp1 (x, y, method, "pp"), xi) == interp1 (x, y, xi, method, "extrap")`.

Duplicate points in *x* specify a discontinuous interpolant. There may be at most 2 consecutive points with the same value. If *x* is increasing, the default discontinuous interpolant is right-continuous. If *x* is decreasing, the default discontinuous interpolant is left-continuous. The continuity condition of the interpolant may be specified by using the options, `"left"` or `"right"`, to select a left-continuous or right-continuous interpolant, respectively. Discontinuous interpolation is only allowed for "nearest" and "linear" methods; in all other cases, the x-values must be unique.

An example of the use of interp1 is

```
xf = [0:0.05:10];
yf = sin (2*pi*xf/5);
xp = [0:10];
yp = sin (2*pi*xp/5);
lin = interp1 (xp, yp, xf);
spl = interp1 (xp, yp, xf, "spline");
cub = interp1 (xp, yp, xf, "cubic");
near = interp1 (xp, yp, xf, "nearest");
plot (xf, yf, "r", xf, lin, "g", xf, spl, "b",
      xf, cub, "c", xf, near, "m", xp, yp, "r*");
legend ("original", "linear", "spline", "cubic", "nearest");
```

See also: [interpft], page 619, [interp2], page 621, [interp3], page 622, [interpn], page 623.

There are some important differences between the various interpolation methods. The "spline" method enforces that both the first and second derivatives of the interpolated values have a continuous derivative, whereas the other methods do not. This means that the results of the "spline" method are generally smoother. If the function to be interpolated is in fact smooth, then "spline" will give excellent results. However, if the function to be evaluated is in some manner discontinuous, then "pchip" interpolation might give better results.

This can be demonstrated by the code

```
t = -2:2;
dt = 1;
ti =-2:0.025:2;
dti = 0.025;
y = sign (t);
ys = interp1 (t,y,ti,"spline");
yp = interp1 (t,y,ti,"pchip");
ddys = diff (diff (ys)./dti) ./ dti;
ddyp = diff (diff (yp)./dti) ./ dti;
figure (1);
plot (ti,ys,"r-", ti,yp,"g-");
legend ("spline", "pchip", 4);
figure (2);
plot (ti,ddys,"r+", ti,ddyp,"g*");
legend ("spline", "pchip");
```

The result of which can be seen in Figure 29.1 and Figure 29.2.

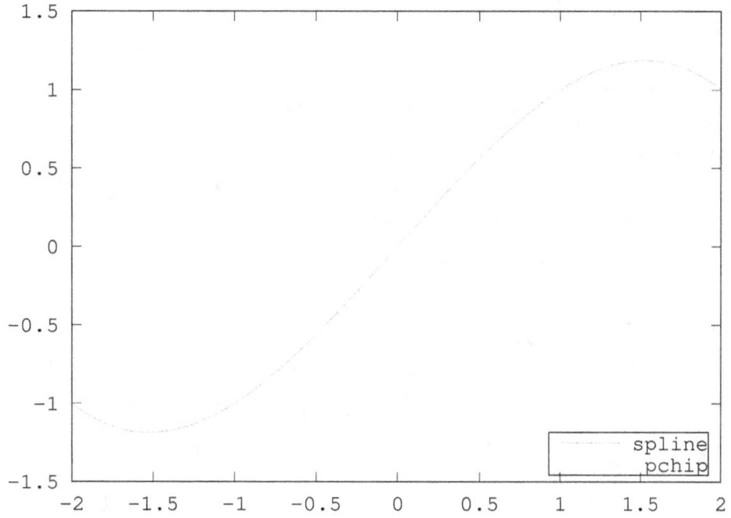

Figure 29.1: Comparison of "`pchip`" and "`spline`" interpolation methods for a step function

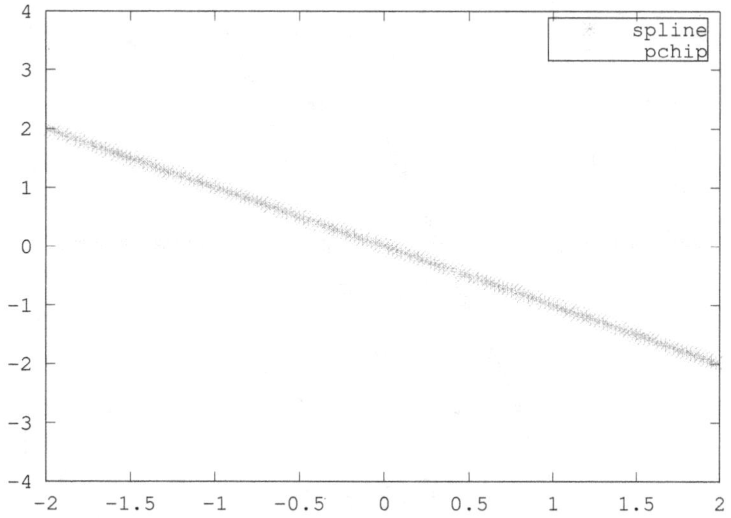

Figure 29.2: Comparison of the second derivative of the "`pchip`" and "`spline`" interpolation methods for a step function

Fourier interpolation, is a resampling technique where a signal is converted to the frequency domain, padded with zeros and then reconverted to the time domain.

interpft (*x*, *n*) [Function File]
interpft (*x*, *n*, *dim*) [Function File]

Fourier interpolation. If *x* is a vector, then *x* is resampled with *n* points. The data in *x* is assumed to be equispaced. If *x* is a matrix or an N-dimensional array, the interpolation is performed on each column of *x*. If *dim* is specified, then interpolate along the dimension *dim*.

interpft assumes that the interpolated function is periodic, and so assumptions are made about the endpoints of the interpolation.

See also: [interp1], page 617.

There are two significant limitations on Fourier interpolation. First, the function signal is assumed to be periodic, and so non-periodic signals will be poorly represented at the edges. Second, both the signal and its interpolation are required to be sampled at equispaced points. An example of the use of `interpft` is

```
t = 0 : 0.3 : pi; dt = t(2)-t(1);
n = length (t); k = 100;
ti = t(1) + [0 : k-1]*dt*n/k;
y = sin (4*t + 0.3) .* cos (3*t - 0.1);
yp = sin (4*ti + 0.3) .* cos (3*ti - 0.1);
plot (ti, yp, "g", ti, interp1 (t, y, ti, "spline"), "b", ...
      ti, interpft (y, k), "c", t, y, "r+");
legend ("sin(4t+0.3)cos(3t-0.1)", "spline", "interpft", "data");
```

which demonstrates the poor behavior of Fourier interpolation for non-periodic functions, as can be seen in Figure 29.3.

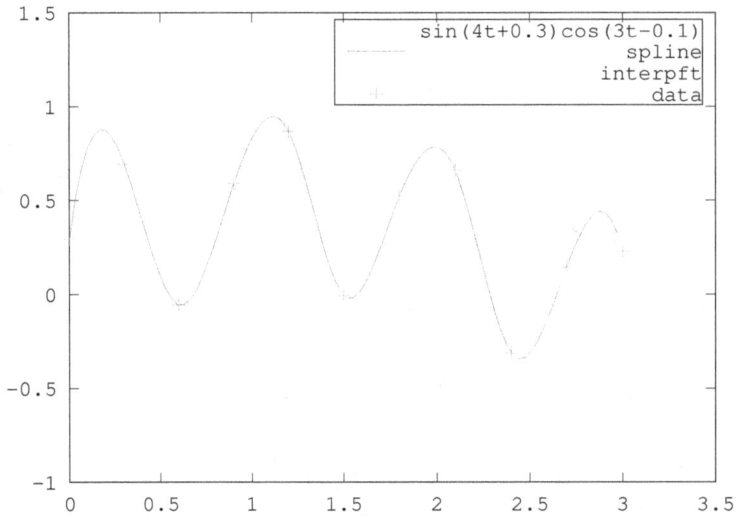

Figure 29.3: Comparison of `interp1` and `interpft` for non-periodic data

In addition, the support functions `spline` and `lookup` that underlie the `interp1` function can be called directly.

pp = `spline` (*x*, *y*) [Function File]
yi = `spline` (*x*, *y*, *xi*) [Function File]

Return the cubic spline interpolant of points *x* and *y*.

When called with two arguments, return the piecewise polynomial *pp* that may be used with `ppval` to evaluate the polynomial at specific points. When called with a third input argument, `spline` evaluates the spline at the points *xi*. The third calling form `spline (x, y, xi)` is equivalent to `ppval (spline (x, y), xi)`.

The variable *x* must be a vector of length *n*. *y* can be either a vector or array. If *y* is a vector it must have a length of either *n* or `n + 2`. If the length of *y* is *n*, then the "not-a-knot" end condition is used. If the length of *y* is `n + 2`, then the first and last values of the vector *y* are the values of the first derivative of the cubic spline at the endpoints.

If *y* is an array, then the size of *y* must have the form

$$[s_1, s_2, \cdots, s_k, n]$$

or

$$[s_1, s_2, \cdots, s_k, n + 2].$$

The array is reshaped internally to a matrix where the leading dimension is given by

$$s_1 s_2 \cdots s_k$$

and each row of this matrix is then treated separately. Note that this is exactly opposite to `interp1` but is done for MATLAB compatibility.

See also: [pchip], page 647, [ppval], page 614, [mkpp], page 614, [unmkpp], page 614.

29.2 Multi-dimensional Interpolation

There are three multi-dimensional interpolation functions in Octave, with similar capabilities. Methods using Delaunay tessellation are described in Section 30.4 [Interpolation on Scattered Data], page 637.

zi = `interp2` (*x, y, z, xi, yi*) [Function File]
zi = `interp2` (*Z, xi, yi*) [Function File]
zi = `interp2` (*Z, n*) [Function File]
zi = `interp2` (..., *method*) [Function File]
zi = `interp2` (..., *method, extrapval*) [Function File]

Two-dimensional interpolation. *x*, *y* and *z* describe a surface function. If *x* and *y* are vectors their length must correspondent to the size of *z*. *x* and *y* must be monotonic. If they are matrices they must have the `meshgrid` format.

`interp2 (x, y, Z, xi, yi, ...)`
: Returns a matrix corresponding to the points described by the matrices *xi*, *yi*.

  If the last argument is a string, the interpolation method can be specified. The method can be `"linear"`, `"nearest"` or `"cubic"`. If it is omitted `"linear"` interpolation is assumed.

`interp2 (z, xi, yi)`
: Assumes `x = 1:rows (z)` and `y = 1:columns (z)`

`interp2 (z, n)`
: Interleaves the matrix *z* n-times. If *n* is omitted a value of `n = 1` is assumed.

The variable *method* defines the method to use for the interpolation. It can take one of the following values

`"nearest"`
: Return the nearest neighbor.

`"linear"` Linear interpolation from nearest neighbors.

`"pchip"` Piecewise cubic Hermite interpolating polynomial.

`"cubic"` Cubic interpolation from four nearest neighbors.

`"spline"` Cubic spline interpolation—smooth first and second derivatives throughout the curve.

If a scalar value *extrapval* is defined as the final value, then values outside the mesh as set to this value. Note that in this case *method* must be defined as well. If *extrapval* is not defined then NA is assumed.

See also: [interp1], page 617.

vi = **interp3** (*x*, *y*, *z*, *v*, *xi*, *yi*, *zi*) [Function File]
vi = **interp3** (*v*, *xi*, *yi*, *zi*) [Function File]
vi = **interp3** (*v*, *m*) [Function File]
vi = **interp3** (*v*) [Function File]
vi = **interp3** (..., *method*) [Function File]
vi = **interp3** (..., *method*, *extrapval*) [Function File]

Perform 3-dimensional interpolation. Each element of the 3-dimensional array *v* represents a value at a location given by the parameters *x*, *y*, and *z*. The parameters *x*, *x*, and *z* are either 3-dimensional arrays of the same size as the array *v* in the `"meshgrid"` format or vectors. The parameters *xi*, etc. respect a similar format to *x*, etc., and they represent the points at which the array *vi* is interpolated.

If *x*, *y*, *z* are omitted, they are assumed to be `x = 1 : size (v, 2)`, `y = 1 : size (v, 1)` and `z = 1 : size (v, 3)`. If *m* is specified, then the interpolation adds a point half way between each of the interpolation points. This process is performed *m* times. If only *v* is specified, then *m* is assumed to be `1`.

Method is one of:

`"nearest"`
: Return the nearest neighbor.

`"linear"` Linear interpolation from nearest neighbors.

`"cubic"` Cubic interpolation from four nearest neighbors (not implemented yet).

`"spline"` Cubic spline interpolation—smooth first and second derivatives throughout the curve.

The default method is `"linear"`.

If *extrap* is the string `"extrap"`, then extrapolate values beyond the endpoints. If *extrap* is a number, replace values beyond the endpoints with that number. If *extrap* is missing, assume NA.

See also: [interp1], page 617, [interp2], page 621, [spline], page 620, [meshgrid], page 302.

vi = **interpn** (*x1*, *x2*, ..., *v*, *y1*, *y2*, ...) [Function File]
vi = **interpn** (*v*, *y1*, *y2*, ...) [Function File]
vi = **interpn** (*v*, *m*) [Function File]
vi = **interpn** (*v*) [Function File]
vi = **interpn** (..., *method*) [Function File]
vi = **interpn** (..., *method*, *extrapval*) [Function File]

Perform n-dimensional interpolation, where n is at least two. Each element of the n-dimensional array *v* represents a value at a location given by the parameters *x1*, *x2*, ..., *xn*. The parameters *x1*, *x2*, ..., *xn* are either n-dimensional arrays of the same size as the array *v* in the `"ndgrid"` format or vectors. The parameters *y1*, etc. respect a similar format to *x1*, etc., and they represent the points at which the array *vi* is interpolated.

If *x1*, ..., *xn* are omitted, they are assumed to be `x1 = 1 : size (v, 1)`, etc. If *m* is specified, then the interpolation adds a point half way between each of the interpolation points. This process is performed *m* times. If only *v* is specified, then *m* is assumed to be `1`.

Method is one of:

`"nearest"`
: Return the nearest neighbor.

`"linear"`
: Linear interpolation from nearest neighbors.

`"cubic"`
: Cubic interpolation from four nearest neighbors (not implemented yet).

`"spline"`
: Cubic spline interpolation—smooth first and second derivatives throughout the curve.

The default method is `"linear"`.

If *extrapval* is the scalar value, use it to replace the values beyond the endpoints with that number. If *extrapval* is missing, assume NA.

See also: [interp1], page 617, [interp2], page 621, [spline], page 620, [ndgrid], page 303.

A significant difference between `interpn` and the other two multi-dimensional interpolation functions is the fashion in which the dimensions are treated. For `interp2` and `interp3`, the y-axis is considered to be the columns of the matrix, whereas the x-axis corresponds to the rows of the array. As Octave indexes arrays in column major order, the first dimension of any array is the columns, and so `interpn` effectively reverses the 'x' and 'y' dimensions. Consider the example,

```
x = y = z = -1:1;
f = @(x,y,z) x.^2 - y - z.^2;
[xx, yy, zz] = meshgrid (x, y, z);
v = f (xx,yy,zz);
xi = yi = zi = -1:0.1:1;
[xxi, yyi, zzi] = meshgrid (xi, yi, zi);
vi = interp3 (x, y, z, v, xxi, yyi, zzi, "spline");
[xxi, yyi, zzi] = ndgrid (xi, yi, zi);
vi2 = interpn (x, y, z, v, xxi, yyi, zzi, "spline");
mesh (zi, yi, squeeze (vi2(1,:,:)));
```

where vi and vi2 are identical. The reversal of the dimensions is treated in the **meshgrid** and **ndgrid** functions respectively. The result of this code can be seen in Figure 29.4.

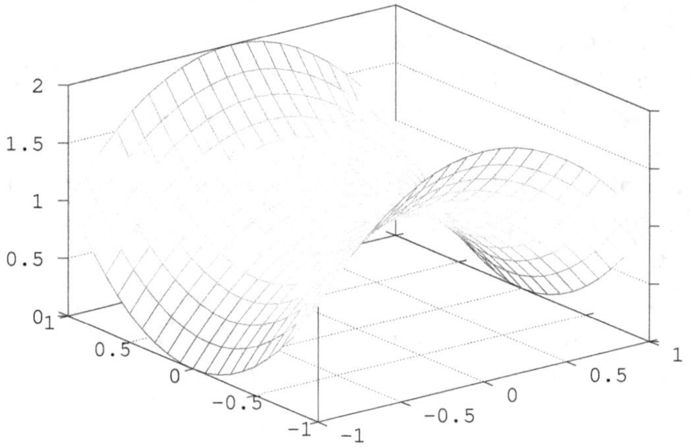

Figure 29.4: Demonstration of the use of **interpn**

In additional the support function **bicubic** that underlies the cubic interpolation of **interp2** function can be called directly.

zi = **bicubic** (*x*, *y*, *z*, *xi*, *yi*, *extrapval*) [Function File]

Return a matrix *zi* corresponding to the bicubic interpolations at *xi* and *yi* of the data supplied as *x*, *y* and *z*. Points outside the grid are set to *extrapval*.

See http://wiki.woodpecker.org.cn/moin/Octave/Bicubic for further information.

See also: [interp2], page 621.

30 Geometry

Much of the geometry code in Octave is based on the Qhull library[1]. Some of the documentation for Qhull, particularly for the options that can be passed to `delaunay`, `voronoi` and `convhull`, etc., is relevant to Octave users.

30.1 Delaunay Triangulation

The Delaunay triangulation is constructed from a set of circum-circles. These circum-circles are chosen so that there are at least three of the points in the set to triangulation on the circumference of the circum-circle. None of the points in the set of points falls within any of the circum-circles.

In general there are only three points on the circumference of any circum-circle. However, in some cases, and in particular for the case of a regular grid, 4 or more points can be on a single circum-circle. In this case the Delaunay triangulation is not unique.

`delaunay` (*x*, *y*) [Function File]
`delaunay` (*x*) [Function File]
`delaunay` (..., *options*) [Function File]
tri = `delaunay` (...) [Function File]

Compute the Delaunay triangulation for a 2-D set of points. The return value *tri* is a set of triangles which satisfies the Delaunay circum-circle criterion, i.e., only a single data point from [*x*, *y*] is within the circum-circle of the defining triangle. The input *x* may also be a matrix with two columns where the first column contains x-data and the second y-data.

The set of triangles *tri* is a matrix of size [n, 3]. Each row defines a triangle and the three columns are the three vertices of the triangle. The value of `tri(i,j)` is an index into *x* and *y* for the location of the j-th vertex of the i-th triangle.

The optional last argument, which must be a string or cell array of strings, contains options passed to the underlying qhull command. See the documentation for the Qhull library for details `http://www.qhull.org/html/qh-quick.htm#options`. The default options are `{"Qt", "Qbb", "Qc", "Qz"}`.

If *options* is not present or `[]` then the default arguments are used. Otherwise, *options* replaces the default argument list. To append user options to the defaults it is necessary to repeat the default arguments in *options*. Use a null string to pass no arguments.

If no output argument is specified the resulting Delaunay triangulation is plotted along with the original set of points.

```
x = rand (1, 10);
y = rand (1, 10);
T = delaunay (x, y);
VX = [ x(T(:,1)); x(T(:,2)); x(T(:,3)); x(T(:,1)) ];
VY = [ y(T(:,1)); y(T(:,2)); y(T(:,3)); y(T(:,1)) ];
axis ([0,1,0,1]);
plot (VX, VY, "b", x, y, "r*");
```

[1] Barber, C.B., Dobkin, D.P., and Huhdanpaa, H.T., *The Quickhull Algorithm for Convex Hulls*, ACM Trans. on Mathematical Software, 22(4):469–483, Dec 1996, `http://www.qhull.org`

See also: [delaunay3], page 626, [delaunayn], page 626, [convhull], page 635, [voronoi], page 632, [triplot], page 627, [trimesh], page 628, [trisurf], page 628.

The 3- and N-dimensional extension of the Delaunay triangulation are given by `delaunay3` and `delaunayn` respectively. `delaunay3` returns a set of tetrahedra that satisfy the Delaunay circum-circle criteria. Similarly, `delaunayn` returns the N-dimensional simplex satisfying the Delaunay circum-circle criteria. The N-dimensional extension of a triangulation is called a tessellation.

tetr = **delaunay3** (*x*, *y*, *z*) [Function File]
tetr = **delaunay3** (*x*, *y*, *z*, *options*) [Function File]

Compute the Delaunay triangulation for a 3-D set of points. The return value *tetr* is a set of tetrahedrons which satisfies the Delaunay circum-circle criterion, i.e., only a single data point from [x, y, z] is within the circum-circle of the defining tetrahedron.

The set of tetrahedrons *tetr* is a matrix of size [n, 4]. Each row defines a tetrahedron and the four columns are the four vertices of the tetrahedron. The value of `tetr(i,j)` is an index into *x*, *y*, *z* for the location of the j-th vertex of the i-th tetrahedron.

An optional fourth argument, which must be a string or cell array of strings, contains options passed to the underlying qhull command. See the documentation for the Qhull library for details `http://www.qhull.org/html/qh-quick.htm#options`. The default options are `{"Qt", "Qbb", "Qc", "Qz"}`.

If *options* is not present or `[]` then the default arguments are used. Otherwise, *options* replaces the default argument list. To append user options to the defaults it is necessary to repeat the default arguments in *options*. Use a null string to pass no arguments.

See also: [delaunay], page 625, [delaunayn], page 626, [convhull], page 635, [voronoi], page 632, [tetramesh], page 629.

T = **delaunayn** (*pts*) [Function File]
T = **delaunayn** (*pts*, *options*) [Function File]

Compute the Delaunay triangulation for an N-dimensional set of points. The Delaunay triangulation is a tessellation of the convex hull of a set of points such that no N-sphere defined by the N-triangles contains any other points from the set.

The input matrix *pts* of size [n, dim] contains n points in a space of dimension dim. The return matrix *T* has size [m, dim+1]. Each row of *T* contains a set of indices back into the original set of points *pts* which describes a simplex of dimension dim. For example, a 2-D simplex is a triangle and 3-D simplex is a tetrahedron.

An optional second argument, which must be a string or cell array of strings, contains options passed to the underlying qhull command. See the documentation for the Qhull library for details `http://www.qhull.org/html/qh-quick.htm#options`. The default options depend on the dimension of the input:

- 2-D and 3-D: *options* = `{"Qt", "Qbb", "Qc", "Qz"}`
- 4-D and higher: *options* = `{"Qt", "Qbb", "Qc", "Qx"}`

If *options* is not present or `[]` then the default arguments are used. Otherwise, *options* replaces the default argument list. To append user options to the defaults it is necessary to repeat the default arguments in *options*. Use a null string to pass no arguments.

See also: [delaunay], page 625, [delaunay3], page 626, [convhulln], page 636, [voronoin], page 633, [trimesh], page 628, [tetramesh], page 629.

An example of a Delaunay triangulation of a set of points is

```
rand ("state", 2);
x = rand (10, 1);
y = rand (10, 1);
T = delaunay (x, y);
X = [ x(T(:,1)); x(T(:,2)); x(T(:,3)); x(T(:,1)) ];
Y = [ y(T(:,1)); y(T(:,2)); y(T(:,3)); y(T(:,1)) ];
axis ([0, 1, 0, 1]);
plot (X, Y, "b", x, y, "r*");
```

The result of which can be seen in Figure 30.1.

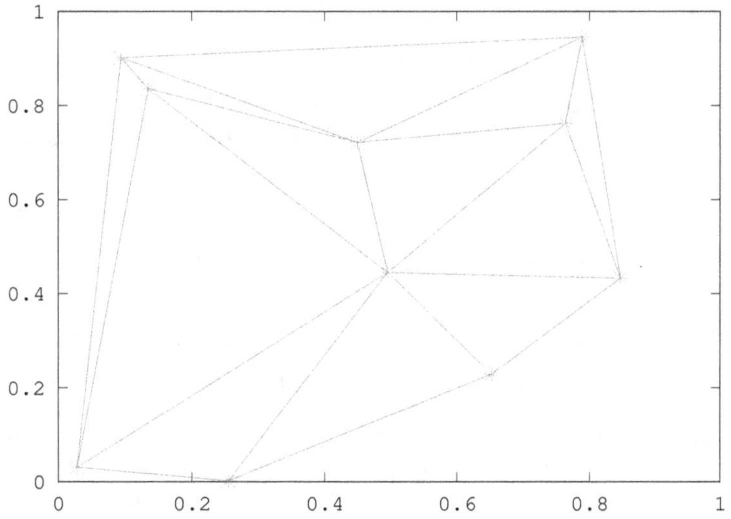

Figure 30.1: Delaunay triangulation of a random set of points

30.1.1 Plotting the Triangulation

Octave has the functions **triplot**, **trimesh**, and **trisurf** to plot the Delaunay triangulation of a 2-dimensional set of points. **tetramesh** will plot the triangulation of a 3-dimensional set of points.

triplot (*tri*, *x*, *y*) [Function File]
triplot (*tri*, *x*, *y*, *linespec*) [Function File]
h = **triplot** (...) [Function File]

Plot a 2-D triangular mesh.

tri is typically the output of a Delaunay triangulation over the grid of *x*, *y*. Every row of *tri* represents one triangle and contains three indices into [*x*, *y*] which are the vertices of the triangles in the x-y plane.

The linestyle to use for the plot can be defined with the argument *linespec* of the same format as the `plot` command.

The optional return value *h* is a graphics handle to the created patch object.

See also: [plot], page 260, [trimesh], page 628, [trisurf], page 628, [delaunay], page 625.

`trimesh` (*tri*, *x*, *y*, *z*, *c*) [Function File]
`trimesh` (*tri*, *x*, *y*, *z*) [Function File]
`trimesh` (*tri*, *x*, *y*) [Function File]
`trimesh` (..., *prop*, *val*, ...) [Function File]
h = `trimesh` (...) [Function File]

Plot a 3-D triangular wireframe mesh.

In contrast to `mesh`, which plots a mesh using rectangles, `trimesh` plots the mesh using triangles.

tri is typically the output of a Delaunay triangulation over the grid of *x*, *y*. Every row of *tri* represents one triangle and contains three indices into [*x*, *y*] which are the vertices of the triangles in the x-y plane. *z* determines the height above the plane of each vertex. If no *z* input is given then the triangles are plotted as a 2-D figure.

The color of the trimesh is computed by linearly scaling the *z* values to fit the range of the current colormap. Use `caxis` and/or change the colormap to control the appearance.

Optionally, the color of the mesh can be specified independently of *z* by supplying a color matrix, *c*. If *z* has N elements, then *c* should be an Nx1 vector for colormap data or an Nx3 matrix for RGB data.

Any property/value pairs are passed directly to the underlying patch object.

The optional return value *h* is a graphics handle to the created patch object.

See also: [mesh], page 292, [tetramesh], page 629, [triplot], page 627, [trisurf], page 628, [delaunay], page 625, [patch], page 337, [hidden], page 294.

`trisurf` (*tri*, *x*, *y*, *z*, *c*) [Function File]
`trisurf` (*tri*, *x*, *y*, *z*) [Function File]
`trisurf` (..., *prop*, *val*, ...) [Function File]
h = `trisurf` (...) [Function File]

Plot a 3-D triangular surface.

In contrast to `surf`, which plots a surface mesh using rectangles, `trisurf` plots the mesh using triangles.

tri is typically the output of a Delaunay triangulation over the grid of *x*, *y*. Every row of *tri* represents one triangle and contains three indices into [*x*, *y*] which are the vertices of the triangles in the x-y plane. *z* determines the height above the plane of each vertex.

The color of the trimesh is computed by linearly scaling the *z* values to fit the range of the current colormap. Use `caxis` and/or change the colormap to control the appearance.

Optionally, the color of the mesh can be specified independently of *z* by supplying a color matrix, *c*. If *z* has N elements, then *c* should be an Nx1 vector for colormap data or an Nx3 matrix for RGB data.

Any property/value pairs are passed directly to the underlying patch object.

The optional return value *h* is a graphics handle to the created patch object.

See also: [surf], page 295, [triplot], page 627, [trimesh], page 628, [delaunay], page 625, [patch], page 337, [shading], page 306.

`tetramesh` (*T*, *X*) [Function File]
`tetramesh` (*T*, *X*, *C*) [Function File]
`tetramesh` (..., *property*, *val*, ...) [Function File]
h = `tetramesh` (...) [Function File]

Display the tetrahedrons defined in the m-by-4 matrix *T* as 3-D patches.

T is typically the output of a Delaunay triangulation of a 3-D set of points. Every row of *T* contains four indices into the n-by-3 matrix *X* of the vertices of a tetrahedron. Every row in *X* represents one point in 3-D space.

The vector *C* specifies the color of each tetrahedron as an index into the current colormap. The default value is 1:m where m is the number of tetrahedrons; the indices are scaled to map to the full range of the colormap. If there are more tetrahedrons than colors in the colormap then the values in *C* are cyclically repeated.

Calling `tetramesh (..., "property", "value", ...)` passes all property/value pairs directly to the patch function as additional arguments.

The optional return value *h* is a vector of patch handles where each handle represents one tetrahedron in the order given by *T*. A typical use case for *h* is to turn the respective patch `"visible"` property `"on"` or `"off"`.

Type `demo tetramesh` to see examples on using `tetramesh`.

See also: [trimesh], page 628, [delaunay3], page 626, [delaunayn], page 626, [patch], page 337.

The difference between `triplot`, and `trimesh` or `triplot`, is that the former only plots the 2-dimensional triangulation itself, whereas the second two plot the value of a function `f (x, y)`. An example of the use of the `triplot` function is

```
rand ("state", 2)
x = rand (20, 1);
y = rand (20, 1);
tri = delaunay (x, y);
triplot (tri, x, y);
```

which plots the Delaunay triangulation of a set of random points in 2-dimensions. The output of the above can be seen in Figure 30.2.

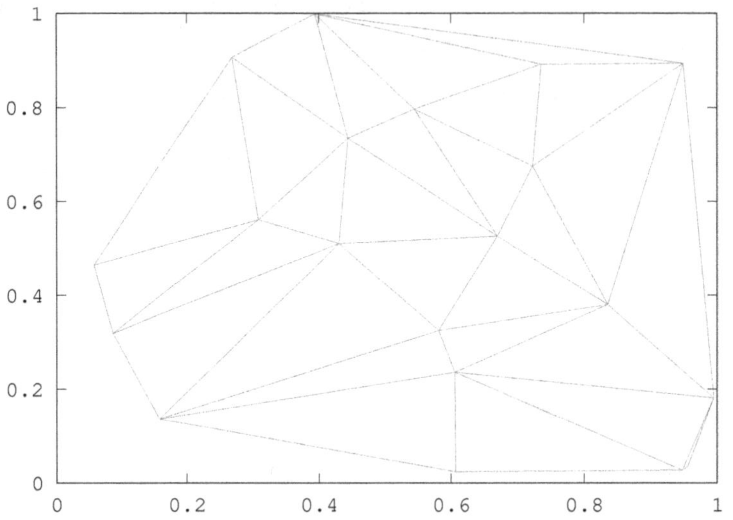

Figure 30.2: Delaunay triangulation of a random set of points

30.1.2 Identifying Points in Triangulation

It is often necessary to identify whether a particular point in the N-dimensional space is within the Delaunay tessellation of a set of points in this N-dimensional space, and if so which N-simplex contains the point and which point in the tessellation is closest to the desired point. The functions `tsearch` and `dsearch` perform this function in a triangulation, and `tsearchn` and `dsearchn` in an N-dimensional tessellation.

To identify whether a particular point represented by a vector *p* falls within one of the simplices of an N-simplex, we can write the Cartesian coordinates of the point in a parametric form with respect to the N-simplex. This parametric form is called the Barycentric Coordinates of the point. If the points defining the N-simplex are given by `N + 1` vectors *t(i,:)*, then the Barycentric coordinates defining the point *p* are given by

```
p = sum (beta(1:N+1) * t(1:N+1),:)
```

where there are `N + 1` values `beta(i)` that together as a vector represent the Barycentric coordinates of the point *p*. To ensure a unique solution for the values of `beta(i)` an additional criteria of

```
sum (beta(1:N+1)) == 1
```

is imposed, and we can therefore write the above as

```
p - t(end, :) = beta(1:end-1) * (t(1:end-1, :)
        - ones (N, 1) * t(end, :)
```

Solving for *beta* we can then write

```
beta(1:end-1) = (p - t(end, :)) / (t(1:end-1, :)
        - ones (N, 1) * t(end, :))
beta(end) = sum (beta(1:end-1))
```

which gives the formula for the conversion of the Cartesian coordinates of the point *p* to the Barycentric coordinates *beta*. An important property of the Barycentric coordinates is that for all points in the N-simplex

```
0 <= beta(i) <= 1
```

Therefore, the test in `tsearch` and `tsearchn` essentially only needs to express each point in terms of the Barycentric coordinates of each of the simplices of the N-simplex and test the values of *beta*. This is exactly the implementation used in `tsearchn`. `tsearch` is optimized for 2-dimensions and the Barycentric coordinates are not explicitly formed.

idx = **tsearch** (*x*, *y*, *t*, *xi*, *yi*) [Loadable Function]

Search for the enclosing Delaunay convex hull. For `t = delaunay (x, y)`, finds the index in *t* containing the points `(xi, yi)`. For points outside the convex hull, *idx* is NaN.

See also: [delaunay], page 625, [delaunayn], page 626.

[*idx*, *p*] = **tsearchn** (*x*, *t*, *xi*) [Function File]

Search for the enclosing Delaunay convex hull. For `t = delaunayn (x)`, finds the index in *t* containing the points *xi*. For points outside the convex hull, *idx* is NaN. If requested `tsearchn` also returns the Barycentric coordinates *p* of the enclosing triangles.

See also: [delaunay], page 625, [delaunayn], page 626.

An example of the use of `tsearch` can be seen with the simple triangulation

```
x = [-1; -1; 1; 1];
y = [-1; 1; -1; 1];
tri = [1, 2, 3; 2, 3, 1];
```

consisting of two triangles defined by *tri*. We can then identify which triangle a point falls in like

```
tsearch (x, y, tri, -0.5, -0.5)
⇒ 1
tsearch (x, y, tri, 0.5, 0.5)
⇒ 2
```

and we can confirm that a point doesn't lie within one of the triangles like

```
tsearch (x, y, tri, 2, 2)
⇒ NaN
```

The `dsearch` and `dsearchn` find the closest point in a tessellation to the desired point. The desired point does not necessarily have to be in the tessellation, and even if it the returned point of the tessellation does not have to be one of the vertexes of the N-simplex within which the desired point is found.

idx = **dsearch** (*x*, *y*, *tri*, *xi*, *yi*) [Function File]
idx = **dsearch** (*x*, *y*, *tri*, *xi*, *yi*, *s*) [Function File]

Return the index *idx* or the closest point in `x, y` to the elements `[xi(:), yi(:)]`. The variable *s* is accepted for compatibility but is ignored.

See also: [dsearchn], page 631, [tsearch], page 631.

idx = **dsearchn** (*x*, *tri*, *xi*) [Function File]
idx = **dsearchn** (*x*, *tri*, *xi*, *outval*) [Function File]
idx = **dsearchn** (*x*, *xi*) [Function File]

`[idx, d] = dsearchn (...)` [Function File]

Return the index *idx* or the closest point in *x* to the elements *xi*. If *outval* is supplied, then the values of *xi* that are not contained within one of the simplices *tri* are set to *outval*. Generally, *tri* is returned from `delaunayn (x)`.

See also: [dsearch], page 631, [tsearch], page 631.

An example of the use of `dsearch`, using the above values of *x*, *y* and *tri* is

```
dsearch (x, y, tri, -2, -2)
⇒ 1
```

If you wish the points that are outside the tessellation to be flagged, then `dsearchn` can be used as

```
dsearchn ([x, y], tri, [-2, -2], NaN)
⇒ NaN
dsearchn ([x, y], tri, [-0.5, -0.5], NaN)
⇒ 1
```

where the point outside the tessellation are then flagged with `NaN`.

30.2 Voronoi Diagrams

A Voronoi diagram or Voronoi tessellation of a set of points *s* in an N-dimensional space, is the tessellation of the N-dimensional space such that all points in `v(p)`, a partitions of the tessellation where *p* is a member of *s*, are closer to *p* than any other point in *s*. The Voronoi diagram is related to the Delaunay triangulation of a set of points, in that the vertexes of the Voronoi tessellation are the centers of the circum-circles of the simplices of the Delaunay tessellation.

`voronoi (x, y)` [Function File]
`voronoi (x, y, options)` [Function File]
`voronoi (..., "linespec")` [Function File]
`voronoi (hax, ...)` [Function File]
`h = voronoi (...)` [Function File]
`[vx, vy] = voronoi (...)` [Function File]

Plot the Voronoi diagram of points `(x, y)`. The Voronoi facets with points at infinity are not drawn.

If `"linespec"` is given it is used to set the color and line style of the plot. If an axis graphics handle *hax* is supplied then the Voronoi diagram is drawn on the specified axis rather than in a new figure.

The *options* argument, which must be a string or cell array of strings, contains options passed to the underlying qhull command. See the documentation for the Qhull library for details `http://www.qhull.org/html/qh-quick.htm#options`.

If a single output argument is requested then the Voronoi diagram will be plotted and a graphics handle *h* to the plot is returned. [*vx*, *vy*] = voronoi (...) returns the Voronoi vertices instead of plotting the diagram.

```
x = rand (10, 1);
y = rand (size (x));
h = convhull (x, y);
[vx, vy] = voronoi (x, y);
plot (vx, vy, "-b", x, y, "o", x(h), y(h), "-g");
legend ("", "points", "hull");
```

See also: [voronoin], page 633, [delaunay], page 625, [convhull], page 635.

`[C, F] = voronoin (pts)` [Function File]
`[C, F] = voronoin (pts, options)` [Function File]

Compute N-dimensional Voronoi facets. The input matrix *pts* of size [n, dim] contains n points in a space of dimension dim. *C* contains the points of the Voronoi facets. The list *F* contains, for each facet, the indices of the Voronoi points.

An optional second argument, which must be a string or cell array of strings, contains options passed to the underlying qhull command. See the documentation for the Qhull library for details `http://www.qhull.org/html/qh-quick.htm#options`.

The default options depend on the dimension of the input:

- 2-D and 3-D: *options* = `{"Qbb"}`
- 4-D and higher: *options* = `{"Qbb", "Qx"}`

If *options* is not present or `[]` then the default arguments are used. Otherwise, *options* replaces the default argument list. To append user options to the defaults it is necessary to repeat the default arguments in *options*. Use a null string to pass no arguments.

See also: [voronoi], page 632, [convhulln], page 636, [delaunayn], page 626.

An example of the use of `voronoi` is

```
rand ("state",9);
x = rand (10,1);
y = rand (10,1);
tri = delaunay (x, y);
[vx, vy] = voronoi (x, y, tri);
triplot (tri, x, y, "b");
hold on;
plot (vx, vy, "r");
```

The result of which can be seen in Figure 30.3. Note that the circum-circle of one of the triangles has been added to this figure, to make the relationship between the Delaunay tessellation and the Voronoi diagram clearer.

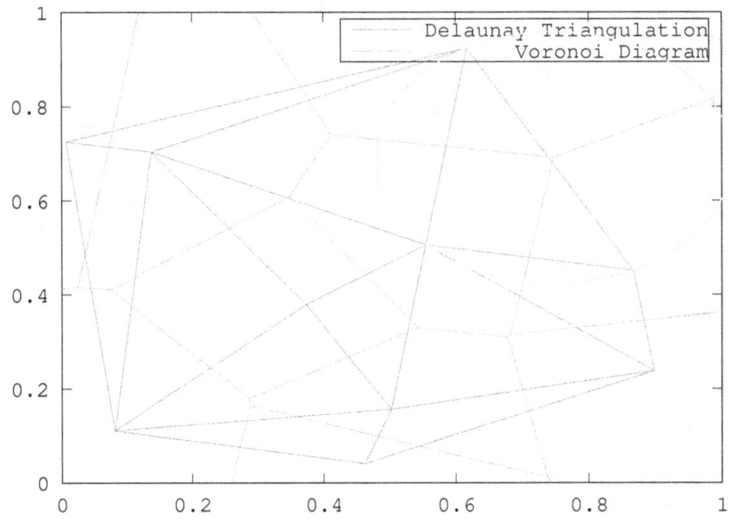

Figure 30.3: Delaunay triangulation and Voronoi diagram of a random set of points

Additional information about the size of the facets of a Voronoi diagram, and which points of a set of points is in a polygon can be had with the **polyarea** and **inpolygon** functions respectively.

polyarea (*x*, *y*) [Function File]
polyarea (*x*, *y*, *dim*) [Function File]

Determine area of a polygon by triangle method. The variables *x* and *y* define the vertex pairs, and must therefore have the same shape. They can be either vectors or arrays. If they are arrays then the columns of *x* and *y* are treated separately and an area returned for each.

If the optional *dim* argument is given, then **polyarea** works along this dimension of the arrays *x* and *y*.

An example of the use of **polyarea** might be

```
rand ("state", 2);
x = rand (10, 1);
y = rand (10, 1);
[c, f] = voronoin ([x, y]);
af = zeros (size (f));
for i = 1 : length (f)
  af(i) = polyarea (c (f {i, :}, 1), c (f {i, :}, 2));
endfor
```

Facets of the Voronoi diagram with a vertex at infinity have infinity area. A simplified version of **polyarea** for rectangles is available with **rectint**

area = **rectint** (*a*, *b*) [Function File]

Compute the area of intersection of rectangles in *a* and rectangles in *b*. Rectangles are defined as [x y width height] where x and y are the minimum values of the two orthogonal dimensions.

If *a* or *b* are matrices, then the output, *area*, is a matrix where the i-th row corresponds to the i-th row of a and the j-th column corresponds to the j-th row of b.

See also: [polyarea], page 634.

`[in, on] = inpolygon (x, y, xv, yv)` [Function File]

For a polygon defined by vertex points `(xv, yv)`, determine if the points `(x, y)` are inside or outside the polygon. The variables *x*, *y*, must have the same dimension. The optional output *on* gives the points that are on the polygon.

An example of the use of `inpolygon` might be

```
randn ("state", 2);
x = randn (100, 1);
y = randn (100, 1);
vx = cos (pi * [-1 : 0.1: 1]);
vy = sin (pi * [-1 : 0.1 : 1]);
in = inpolygon (x, y, vx, vy);
plot (vx, vy, x(in), y(in), "r+", x(!in), y(!in), "bo");
axis ([-2, 2, -2, 2]);
```

The result of which can be seen in Figure 30.4.

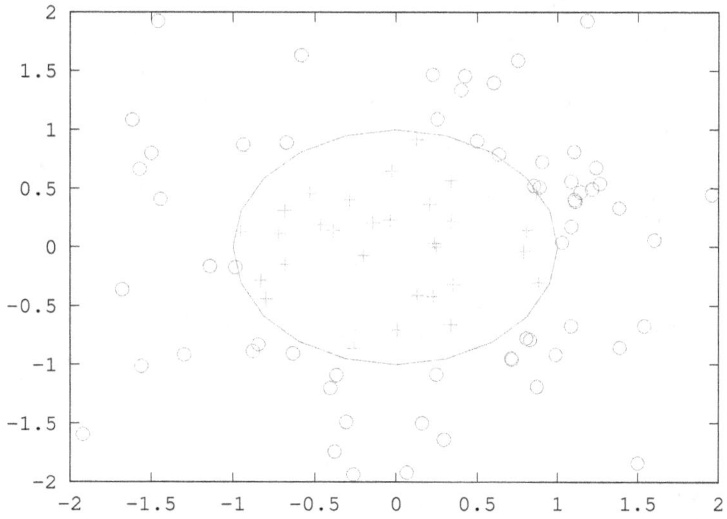

Figure 30.4: Demonstration of the `inpolygon` function to determine the points inside a polygon

30.3 Convex Hull

The convex hull of a set of points is the minimum convex envelope containing all of the points. Octave has the functions `convhull` and `convhulln` to calculate the convex hull of 2-dimensional and N-dimensional sets of points.

H = `convhull` (*x*, *y*) [Function File]
H = `convhull` (*x*, *y*, *options*) [Function File]

Compute the convex hull of the set of points defined by the arrays *x* and *y*. The hull *H* is an index vector into the set of points and specifies which points form the enclosing hull.

An optional third argument, which must be a string or cell array of strings, contains options passed to the underlying qhull command. See the documentation for the Qhull library for details `http://www.qhull.org/html/qh-quick.htm#options`. The default option is `{"Qt"}`.

If *options* is not present or `[]` then the default arguments are used. Otherwise, *options* replaces the default argument list. To append user options to the defaults it is necessary to repeat the default arguments in *options*. Use a null string to pass no arguments.

See also: [convhulln], page 636, [delaunay], page 625, [voronoi], page 632.

h = `convhulln` (*pts*) [Loadable Function]
h = `convhulln` (*pts*, *options*) [Loadable Function]
[*h*, *v*] = `convhulln` (...) [Loadable Function]

Compute the convex hull of the set of points *pts* which is a matrix of size [n, dim] containing n points in a space of dimension dim. The hull *h* is an index vector into the set of points and specifies which points form the enclosing hull.

An optional second argument, which must be a string or cell array of strings, contains options passed to the underlying qhull command. See the documentation for the Qhull library for details `http://www.qhull.org/html/qh-quick.htm#options`. The default options depend on the dimension of the input:

- 2D, 3D, 4D: *options* = `{"Qt"}`
- 5D and higher: *options* = `{"Qt", "Qx"}`

If *options* is not present or `[]` then the default arguments are used. Otherwise, *options* replaces the default argument list. To append user options to the defaults it is necessary to repeat the default arguments in *options*. Use a null string to pass no arguments.

If the second output *v* is requested the volume of the enclosing convex hull is calculated.

See also: [convhull], page 635, [delaunayn], page 626, [voronoin], page 633.

An example of the use of `convhull` is

```
x = -3:0.05:3;
y = abs (sin (x));
k = convhull (x, y);
plot (x(k), y(k), "r-", x, y, "b+");
axis ([-3.05, 3.05, -0.05, 1.05]);
```

The output of the above can be seen in Figure 30.5.

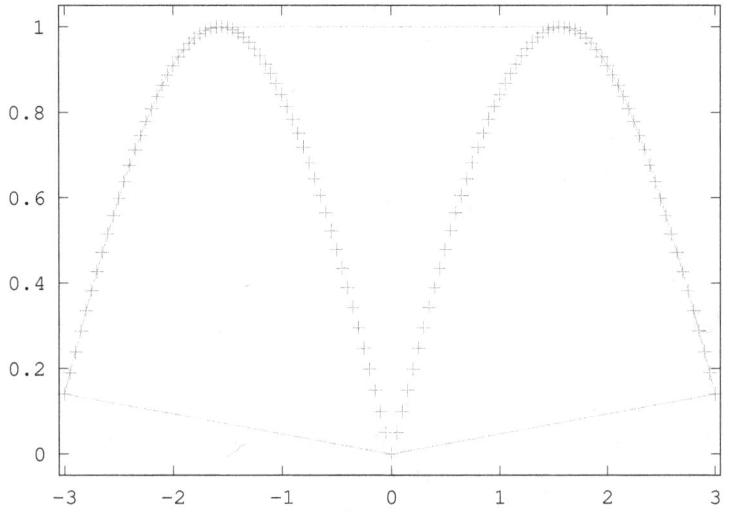

Figure 30.5: The convex hull of a simple set of points

30.4 Interpolation on Scattered Data

An important use of the Delaunay tessellation is that it can be used to interpolate from scattered data to an arbitrary set of points. To do this the N-simplex of the known set of points is calculated with `delaunay`, `delaunay3` or `delaunayn`. Then the simplices in to which the desired points are found are identified. Finally the vertices of the simplices are used to interpolate to the desired points. The functions that perform this interpolation are `griddata`, `griddata3` and `griddatan`.

zi = **griddata** (*x*, *y*, *z*, *xi*, *yi*) [Function File]
zi = **griddata** (*x*, *y*, *z*, *xi*, *yi*, *method*) [Function File]
[*xi*, *yi*, *zi*] = **griddata** (...) [Function File]

Generate a regular mesh from irregular data using interpolation. The function is defined by `z = f (x, y)`. Inputs `x, y, z` are vectors of the same length or `x, y` are vectors and `z` is matrix.

The interpolation points are all `(xi, yi)`. If *xi*, *yi* are vectors then they are made into a 2-D mesh.

The interpolation method can be `"nearest"`, `"cubic"` or `"linear"`. If method is omitted it defaults to `"linear"`.

See also: [griddata3], page 637, [griddatan], page 638, [delaunay], page 625.

vi = **griddata3** (*x*, *y*, *z*, *v*, *xi*, *yi*, *zi*) [Function File]
vi = **griddata3** (*x*, *y*, *z*, *v*, *xi*, *yi*, *zi*, *method*) [Function File]
vi = **griddata3** (*x*, *y*, *z*, *v*, *xi*, *yi*, *zi*, *method*, *options*) [Function File]

Generate a regular mesh from irregular data using interpolation. The function is defined by `v = f (x, y, z)`. The interpolation points are specified by *xi*, *yi*, *zi*.

The interpolation method can be `"nearest"` or `"linear"`. If method is omitted it defaults to `"linear"`.

The optional argument *options* is passed directly to Qhull when computing the Delaunay triangulation used for interpolation. See **delaunayn** for information on the defaults and how to pass different values.

See also: [griddata], page 637, [griddatan], page 638, [delaunayn], page 626.

yi = **griddatan** (*x*, *y*, *xi*) [Function File]
yi = **griddatan** (*x*, *y*, *xi*, *method*) [Function File]
yi = **griddatan** (*x*, *y*, *xi*, *method*, *options*) [Function File]

Generate a regular mesh from irregular data using interpolation. The function is defined by `y = f (x)`. The interpolation points are all *xi*.

The interpolation method can be `"nearest"` or `"linear"`. If method is omitted it defaults to `"linear"`.

The optional argument *options* is passed directly to Qhull when computing the Delaunay triangulation used for interpolation. See **delaunayn** for information on the defaults and how to pass different values.

See also: [griddata], page 637, [griddata3], page 637, [delaunayn], page 626.

An example of the use of the `griddata` function is

```
rand ("state", 1);
x = 2*rand (1000,1) - 1;
y = 2*rand (size (x)) - 1;
z = sin (2*(x.^2+y.^2));
[xx,yy] = meshgrid (linspace (-1,1,32));
griddata (x,y,z,xx,yy);
```

that interpolates from a random scattering of points, to a uniform grid. The output of the above can be seen in Figure 30.6.

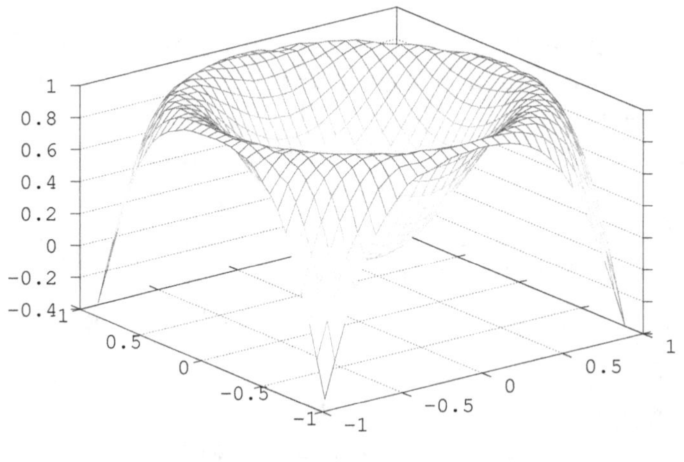

Figure 30.6: Interpolation from a scattered data to a regular grid

31 Signal Processing

This chapter describes the signal processing and fast Fourier transform functions available in Octave. Fast Fourier transforms are computed with the FFTW or FFTPACK libraries depending on how Octave is built.

`fft (x)` [Built-in Function]
`fft (x, n)` [Built-in Function]
`fft (x, n, dim)` [Built-in Function]

Compute the discrete Fourier transform of *A* using a Fast Fourier Transform (FFT) algorithm.

The FFT is calculated along the first non-singleton dimension of the array. Thus if *x* is a matrix, `fft (x)` computes the FFT for each column of *x*.

If called with two arguments, *n* is expected to be an integer specifying the number of elements of *x* to use, or an empty matrix to specify that its value should be ignored. If *n* is larger than the dimension along which the FFT is calculated, then *x* is resized and padded with zeros. Otherwise, if *n* is smaller than the dimension along which the FFT is calculated, then *x* is truncated.

If called with three arguments, *dim* is an integer specifying the dimension of the matrix along which the FFT is performed

See also: [ifft], page 639, [fft2], page 639, [fftn], page 640, [fftw], page 640.

`ifft (x)` [Built-in Function]
`ifft (x, n)` [Built-in Function]
`ifft (x, n, dim)` [Built-in Function]

Compute the inverse discrete Fourier transform of *A* using a Fast Fourier Transform (FFT) algorithm.

The inverse FFT is calculated along the first non-singleton dimension of the array. Thus if *x* is a matrix, `fft (x)` computes the inverse FFT for each column of *x*.

If called with two arguments, *n* is expected to be an integer specifying the number of elements of *x* to use, or an empty matrix to specify that its value should be ignored. If *n* is larger than the dimension along which the inverse FFT is calculated, then *x* is resized and padded with zeros. Otherwise, if *n* is smaller than the dimension along which the inverse FFT is calculated, then *x* is truncated.

If called with three arguments, *dim* is an integer specifying the dimension of the matrix along which the inverse FFT is performed

See also: [fft], page 639, [ifft2], page 640, [ifftn], page 640, [fftw], page 640.

`fft2 (A)` [Built-in Function]
`fft2 (A, m, n)` [Built-in Function]

Compute the two-dimensional discrete Fourier transform of *A* using a Fast Fourier Transform (FFT) algorithm.

The optional arguments *m* and *n* may be used specify the number of rows and columns of *A* to use. If either of these is larger than the size of *A*, *A* is resized and padded with zeros.

If *A* is a multi-dimensional matrix, each two-dimensional sub-matrix of *A* is treated separately.

See also: ifft2, fft, fftn, fftw.

`ifft2 (A)` [Built-in Function]
`ifft2 (A, m, n)` [Built-in Function]

Compute the inverse two-dimensional discrete Fourier transform of *A* using a Fast Fourier Transform (FFT) algorithm.

The optional arguments *m* and *n* may be used specify the number of rows and columns of *A* to use. If either of these is larger than the size of *A*, *A* is resized and padded with zeros.

If *A* is a multi-dimensional matrix, each two-dimensional sub-matrix of *A* is treated separately

See also: fft2, ifft, ifftn, fftw.

`fftn (A)` [Built-in Function]
`fftn (A, size)` [Built-in Function]

Compute the N-dimensional discrete Fourier transform of *A* using a Fast Fourier Transform (FFT) algorithm.

The optional vector argument *size* may be used specify the dimensions of the array to be used. If an element of *size* is smaller than the corresponding dimension of *A*, then the dimension of *A* is truncated prior to performing the FFT. Otherwise, if an element of *size* is larger than the corresponding dimension then *A* is resized and padded with zeros.

See also: [ifftn], page 640, [fft], page 639, [fft2], page 639, [fftw], page 640.

`ifftn (A)` [Built-in Function]
`ifftn (A, size)` [Built-in Function]

Compute the inverse N-dimensional discrete Fourier transform of *A* using a Fast Fourier Transform (FFT) algorithm.

The optional vector argument *size* may be used specify the dimensions of the array to be used. If an element of *size* is smaller than the corresponding dimension of *A*, then the dimension of *A* is truncated prior to performing the inverse FFT. Otherwise, if an element of *size* is larger than the corresponding dimension then *A* is resized and padded with zeros.

See also: [fftn], page 640, [ifft], page 639, [ifft2], page 640, [fftw], page 640.

Octave uses the FFTW libraries to perform FFT computations. When Octave starts up and initializes the FFTW libraries, they read a system wide file (on a Unix system, it is typically '`/etc/fftw/wisdom`') that contains information useful to speed up FFT computations. This information is called the *wisdom*. The system-wide file allows wisdom to be shared between all applications using the FFTW libraries.

Use the `fftw` function to generate and save wisdom. Using the utilities provided together with the FFTW libraries (`fftw-wisdom` on Unix systems), you can even add wisdom generated by Octave to the system-wide wisdom file.

method = `fftw` ("*planner*") [Loadable Function]
`fftw` ("*planner*", ***method***) [Loadable Function]
wisdom = `fftw` ("*dwisdom*") [Loadable Function]
`fftw` ("*dwisdom*", ***wisdom***) [Loadable Function]
`fftw` ("*threads*", ***nthreads***) [Loadable Function]
nthreads = `fftw` ("*threads*") [Loadable Function]

Manage FFTW wisdom data. Wisdom data can be used to significantly accelerate the calculation of the FFTs, but implies an initial cost in its calculation. When the FFTW libraries are initialized, they read a system wide wisdom file (typically in '`/etc/fftw/wisdom`'), allowing wisdom to be shared between applications other than Octave. Alternatively, the `fftw` function can be used to import wisdom. For example,

```
wisdom = fftw ("dwisdom")
```

will save the existing wisdom used by Octave to the string *wisdom*. This string can then be saved to a file and restored using the `save` and `load` commands respectively. This existing wisdom can be re-imported as follows

```
fftw ("dwisdom", wisdom)
```

If *wisdom* is an empty string, then the wisdom used is cleared.

During the calculation of Fourier transforms further wisdom is generated. The fashion in which this wisdom is generated is also controlled by the `fftw` function. There are five different manners in which the wisdom can be treated:

`"estimate"`
: Specifies that no run-time measurement of the optimal means of calculating a particular is performed, and a simple heuristic is used to pick a (probably sub-optimal) plan. The advantage of this method is that there is little or no overhead in the generation of the plan, which is appropriate for a Fourier transform that will be calculated once.

`"measure"`
: In this case a range of algorithms to perform the transform is considered and the best is selected based on their execution time.

`"patient"`
: Similar to `"measure"`, but a wider range of algorithms is considered.

`"exhaustive"`
: Like `"measure"`, but all possible algorithms that may be used to treat the transform are considered.

`"hybrid"`
: As run-time measurement of the algorithm can be expensive, this is a compromise where `"measure"` is used for transforms up to the size of 8192 and beyond that the `"estimate"` method is used.

The default method is `"estimate"`. The current method can be queried with

```
method = fftw ("planner")
```

or set by using

```
fftw ("planner", method)
```

Note that calculated wisdom will be lost when restarting Octave. However, the wisdom data can be reloaded if it is saved to a file as described above. Saved wisdom files should not be used on different platforms since they will not be efficient and the point of calculating the wisdom is lost.

The number of threads used for computing the plans and executing the transforms can be set with

```
fftw ("threads", NTHREADS)
```

Note that octave must be compiled with multi-threaded FFTW support for this feature. The number of processors available to the current process is used per default.

See also: [fft], page 639, [ifft], page 639, [fft2], page 639, [ifft2], page 640, [fftn], page 640, [ifftn], page 640.

`fftconv (x, y)` [Function File]
`fftconv (x, y, n)` [Function File]

Convolve two vectors using the FFT for computation.

`c = fftconv (x, y)` returns a vector of length equal to `length (x) + length (y) - 1`. If *x* and *y* are the coefficient vectors of two polynomials, the returned value is the coefficient vector of the product polynomial.

The computation uses the FFT by calling the function `fftfilt`. If the optional argument *n* is specified, an N-point FFT is used.

See also: [deconv], page 604, [conv], page 603, [conv2], page 604.

`fftfilt (b, x)` [Function File]
`fftfilt (b, x, n)` [Function File]

With two arguments, `fftfilt` filters *x* with the FIR filter *b* using the FFT.

Given the optional third argument, *n*, `fftfilt` uses the overlap-add method to filter *x* with *b* using an N-point FFT. The FFT size must be an even power of 2 and must be greater than or equal to the length of *b*. If the specified *n* does not meet these criteria, it is automatically adjusted to the nearest value that does.

If *x* is a matrix, filter each column of the matrix.

See also: [filter], page 642, [filter2], page 643.

`y = filter (b, a, x)` [Built-in Function]
`[y, sf] = filter (b, a, x, si)` [Built-in Function]
`[y, sf] = filter (b, a, x, [], dim)` [Built-in Function]
`[y, sf] = filter (b, a, x, si, dim)` [Built-in Function]

Return the solution to the following linear, time-invariant difference equation:

$$\sum_{k=0}^{N} a_{k+1} y_{n-k} = \sum_{k=0}^{M} b_{k+1} x_{n-k}, \qquad 1 \le n \le P$$

where $a \in \Re^{N-1}$, $b \in \Re^{M-1}$, and $x \in \Re^{P}$. The result is calculated over the first non-singleton dimension of *x* or over *dim* if supplied.

An equivalent form of the equation is:

$$y_n = -\sum_{k=1}^{N} c_{k+1} y_{n-k} + \sum_{k=0}^{M} d_{k+1} x_{n-k}, \qquad 1 \le n \le P$$

where $c = a/a_1$ and $d = b/a_1$.

If the fourth argument *si* is provided, it is taken as the initial state of the system and the final state is returned as *sf*. The state vector is a column vector whose length is equal to the length of the longest coefficient vector minus one. If *si* is not supplied, the initial state vector is set to all zeros.

In terms of the Z Transform, y is the result of passing the discrete- time signal x through a system characterized by the following rational system function:

$$H(z) = \frac{\sum_{k=0}^{M} d_{k+1} z^{-k}}{1 + \sum_{k+1}^{N} c_{k+1} z^{-k}}$$

See also: [filter2], page 643, [fftfilt], page 642, [freqz], page 643.

`y = filter2 (b, x)` [Function File]
`y = filter2 (b, x, shape)` [Function File]

Apply the 2-D FIR filter *b* to *x*. If the argument *shape* is specified, return an array of the desired shape. Possible values are:

`"full"` pad *x* with zeros on all sides before filtering.

`"same"` unpadded *x* (default)

`"valid"` trim *x* after filtering so edge effects are no included.

Note this is just a variation on convolution, with the parameters reversed and *b* rotated 180 degrees.

See also: [conv2], page 604.

`[h, w] = freqz (b, a, n, "whole")` [Function File]
`[h, w] = freqz (b)` [Function File]
`[h, w] = freqz (b, a)` [Function File]
`[h, w] = freqz (b, a, n)` [Function File]
`h = freqz (b, a, w)` [Function File]
`[...] = freqz (..., Fs)` [Function File]
`freqz (...)` [Function File]

Return the complex frequency response *h* of the rational IIR filter whose numerator and denominator coefficients are *b* and *a*, respectively. The response is evaluated at *n* angular frequencies between 0 and 2π.

The output value *w* is a vector of the frequencies.

If *a* is omitted, the denominator is assumed to be 1 (this corresponds to a simple FIR filter).

If *n* is omitted, a value of 512 is assumed. For fastest computation, *n* should factor into a small number of small primes.

If the fourth argument, `"whole"`, is omitted the response is evaluated at frequencies between 0 and π.

`freqz (b, a, w)`

Evaluate the response at the specific frequencies in the vector *w*. The values for *w* are measured in radians.

`[...] = freqz (..., Fs)`

Return frequencies in Hz instead of radians assuming a sampling rate *Fs*. If you are evaluating the response at specific frequencies *w*, those frequencies should be requested in Hz rather than radians.

`freqz (...)`

Plot the magnitude and phase response of *h* rather than returning them.

See also: [freqz_plot], page 644.

`freqz_plot (w, h)` [Function File]
`freqz_plot (w, h, freq_norm)` [Function File]

Plot the magnitude and phase response of *h*.

If the optional *freq_norm* argument is true, the frequency vector *w* is in units of normalized radians. If *freq_norm* is false, or not given, then *w* is measured in Hertz.

See also: [freqz], page 643.

`sinc (x)` [Function File]

Return $\sin(\pi x)/(\pi x)$.

`b = unwrap (x)` [Function File]
`b = unwrap (x, tol)` [Function File]
`b = unwrap (x, tol, dim)` [Function File]

Unwrap radian phases by adding multiples of 2*pi as appropriate to remove jumps greater than *tol*. *tol* defaults to pi.

Unwrap will work along the dimension *dim*. If *dim* is unspecified it defaults to the first non-singleton dimension.

`[a, b] = arch_fit (y, x, p, iter, gamma, a0, b0)` [Function File]

Fit an ARCH regression model to the time series *y* using the scoring algorithm in Engle's original ARCH paper. The model is

```
y(t) = b(1) * x(t,1) + ... + b(k) * x(t,k) + e(t),
h(t) = a(1) + a(2) * e(t-1)^2 + ... + a(p+1) * e(t-p)^2
```

in which $e(t)$ is $N(0, h(t))$, given a time-series vector *y* up to time $t-1$ and a matrix of (ordinary) regressors *x* up to t. The order of the regression of the residual variance is specified by *p*.

If invoked as `arch_fit (y, k, p)` with a positive integer *k*, fit an ARCH(*k*, *p*) process, i.e., do the above with the t-th row of *x* given by

```
[1, y(t-1), ..., y(t-k)]
```

Optionally, one can specify the number of iterations *iter*, the updating factor *gamma*, and initial values *a0* and *b0* for the scoring algorithm.

`arch_rnd (a, b, t)` [Function File]

Simulate an ARCH sequence of length *t* with AR coefficients *b* and CH coefficients *a*. I.e., the result $y(t)$ follows the model

```
y(t) = b(1) + b(2) * y(t-1) + ... + b(lb) * y(t-lb+1) + e(t),
```

where $e(t)$, given y up to time $t-1$, is $N(0, h(t))$, with

```
h(t) = a(1) + a(2) * e(t-1)^2 + ... + a(la) * e(t-la+1)^2
```

[*pval*, *lm*] = **arch_test** (*y*, *x*, *p*) [Function File]

For a linear regression model

```
y = x * b + e
```

perform a Lagrange Multiplier (LM) test of the null hypothesis of no conditional heteroscedascity against the alternative of CH(p).

I.e., the model is

```
y(t) = b(1) * x(t,1) + ... + b(k) * x(t,k) + e(t),
```

given y up to $t-1$ and x up to t, e(t) is $N(0, h(t))$ with

```
h(t) = v + a(1) * e(t-1)^2 + ... + a(p) * e(t-p)^2,
```

and the null is $a(1) == \ldots == a(p) == 0$.

If the second argument is a scalar integer, k, perform the same test in a linear autoregression model of order k, i.e., with

```
[1, y(t-1), ..., y(t-k)]
```

as the t-th row of x.

Under the null, LM approximately has a chisquare distribution with *p* degrees of freedom and *pval* is the *p*-value (1 minus the CDF of this distribution at LM) of the test.

If no output argument is given, the *p*-value is displayed.

arma_rnd (*a*, *b*, *v*, *t*, *n*) [Function File]

Return a simulation of the ARMA model

```
x(n) = a(1) * x(n-1) + ... + a(k) * x(n-k)
     + e(n) + b(1) * e(n-1) + ... + b(l) * e(n-l)
```

in which *k* is the length of vector *a*, *l* is the length of vector *b* and *e* is Gaussian white noise with variance *v*. The function returns a vector of length *t*.

The optional parameter *n* gives the number of dummy *x*(*i*) used for initialization, i.e., a sequence of length *t*+*n* is generated and *x*(*n*+1:*t*+*n*) is returned. If *n* is omitted, *n* = 100 is used.

autoreg_matrix (*y*, *k*) [Function File]

Given a time series (vector) *y*, return a matrix with ones in the first column and the first *k* lagged values of *y* in the other columns. I.e., for *t* > *k*, `[1, y(t-1), ..., y(t-k)]` is the t-th row of the result. The resulting matrix may be used as a regressor matrix in autoregressions.

bartlett (*m*) [Function File]

Return the filter coefficients of a Bartlett (triangular) window of length *m*.

For a definition of the Bartlett window, see e.g., A. V. Oppenheim & R. W. Schafer, *Discrete-Time Signal Processing*.

`blackman (m)` [Function File]

Return the filter coefficients of a Blackman window of length *m*.

For a definition of the Blackman window, see e.g., A. V. Oppenheim & R. W. Schafer, *Discrete-Time Signal Processing*.

`detrend (x, p)` [Function File]

If *x* is a vector, `detrend (x, p)` removes the best fit of a polynomial of order *p* from the data *x*.

If *x* is a matrix, `detrend (x, p)` does the same for each column in *x*.

The second argument is optional. If it is not specified, a value of 1 is assumed. This corresponds to removing a linear trend.

The order of the polynomial can also be given as a string, in which case *p* must be either `"constant"` (corresponds to `p=0`) or `"linear"` (corresponds to `p=1`).

See also: [polyfit], page 607.

`[d, dd] = diffpara (x, a, b)` [Function File]

Return the estimator *d* for the differencing parameter of an integrated time series.

The frequencies from $[2 * pi * a/t, 2 * pi * b/T]$ are used for the estimation. If *b* is omitted, the interval $[2 * pi/T, 2 * pi * a/T]$ is used. If both *b* and *a* are omitted then $a = 0.5 * sqrt(T)$ and $b = 1.5 * sqrt(T)$ is used, where T is the sample size. If *x* is a matrix, the differencing parameter of each column is estimated.

The estimators for all frequencies in the intervals described above is returned in *dd*. The value of *d* is simply the mean of *dd*.

Reference: P.J. Brockwell & R.A. Davis. *Time Series: Theory and Methods.* Springer 1987.

`durbinlevinson (c, oldphi, oldv)` [Function File]

Perform one step of the Durbin-Levinson algorithm.

The vector *c* specifies the autocovariances `[gamma_0, ..., gamma_t]` from lag 0 to *t*, *oldphi* specifies the coefficients based on *c*(*t*-1) and *oldv* specifies the corresponding error.

If *oldphi* and *oldv* are omitted, all steps from 1 to *t* of the algorithm are performed.

`fftshift (x)` [Function File]
`fftshift (x, dim)` [Function File]

Perform a shift of the vector *x*, for use with the `fft` and `ifft` functions, in order the move the frequency 0 to the center of the vector or matrix.

If *x* is a vector of N elements corresponding to N time samples spaced by *dt*, then `fftshift (fft (x))` corresponds to frequencies

```
f = [ -(ceil((N-1)/2):-1:1)*df 0 (1:floor((N-1)/2))*df ]
```

where $df = 1 / dt$.

If *x* is a matrix, the same holds for rows and columns. If *x* is an array, then the same holds along each dimension.

The optional *dim* argument can be used to limit the dimension along which the permutation occurs.

`ifftshift (x)` [Function File]
`ifftshift (x, dim)` [Function File]

Undo the action of the `fftshift` function. For even length *x*, `fftshift` is its own inverse, but odd lengths differ slightly.

`fractdiff (x, d)` [Function File]

Compute the fractional differences $(1 - L)^d x$ where L denotes the lag-operator and d is greater than -1.

`hamming (m)` [Function File]

Return the filter coefficients of a Hamming window of length *m*.

For a definition of the Hamming window, see e.g., A. V. Oppenheim & R. W. Schafer, *Discrete-Time Signal Processing.*

`hanning (m)` [Function File]

Return the filter coefficients of a Hanning window of length *m*.

For a definition of this window type, see e.g., A. V. Oppenheim & R. W. Schafer, *Discrete-Time Signal Processing.*

`hurst (x)` [Function File]

Estimate the Hurst parameter of sample *x* via the rescaled range statistic. If *x* is a matrix, the parameter is estimated for every single column.

`pp = pchip (x, y)` [Function File]
`yi = pchip (x, y, xi)` [Function File]

Return the Piecewise Cubic Hermite Interpolating Polynomial (pchip) of points *x* and *y*.

If called with two arguments, return the piecewise polynomial *pp* that may be used with `ppval` to evaluate the polynomial at specific points. When called with a third input argument, `pchip` evaluates the pchip polynomial at the points *xi*. The third calling form is equivalent to `ppval (pchip (x, y), xi)`.

The variable *x* must be a strictly monotonic vector (either increasing or decreasing) of length *n*. *y* can be either a vector or array. If *y* is a vector then it must be the same length *n* as x. If *y* is an array then the size of *y* must have the form

$$[s_1, s_2, \cdots, s_k, n]$$

The array is reshaped internally to a matrix where the leading dimension is given by

$$s_1 s_2 \cdots s_k$$

and each row of this matrix is then treated separately. Note that this is exactly opposite to `interp1` but is done for MATLAB compatibility.

See also: [spline], page 620, [ppval], page 614, [mkpp], page 614, [unmkpp], page 614.

`[Pxx, w] = periodogram (x)` [Function File]

For a data matrix *x* from a sample of size *n*, return the periodogram. The angular frequency is returned in *w*.

[Pxx,w] = periodogram (*x*).

[Pxx,w] = periodogram (x,win).

[Pxx,w] = periodogram (x,win,nfft).

[Pxx,f] = periodogram (x,win,nfft,Fs).

[Pxx,f] = periodogram (x,win,nfft,Fs,"range").

- x: data; if real-valued a one-sided spectrum is estimated, if complex-valued or range indicates `"twosided"`, the full spectrum is estimated.
- win: weight data with window, x.*win is used for further computation, if window is empty, a rectangular window is used.
- nfft: number of frequency bins, default max (256, 2.^ceil (log2 (length (x)))).
- Fs: sampling rate, default 1.
- range: `"onesided"` computes spectrum from [0..nfft/2+1]. `"twosided"` computes spectrum from [0..nfft-1]. These strings can appear at any position in the list input arguments after window.
- Pxx: one-, or two-sided power spectrum.
- w: angular frequency [0..2*pi) (two-sided) or [0..pi] one-sided.
- f: frequency [0..Fs) (two-sided) or [0..Fs/2] one-sided.

`sinetone (`*`freq`*`, `*`rate`*`, `*`sec`*`, `*`ampl`*`)` [Function File]

Return a sinetone of frequency *freq* with length of *sec* seconds at sampling rate *rate* and with amplitude *ampl*. The arguments *freq* and *ampl* may be vectors of common size.

Defaults are *rate* = 8000, *sec* = 1 and *ampl* = 64.

`sinewave (`*`m`*`, `*`n`*`, `*`d`*`)` [Function File]

Return an *m*-element vector with *i*-th element given by `sin (2 * pi * (`*`i`*`+`*`d`*`-1) / `*`n`*`)`.

The default value for *d* is 0 and the default value for *n* is *m*.

`spectral_adf (`*`c`*`)` [Function File]
`spectral_adf (`*`c`*`, `*`win`*`)` [Function File]
`spectral_adf (`*`c`*`, `*`win`*`, `*`b`*`)` [Function File]

Return the spectral density estimator given a vector of autocovariances *c*, window name *win*, and bandwidth, *b*.

The window name, e.g., `"triangle"` or `"rectangle"` is used to search for a function called `win_lw`.

If *win* is omitted, the triangle window is used. If *b* is omitted, `1 / sqrt (length (`*`x`*`))` is used.

See also: [spectral_xdf], page 648.

`spectral_xdf (`*`x`*`)` [Function File]
`spectral_xdf (`*`x`*`, `*`win`*`)` [Function File]
`spectral_xdf (`*`x`*`, `*`win`*`, `*`b`*`)` [Function File]

Return the spectral density estimator given a data vector *x*, window name *win*, and bandwidth, *b*.

The window name, e.g., `"triangle"` or `"rectangle"` is used to search for a function called `win_sw`.

If *win* is omitted, the triangle window is used. If *b* is omitted, `1 / sqrt (length (x))` is used.

See also: [spectral_adf], page 648.

`spencer (x)` [Function File]

Return Spencer's 15 point moving average of each column of *x*.

`y = stft (x)` [Function File]
`y = stft (x, win_size)` [Function File]
`y = stft (x, win_size, inc)` [Function File]
`y = stft (x, win_size, inc, num_coef)` [Function File]
`y = stft (x, win_size, inc, num_coef, win_type)` [Function File]
`[y, c] = stft (...)` [Function File]

Compute the short-time Fourier transform of the vector *x* with *num_coef* coefficients by applying a window of *win_size* data points and an increment of *inc* points.

Before computing the Fourier transform, one of the following windows is applied:

`"hanning"`
: win_type = 1

`"hamming"`
: win_type = 2

`"rectangle"`
: win_type = 3

The window names can be passed as strings or by the *win_type* number.

The following defaults are used for unspecified arguments: *win_size* = 80, *inc* = 24, *num_coef* = 64, and *win_type* = 1.

`y = stft (x, ...)` returns the absolute values of the Fourier coefficients according to the *num_coef* positive frequencies.

`[y, c] = stft (x, ...)` returns the entire STFT-matrix *y* and a 3-element vector *c* containing the window size, increment, and window type, which is needed by the `synthesis` function.

See also: [synthesis], page 649.

`x = synthesis (y, c)` [Function File]

Compute a signal from its short-time Fourier transform *y* and a 3-element vector *c* specifying window size, increment, and window type.

The values *y* and *c* can be derived by

```
[y, c] = stft (x , ...)
```

See also: [stft], page 649.

`[a, v] = yulewalker (c)` [Function File]

Fit an AR (p)-model with Yule-Walker estimates given a vector *c* of autocovariances `[gamma_0, ..., gamma_p]`.

Returns the AR coefficients, *a*, and the variance of white noise, *v*.

32 Image Processing

Since an image is basically a matrix, Octave is a very powerful environment for processing and analyzing images. To illustrate how easy it is to do image processing in Octave, the following example will load an image, smooth it by a 5-by-5 averaging filter, and compute the gradient of the smoothed image.

```
I = imread ("myimage.jpg");
S = conv2 (I, ones (5, 5) / 25, "same");
[Dx, Dy] = gradient (S);
```

In this example `S` contains the smoothed image, and `Dx` and `Dy` contains the partial spatial derivatives of the image.

32.1 Loading and Saving Images

The first step in most image processing tasks is to load an image into Octave which is done with the `imread` function. The `imwrite` function is the corresponding function for writing images to the disk.

In summary, most image processing code will follow the structure of this code

```
I = imread ("my_input_image.img");
J = process_my_image (I);
imwrite (J, "my_output_image.img");
```

[*img*, *map*, *alpha*] = imread (*filename*) [Function File]
[...] = imread (*filename*, *ext*) [Function File]
[...] = imread (*url*) [Function File]
[...] = imread (..., *idx*) [Function File]
[...] = imread (..., *param1*, *val1*, ...) [Function File]

Read images from various file formats.

Reads an image as a matrix from the file *filename*. If there is no file *filename*, and *ext* was specified, it will look for a file named *filename* and extension *ext*, i.e., a file named *filename.ext*.

The size and class of the output depends on the format of the image. A color image is returned as an MxNx3 matrix. Gray-level and black-and-white images are of size MxN. Multipage images will have an additional 4th dimension.

The bit depth of the image determines the class of the output: `"uint8"`, `"uint16"` or `"single"` for gray and color, and `"logical"` for black and white. Note that indexed images always return the indexes for a colormap, independent if *map* is a requested output. To obtain the actual RGB image, use `ind2rgb`. When more than one indexed image is being read, *map* is obtained from the first. In some rare cases this may be incorrect and `imfinfo` can be used to obtain the colormap of each image.

See the Octave manual for more information in representing images.

Some file formats, such as TIFF and GIF, are able to store multiple images in a single file. *idx* can be a scalar or vector specifying the index of the images to read. By default, Octave will only read the first page.

Depending on the file format, it is possible to configure the reading of images with *param*, *val* pairs. The following options are supported:

'"Frames" or "Index"'
: This is an alternative method to specify *idx*. When specifying it in this way, its value can also be the string "all".

'"Info"'
: This option exists for MATLAB compatibility and has no effect. For maximum performance while reading multiple images from a single file, use the Index option.

'"PixelRegion"'
: Controls the image region that is read. Takes as value a cell array with two arrays of 3 elements {*rows cols*}. The elements in the array are the start, increment and end pixel to be read. If the increment value is omitted, defaults to 1.

See also: [imwrite], page 652, [imfinfo], page 653, [imformats], page 656.

`imwrite` (*img*, *filename*) [Function File]
`imwrite` (*img*, *filename*, *ext*) [Function File]
`imwrite` (*img*, *map*, *filename*) [Function File]
`imwrite` (..., *param1*, *val1*, ...) [Function File]

Write images in various file formats.

The image *img* can be a binary, grayscale, RGB, or multi-dimensional image. The size and class of *img* should be the same as what should be expected when reading it with `imread`: the 3rd and 4th dimensions reserved for color space, and multiple pages respectively. If it's an indexed image, the colormap *map* must also be specified.

If *ext* is not supplied, the file extension of *filename* is used to determine the format. The actual supported formats are dependent on options made during the build of Octave. Use `imformats` to check the support of the different image formats.

Depending on the file format, it is possible to configure the writing of images with *param*, *val* pairs. The following options are supported:

'Alpha'
: Alpha (transparency) channel for the image. This must be a matrix with same class, and number of rows and columns of *img*. In case of a multipage image, the size of the 4th dimension must also match and the third dimension must be a singleton. By default, image will be completely opaque.

'DelayTime'
: For formats that accept animations (such as GIF), controls for how long a frame is displayed until it moves to the next one. The value must be scalar (which will applied to all frames in *img*), or a vector of length equal to the number of frames in *im*. The value is in seconds, must be between 0 and 655.35, and defaults to 0.5.

'DisposalMethod'
: For formats that accept animations (such as GIF), controls what happens to a frame before drawing the next one. Its value can be one of the following strings: "doNotSpecify" (default); "leaveInPlace"; "restoreBG"; and "restorePrevious", or a cell array of those string with length equal to the number of frames in *img*.

'LoopCount'
: For formats that accept animations (such as GIF), controls how many times the sequence is repeated. A value of Inf means an infinite loop (default), a value of 0 or 1 that the sequence is played only once (loops zero times), while a value of 2 or above loops that number of times (looping twice means it plays the complete sequence 3 times). This option is ignored when there is only a single image at the end of writing the file.

'Quality'
: Set the quality of the compression. The value should be an integer between 0 and 100, with larger values indicating higher visual quality and lower compression. Defaults to 75.

'WriteMode'
: Some file formats, such as TIFF and GIF, are able to store multiple images in a single file. This option specifies if *img* should be appended to the file (if it exists) or if a new file should be created for it (possibly overwriting an existing file). The value should be the string `"Overwrite"` (default), or `"Append"`.

  Despite this option, the most efficient method of writing a multipage image is to pass a 4 dimensional *img* to `imwrite`, the same matrix that could be expected when using `imread` with the option `"Index"` set to `"all"`.

See also: [imread], page 651, [imfinfo], page 653, [imformats], page 656.

val = IMAGE_PATH () [Built-in Function]
old_val = IMAGE_PATH (*new_val*) [Built-in Function]
IMAGE_PATH (*new_val*, "*local*") [Built-in Function]

Query or set the internal variable that specifies a colon separated list of directories in which to search for image files.

When called from inside a function with the `"local"` option, the variable is changed locally for the function and any subroutines it calls. The original variable value is restored when exiting the function.

See also: [EXEC_PATH], page 724, [OCTAVE_HOME], page 733.

It is possible to get information about an image file on disk, without actually reading it into Octave. This is done using the `imfinfo` function which provides read access to many of the parameters stored in the header of the image file.

info = imfinfo (*filename*) [Function File]
info = imfinfo (*filename*, *ext*) [Function File]
info = imfinfo (*url*) [Function File]

Read image information from a file.

`imfinfo` returns a structure containing information about the image stored in the file *filename*. If there is no file *filename*, and *ext* was specified, it will look for a file named *filename* and extension *ext*, i.e., a file named *filename.ext*.

The output structure *info* contains the following fields:

'Filename'
: The full name of the image file.

'FileModDate'
: Date of last modification to the file.

'FileSize'
: Number of bytes of the image on disk

'Format'
: Image format (e.g., `"jpeg"`).

'Height'
: Image height in pixels.

'Width'
: Image Width in pixels.

'BitDepth'
: Number of bits per channel per pixel.

'ColorType'
: Image type. Value is `"grayscale"`, `"indexed"`, `"truecolor"`, `"CMYK"`, or `"undefined"`.

'XResolution'
: X resolution of the image.

'YResolution'
: Y resolution of the image.

'ResolutionUnit'
: Units of image resolution. Value is `"Inch"`, `"Centimeter"`, or `"undefined"`.

'DelayTime'
: Time in 1/100ths of a second (0 to 65535) which must expire before displaying the next image in an animated sequence.

'LoopCount'
: Number of iterations to loop an animation.

'ByteOrder'
: Endian option for formats that support it. Value is `"little-endian"`, `"big-endian"`, or `"undefined"`.

'Gamma'
: Gamma level of the image. The same color image displayed on two different workstations may look different due to differences in the display monitor.

'Quality'
: JPEG/MIFF/PNG compression level. Value is an integer in the range [0 100].

'DisposalMethod'
: Only valid for GIF images, control how successive frames are rendered (how the preceding frame is disposed of) when creating a GIF animation. Values can be `"doNotSpecify"`, `"leaveInPlace"`, `"restoreBG"`, or `"restorePrevious"`. For non-GIF files, value is an empty string.

'Chromaticities'
: Value is a 1x8 Matrix with the x,y chromaticity values for white, red, green, and blue points, in that order.

'Comment' Image comment.

'Compression'
: Compression type. Value can be "none", "bzip", "fax3", "fax4", "jpeg", "lzw", "rle", "deflate", "lzma", "jpeg2000", "jbig2", "jbig2", or "undefined".

'Colormap'
: Colormap for each image.

'Orientation'
: The orientation of the image with respect to the rows and columns. Value is an integer between 1 and 8 as defined in the TIFF 6 specifications, and for MATLAB compatibility.

'Software'
: Name and version of the software or firmware of the camera or image input device used to generate the image.

'Make' The manufacturer of the recording equipment. This is the manufacture of the DSC, scanner, video digitizer or other equipment that generated the image.

'Model' The model name or model number of the recording equipment as mentioned on the field "Make".

'DateTime'
: The date and time of image creation as defined by the Exif standard, i.e., it is the date and time the file was changed.

'ImageDescription'
: The title of the image as defined by the Exif standard.

'Artist' Name of the camera owner, photographer or image creator.

'Copyright'
: Copyright notice of the person or organization claiming rights to the image.

'DigitalCamera'
: A struct with information retrieved from the Exif tag.

'GPSInfo' A struct with geotagging information retrieved from the Exif tag.

See also: [imread], page 651, [imwrite], page 652, [imshow], page 657, [imformats], page 656.

By default, Octave's image IO functions (imread, imwrite, and imfinfo) use the GraphicsMagick library for their operations. This means a vast number of image formats is supported but considering the large amount of image formats in science and its commonly closed nature, it is impossible to have a library capable of reading them all. Because of this, the function `imformats` keeps a configurable list of available formats, their extensions, and what functions should the image IO functions use. This allows to expand Octave's image IO capabilities by creating functions aimed at acting on specific file formats.

While it would be possible to call the extra functions directly, properly configuring Octave with `imformats` allows to keep a consistent code that is abstracted from file formats.

It is important to note that a file format is not actually defined by its file extension and that `GraphicsMagick` is capable to read and write more file formats than the ones listed by `imformats`. What this means is that even with an incorrect or missing extension the image may still be read correctly, and that even unlisted formats are not necessarily unsupported.

`imformats ()` [Function File]
formats = `imformats` (*ext*) [Function File]
formats = `imformats` (*format*) [Function File]
formats = `imformats` ("*add*", *format*) [Function File]
formats = `imformats` ("*remove*", *ext*) [Function File]
formats = `imformats` ("*update*", *ext*, *format*) [Function File]
formats = `imformats` ("*factory*") [Function File]

Manage supported image formats.

formats is a structure with information about each supported file format, or from a specific format *ext*, the value displayed on the field `ext`. It contains the following fields:

ext — The name of the file format. This may match the file extension but Octave will automatically detect the file format.

description — A long description of the file format.

isa — A function handle to confirm if a file is of the specified format.

write — A function handle to write if a file is of the specified format.

read — A function handle to open files the specified format.

info — A function handle to obtain image information of the specified format.

alpha — Logical value if format supports alpha channel (transparency or matte).

multipage — Logical value if format supports multipage (multiple images per file).

It is possible to change the way Octave manages file formats with the options `"add"`, `"remove"`, and `"update"`, and supplying a structure *format* with the required fields. The option `"factory"` resets the configuration to the default.

This can be used by Octave packages to extend the image reading capabilities Octave, through use of the PKG_ADD and PKG_DEL commands.

See also: [imfinfo], page 653, [imread], page 651, [imwrite], page 652.

32.2 Displaying Images

A natural part of image processing is visualization of an image. The most basic function for this is the `imshow` function that shows the image given in the first input argument.

`imshow (`*im*`)` [Function File]
`imshow (`*im*, *limits*`)` [Function File]
`imshow (`*im*, *map*`)` [Function File]
`imshow (`*rgb*, ...`)` [Function File]
`imshow (`*filename*`)` [Function File]
`imshow (`..., *string_param1*, *value1*, ...`)` [Function File]
h `= imshow (`...`)` [Function File]

Display the image *im*, where *im* can be a 2-dimensional (grayscale image) or a 3-dimensional (RGB image) matrix.

If *limits* is a 2-element vector `[low, high]`, the image is shown using a display range between *low* and *high*. If an empty matrix is passed for *limits*, the display range is computed as the range between the minimal and the maximal value in the image.

If *map* is a valid color map, the image will be shown as an indexed image using the supplied color map.

If a file name is given instead of an image, the file will be read and shown.

If given, the parameter *string_param1* has value *value1*. *string_param1* can be any of the following:

`"displayrange"`
value1 is the display range as described above.

`"xdata"` If *value1* is a two element vector, it must contain horizontal axis limits in the form [xmin xmax]; Otherwise *value1* must be a vector and only the first and last elements will be used for xmin and xmax respectively.

`"ydata"` If *value1* is a two element vector, it must contain vertical axis limits in the form [ymin ymax]; Otherwise *value1* must be a vector and only the first and last elements will be used for ymin and ymax respectively.

The optional return value *h* is a graphics handle to the image.

See also: [image], page 657, [imagesc], page 658, [colormap], page 660, [gray2ind], page 659, [rgb2ind], page 659.

`image (`*img*`)` [Function File]
`image (`*x*, *y*, *img*`)` [Function File]
`image (`..., `"`*prop*`"`, *val*, ...`)` [Function File]
`image ("`*prop1*`"`, *val1*, ...`)` [Function File]
h `= image (`...`)` [Function File]

Display a matrix as an indexed color image.

The elements of *img* are indices into the current colormap. *x* and *y* are optional 2-element vectors, `[min, max]`, which specify the range for the axis labels. If a range is specified as `[max, min]` then the image will be reversed along that axis. For convenience, *x* and *y* may be specified as N-element vectors matching the length of the data in *img*. However, only the first and last elements will be used to determine the axis limits. **Warning:** *x* and *y* are ignored when using gnuplot 4.0 or earlier.

Multiple property/value pairs may be specified for the image object, but they must appear in pairs.

The optional return value *h* is a graphics handle to the image.

Implementation Note: The origin (0, 0) for images is located in the upper left. For ordinary plots, the origin is located in the lower left. Octave handles this inversion by plotting the data normally, and then reversing the direction of the y-axis by setting the `ydir` property to `"reverse"`. This has implications whenever an image and an ordinary plot need to be overlaid. The recommended solution is to display the image and then plot the reversed ydata using, for example, `flipud (ydata)`.

Calling Forms: The `image` function can be called in two forms: High-Level and Low-Level. When invoked with normal options, the High-Level form is used which first calls `newplot` to prepare the graphic figure and axes. When the only inputs to `image` are property/value pairs the Low-Level form is used which creates a new instance of an image object and inserts it in the current axes.

See also: [imshow], page 657, [imagesc], page 658, [colormap], page 660.

`imagesc (img)` [Function File]
`imagesc (x, y, img)` [Function File]
`imagesc (..., climits)` [Function File]
`imagesc (..., "prop", val, ...)` [Function File]
`imagesc ("prop1", val1, ...)` [Function File]
`imagesc (hax, ...)` [Function File]
`h = imagesc (...)` [Function File]

Display a scaled version of the matrix *img* as a color image. The colormap is scaled so that the entries of the matrix occupy the entire colormap. If `climits = [lo, hi]` is given, then that range is set to the `"clim"` of the current axes.

The axis values corresponding to the matrix elements are specified in *x* and *y*, either as pairs giving the minimum and maximum values for the respective axes, or as values for each row and column of the matrix *img*.

The optional return value *h* is a graphics handle to the image.

Calling Forms: The `imagesc` function can be called in two forms: High-Level and Low-Level. When invoked with normal options, the High-Level form is used which first calls `newplot` to prepare the graphic figure and axes. When the only inputs to `image` are property/value pairs the Low-Level form is used which creates a new instance of an image object and inserts it in the current axes.

See also: [image], page 657, [imshow], page 657, [caxis], page 287.

32.3 Representing Images

In general Octave supports four different kinds of images, grayscale images, RGB images, binary images, and indexed images. A grayscale image is represented with an M-by-N matrix in which each element corresponds to the intensity of a pixel. An RGB image is represented with an M-by-N-by-3 array where each 3-vector corresponds to the red, green, and blue intensities of each pixel.

The actual meaning of the value of a pixel in a grayscale or RGB image depends on the class of the matrix. If the matrix is of class `double` pixel intensities are between 0 and 1, if it is of class `uint8` intensities are between 0 and 255, and if it is of class `uint16` intensities are between 0 and 65535.

A binary image is an M-by-N matrix of class `logical`. A pixel in a binary image is black if it is `false` and white if it is `true`.

An indexed image consists of an M-by-N matrix of integers and a C-by-3 color map. Each integer corresponds to an index in the color map, and each row in the color map corresponds to an RGB color. The color map must be of class `double` with values between 0 and 1.

`iscolormap (`*cmap*`)` [Function File]

Return true if *cmap* is a colormap.

A colormap is a real matrix with *n* rows and 3 columns. Each row represents a single color. The columns contain red, green, and blue intensities respectively. All entries must be between 0 and 1 inclusive.

See also: [colormap], page 660, [rgbplot], page 661.

img `= gray2ind (`*I*`)` [Function File]
img `= gray2ind (`*I*, *n*`)` [Function File]
img `= gray2ind (`*BW*`)` [Function File]
img `= gray2ind (`*BW*, *n*`)` [Function File]
`[`*img*, *map*`] = gray2ind (...)` [Function File]

Convert a grayscale or binary intensity image to an indexed image.

The indexed image will consist of *n* different intensity values. If not given *n* defaults to 64 for grayscale images or 2 for binary black and white images.

The output *img* is of class uint8 if *n* is less than or equal to 256; Otherwise the return class is uint16.

See also: [ind2gray], page 659, [rgb2ind], page 659.

I `= ind2gray (`*x*, *map*`)` [Function File]

Convert a color indexed image to a grayscale intensity image.

The image *x* must be an indexed image which will be converted using the colormap *cmap*. If *cmap* does not contain enough colors for the image, pixels in *x* outside the range are mapped to the last color in the map before conversion to grayscale.

The output *I* is of the same class as the input *x* and may be one of `uint8`, `uint16`, `single`, or `double`.

Implementation Note: There are several ways of converting colors to grayscale intensities. This functions uses the luminance value obtained from `rgb2ntsc` which is `I = 0.299*R + 0.587*G + 0.114*B`. Other possibilities include the value component from `rgb2hsv` or using a single color channel from `ind2rgb`.

See also: [gray2ind], page 659, [ind2rgb], page 660.

`[`*x*, *map*`] = rgb2ind (`*rgb*`)` [Function File]
`[`*x*, *map*`] = rgb2ind (`*R*, *G*, *B*`)` [Function File]

Convert an image in red-green-blue (RGB) color space to an indexed image.

The input image *rgb* can be specified as a single matrix of size MxNx3, or as three separate variables, *R*, *G*, and *B*, its three colour channels, red, green, and blue.

It outputs an indexed image *x* and a colormap *map* to interpret an image exactly the same as the input. No dithering or other form of color quantization is performed. The output class of the indexed image *x* can be uint8, uint16 or double, whichever is required to specify the number of unique colors in the image (which will be equal to the number of rows in *map*) in order

Multi-dimensional indexed images (of size MxNx3xK) are also supported, both via a single input (*rgb*) or its three colour channels as separate variables.

See also: [ind2rgb], page 660, [rgb2hsv], page 666, [rgb2ntsc], page 667.

`rgb = ind2rgb (x, map)` [Function File]
`[R, G, B] = ind2rgb (x, map)` [Function File]

Convert an indexed image to red, green, and blue color components.

The image *x* must be an indexed image which will be converted using the colormap *map*. If *map* does not contain enough colors for the image, pixels in *x* outside the range are mapped to the last color in the map.

The output may be a single RGB image (MxNx3 matrix where M and N are the original image *x* dimensions, one for each of the red, green and blue channels). Alternatively, the individual red, green, and blue color matrices of size MxN may be returned.

Multi-dimensional indexed images (of size MxNx1xK) are also supported.

See also: [rgb2ind], page 659, [ind2gray], page 659, [hsv2rgb], page 667, [ntsc2rgb], page 667.

`cmap = colormap ()` [Function File]
`cmap = colormap (map)` [Function File]
`cmap = colormap ("default")` [Function File]
`cmap = colormap ("map_name")` [Function File]
`cmap = colormap (hax, ...)` [Function File]
`colormap map_name` [Command]
`cmaps = colormap ("list")` [Function File]
`colormap ("register", "name")` [Function File]
`colormap ("unregister", "name")` [Function File]

Query or set the current colormap.

With no input arguments, `colormap` returns the current color map.

`colormap (map)` sets the current colormap to *map*. The colormap should be an *n* row by 3 column matrix. The columns contain red, green, and blue intensities respectively. All entries must be between 0 and 1 inclusive. The new colormap is returned.

`colormap ("default")` restores the default colormap (the `jet` map with 64 entries). The default colormap is returned.

The map may also be specified by a string, `"map_name"`, where *map_name* is the name of a function that returns a colormap.

If the first argument *hax* is an axes handle, then the colormap for the parent figure of *hax* is queried or set.

For convenience, it is also possible to use this function with the command form, `colormap` *map_name*.

`colormap ("list")` returns a cell array with all of the available colormaps. The options `"register"` and `"unregister"` add or remove the colormap *name* from this list.

See also: [jet], page 663.

`rgbplot (`*cmap*`)` [Function File]
`rgbplot (`*cmap*, *style*`)` [Function File]
h `= rgbplot (...)` [Function File]

Plot the components of a colormap.

Two different *style*s are available for displaying the *cmap*:

profile (default)
: Plot the RGB line profile of the colormap for each of the channels (red, green and blue) with the plot lines colored appropriately. Each line represents the intensity of each RGB components across the colormap.

composite
: Draw the colormap across the X-axis so that the actual index colors are visible rather than the individual color components.

The optional return value *h* is a graphics handle to the created plot.

Run `demo rgbplot` to see an example of `rgbplot` and each style option.

See also: [colormap], page 660.

map `= autumn ()` [Function File]
map `= autumn (`*n*`)` [Function File]

Create color colormap. This colormap ranges from red through orange to yellow. The argument *n* must be a scalar. If unspecified, the length of the current colormap, or 64, is used.

See also: [colormap], page 660.

map `= bone ()` [Function File]
map `= bone (`*n*`)` [Function File]

Create color colormap. This colormap varies from black to white with gray-blue shades. The argument *n* must be a scalar. If unspecified, the length of the current colormap, or 64, is used.

See also: [colormap], page 660.

map `= colorcube ()` [Function File]
map `= colorcube (`*n*`)` [Function File]

Create color colormap. This colormap is composed of as many equally spaced colors (not grays) in the RGB color space as possible. If there are not a perfect number *n* of regularly spaced colors then the remaining entries in the colormap are gradients of pure red, green, blue, and gray. The argument *n* must be a scalar. If unspecified, the length of the current colormap, or 64, is used.

See also: [colormap], page 660.

map = `cool ()` [Function File]
map = `cool (`*n*`)` [Function File]

Create color colormap. The colormap varies from cyan to magenta. The argument *n* must be a scalar. If unspecified, the length of the current colormap, or 64, is used.

See also: [colormap], page 660.

map = `copper ()` [Function File]
map = `copper (`*n*`)` [Function File]

Create color colormap. This colormap varies from black to a light copper tone. The argument *n* must be a scalar. If unspecified, the length of the current colormap, or 64, is used.

See also: [colormap], page 660.

map = `flag ()` [Function File]
map = `flag (`*n*`)` [Function File]

Create color colormap. This colormap cycles through red, white, blue, and black with each index change. The argument *n* must be a scalar. If unspecified, the length of the current colormap, or 64, is used.

See also: [colormap], page 660.

map = `gray ()` [Function File]
map = `gray (`*n*`)` [Function File]

Create gray colormap. This colormap varies from black to white with shades of gray. The argument *n* must be a scalar. If unspecified, the length of the current colormap, or 64, is used.

See also: [colormap], page 660.

map = `hot ()` [Function File]
map = `hot (`*n*`)` [Function File]

Create color colormap. This colormap ranges from black through dark red, red, orange, yellow, to white. The argument *n* must be a scalar. If unspecified, the length of the current colormap, or 64, is used.

See also: [colormap], page 660.

`hsv (`*n*`)` [Function File]

Create color colormap. This colormap begins with red, changes through yellow, green, cyan, blue, and magenta, before returning to red. It is useful for displaying periodic functions. The map is obtained by linearly varying the hue through all possible values while keeping constant maximum saturation and value. The equivalent code is `hsv2rgb ([(0:N-1)'/N, ones(N,2)])`.

The argument *n* must be a scalar. If unspecified, the length of the current colormap, or 64, is used.

See also: [colormap], page 660.

map = `jet` () [Function File]
map = `jet` (*n*) [Function File]

Create color colormap. This colormap ranges from dark blue through blue, cyan, green, yellow, red, to dark red. The argument *n* must be a scalar. If unspecified, the length of the current colormap, or 64, is used.

See also: [colormap], page 660.

map = `lines` () [Function File]
map = `lines` (*n*) [Function File]

Create color colormap. This colormap is composed of the list of colors in the current axes `"ColorOrder"` property. The default is blue, green, red, cyan, pink, yellow, and gray. The argument *n* must be a scalar. If unspecified, the length of the current colormap, or 64, is used.

See also: [colormap], page 660.

map = `ocean` () [Function File]
map = `ocean` (*n*) [Function File]

Create color colormap. This colormap varies from black to white with shades of blue. The argument *n* must be a scalar. If unspecified, the length of the current colormap, or 64, is used.

See also: [colormap], page 660.

map = `pink` () [Function File]
map = `pink` (*n*) [Function File]

Create color colormap. This colormap varies from black to white with shades of gray-pink. It gives a sepia tone when used on grayscale images. The argument *n* must be a scalar. If unspecified, the length of the current colormap, or 64, is used.

See also: [colormap], page 660.

map = `prism` () [Function File]
map = `prism` (*n*) [Function File]

Create color colormap. This colormap cycles through red, orange, yellow, green, blue and violet with each index change. The argument *n* must be a scalar. If unspecified, the length of the current colormap, or 64, is used.

See also: [colormap], page 660.

map = `rainbow` () [Function File]
map = `rainbow` (*n*) [Function File]

Create color colormap. This colormap ranges from red through orange, yellow, green, blue, to violet. The argument *n* must be a scalar. If unspecified, the length of the current colormap, or 64, is used.

See also: [colormap], page 660.

map = `spring` () [Function File]
map = `spring` (*n*) [Function File]

Create color colormap. This colormap varies from magenta to yellow. The argument *n* must be a scalar. If unspecified, the length of the current colormap, or 64, is used.

See also: [colormap], page 660.

map = `summer` () [Function File]
map = `summer` (*n*) [Function File]

Create color colormap. This colormap varies from green to yellow. The argument *n* must be a scalar. If unspecified, the length of the current colormap, or 64, is used.

See also: [colormap], page 660.

map = `white` () [Function File]
map = `white` (*n*) [Function File]

Create color colormap. This colormap is completely white. The argument *n* should be a scalar. If it is omitted, the length of the current colormap or 64 is assumed.

See also: [colormap], page 660.

map = `winter` () [Function File]
map = `winter` (*n*) [Function File]

Create color colormap. This colormap varies from blue to green. The argument *n* must be a scalar. If unspecified, the length of the current colormap, or 64, is used.

See also: [colormap], page 660.

cmap = `contrast` (*x*) [Function File]
cmap = `contrast` (*x*, *n*) [Function File]

Return a gray colormap that maximizes the contrast in an image. The returned colormap will have *n* rows. If *n* is not defined then the size of the current colormap is used.

See also: [colormap], page 660, [brighten], page 664.

An additional colormap is `gmap40`. This code map contains only colors with integer values of the red, green and blue components. This is a workaround for a limitation of gnuplot 4.0, that does not allow the color of line or patch objects to be set. `gmap40` is chiefly useful to gnuplot 4.0 users, and particularly in conjunction with the *bar*, *surf*, and *contour* functions.

map = `gmap40` () [Function File]
map = `gmap40` (*n*) [Function File]

Create color colormap. The colormap consists of red, green, blue, yellow, magenta and cyan. This colormap is specifically designed for users of gnuplot 4.0 where these 6 colors are the allowable ones for patch objects. The argument *n* must be a scalar. If unspecified, a length of 6 is assumed. Larger values of *n* result in a repetition of the above colors.

See also: [colormap], page 660.

The following three functions modify the existing colormap rather than replace it.

map_out = `brighten` (*map*, *beta*) [Function File]
map_out = `brighten` (*beta*) [Function File]
map_out = `brighten` (*h*, *beta*) [Function File]

Brighten or darken a colormap. If the *map* argument is omitted, the function is applied to the current colormap. The first argument can also be a valid graphics handle *h*, in which case `brighten` is applied to the colormap associated with this handle.

The argument *beta* must be a scalar between -1 and 1, where a negative value darkens and a positive value brightens the colormap.

If no output is specified then the result is written to the current colormap.

See also: [colormap], page 660, [contrast], page 664.

`spinmap ()` [Function File]
`spinmap (`*t*`)` [Function File]
`spinmap (`*t*`, `*inc*`)` [Function File]
`spinmap ("`*inf*`")` [Function File]

Cycle the colormap for *t* seconds with a color increment of *inc*. Both parameters are optional. The default cycle time is 5 seconds and the default increment is 2. If the option `"inf"` is given then cycle continuously until `Control-C` is pressed.

When rotating the original color 1 becomes color 2, color 2 becomes color 3, etc. A positive or negative increment is allowed and a higher value of *inc* will cause faster cycling through the colormap.

See also: [colormap], page 660.

`whitebg ()` [Function File]
`whitebg (`*color*`)` [Function File]
`whitebg ("`*none*`")` [Function File]
`whitebg (`*hfig*`, ...)` [Function File]

Invert the colors in the current color scheme.

The root properties are also inverted such that all subsequent plot use the new color scheme.

If the optional argument *color* is present then the background color is set to *color* rather than inverted. *color* may be a string representing one of the eight known colors or an RGB triplet. The special string argument `"none"` restores the plot to the default colors.

If the first argument *hfig* is a figure handle, then operate on this figure rather than the current figure returned by `gcf`. The root properties will not be changed.

See also: [reset], page 361, [get], page 341, [set], page 342.

The following functions can be used to manipulate colormaps.

`[`*Y*`, `*newmap*`] = cmunique (`*X*`, `*map*`)` [Function File]
`[`*Y*`, `*newmap*`] = cmunique (`*RGB*`)` [Function File]
`[`*Y*`, `*newmap*`] = cmunique (`*I*`)` [Function File]

Convert an input image *X* to an ouput indexed image *Y* which uses the smallest colormap possible *newmap*.

When the input is an indexed image (*X* with colormap *map*) the output is a colormap *newmap* from which any repeated rows have been eliminated. The output image, *Y*, is the original input image with the indices adjusted to match the new, possibly smaller, colormap.

When the input is an RGB image (an MxNx3 array), the output colormap will contain one entry for every unique color in the original image. In the worst case the new map could have as many rows as the number of pixels in the original image.

When the input is a grayscale image *I*, the output colormap will contain one entry for every unique intensity value in the original image. In the worst case the new map could have as many rows as the number of pixels in the original image.

Implementation Details:

newmap is always an Mx3 matrix, even if the input image is an intensity grayscale image *I* (all three RGB planes are assigned the same value).

The output image is of class uint8 if the size of the new colormap is less than or equal to 256. Otherwise, the output image is of class double.

See also: [rgb2ind], page 659, [gray2ind], page 659.

`[Y, newmap] = cmpermute (X, map)` [Function File]
`[Y, newmap] = cmpermute (X, map, index)` [Function File]

Reorder colors in a colormap.

When called with only two arguments, `cmpermute` randomly rearranges the colormap *map* and returns a new colormap *newmap*. It also returns the indexed image *Y* which is the equivalent of the original input image *X* when displayed using *newmap*.

When called with an optional third argument the order of colors in the new colormap is defined by *index*.

Caution: `index` should not have repeated elements or the function will fail.

32.4 Plotting on top of Images

If gnuplot is being used to display images it is possible to plot on top of images. Since an image is a matrix it is indexed by row and column values. The plotting system is, however, based on the traditional (x, y) system. To minimize the difference between the two systems Octave places the origin of the coordinate system in the point corresponding to the pixel at $(1, 1)$. So, to plot points given by row and column values on top of an image, one should simply call `plot` with the column values as the first argument and the row values as the second. As an example the following code generates an image with random intensities between 0 and 1, and shows the image with red circles over pixels with an intensity above 0.99.

```
I = rand (100, 100);
[row, col] = find (I > 0.99);
hold ("on");
imshow (I);
plot (col, row, "ro");
hold ("off");
```

32.5 Color Conversion

Octave supports conversion from the RGB color system to NTSC and HSV and vice versa.

hsv_map = rgb2hsv (*rgb*) [Function File]
hsv_map = rgb2hsv (*rgb*) [Function File]

Transform a colormap or image from red-green-blue (RGB) space to hue-saturation-value (HSV) space.

A color in the RGB space consists of red, green, and blue intensities.

A color in HSV space is represented by hue, saturation, and value (brightness) levels. Value gives the amount of light in the color. Hue describes the dominant wavelength. Saturation is the amount of hue mixed into the color.

See also: [hsv2rgb], page 667, [rgb2ind], page 659, [rgb2ntsc], page 667.

rgb_map = hsv2rgb (*hsv_map*) [Function File]
rgb_img = hsv2rgb (*hsv_img*) [Function File]

Transform a colormap or image from hue-saturation-value (HSV) space to red-green-blue (RGB) space.

A color in HSV space is represented by hue, saturation and value (brightness) levels. Value gives the amount of light in the color. Hue describes the dominant wavelength. Saturation is the amount of hue mixed into the color.

A color in the RGB space consists of red, green, and blue intensities.

See also: [rgb2hsv], page 666, [ind2rgb], page 660, [ntsc2rgb], page 667.

yiq_map = rgb2ntsc (*rgb_map*) [Function File]
yiq_img = rgb2ntsc (*rgb_img*) [Function File]

Transform a colormap or image from red-green-blue (RGB) color space to luminance-chrominance (NTSC) space. The input may be of class uint8, uint16, single, or double. The output is of class double.

Implementation Note: The reference matrix for the transformation is

```
/Y\     0.299  0.587  0.114  /R\
|I|  =  0.596 -0.274 -0.322  |G|
\Q/     0.211 -0.523  0.312  \B/
```

as documented in `http://en.wikipedia.org/wiki/YIQ` and truncated to 3 significant figures. Note: The FCC version of NTSC uses only 2 significant digits and is slightly different.

See also: [ntsc2rgb], page 667, [rgb2hsv], page 666, [rgb2ind], page 659.

rgb_map = ntsc2rgb (*yiq_map*) [Function File]
rgb_img = ntsc2rgb (*yiq_img*) [Function File]

Transform a colormap or image from luminance-chrominance (NTSC) space to red-green-blue (RGB) color space.

Implementation Note: The conversion matrix is chosen to be the inverse of the matrix used for rgb2ntsc such that

```
x == ntsc2rgb (rgb2ntsc (x))
```

MATLAB uses a slightly different matrix where rounding means the equality above does not hold.

See also: [rgb2ntsc], page 667, [hsv2rgb], page 667, [ind2rgb], page 660.

33 Audio Processing

Octave provides a few functions for dealing with audio data. An audio 'sample' is a single output value from an A/D converter, i.e., a small integer number (usually 8 or 16 bits), and audio data is just a series of such samples. It can be characterized by three parameters: the sampling rate (measured in samples per second or Hz, e.g., 8000 or 44100), the number of bits per sample (e.g., 8 or 16), and the number of channels (1 for mono, 2 for stereo, etc.).

There are many different formats for representing such data. Currently, only the two most popular, *linear encoding* and *mu-law encoding*, are supported by Octave. There is an excellent FAQ on audio formats by Guido van Rossum `guido@cwi.nl` which can be found at any FAQ ftp site, in particular in the directory '`/pub/usenet/news.answers/audio-fmts`' of the archive site `rtfm.mit.edu`.

Octave simply treats audio data as vectors of samples (non-mono data are not supported yet). It is assumed that audio files using linear encoding have one of the extensions '`lin`' or '`raw`', and that files holding data in mu-law encoding end in '`au`', '`mu`', or '`snd`'.

`lin2mu` (*x*, *n*) [Function File]

Convert audio data from linear to mu-law. Mu-law values use 8-bit unsigned integers. Linear values use *n*-bit signed integers or floating point values in the range $-1 \le x \le 1$ if *n* is 0.

If *n* is not specified it defaults to 0, 8, or 16 depending on the range of values in *x*.

See also: [mu2lin], page 669, [loadaudio], page 669, [saveaudio], page 669.

`mu2lin` (*x*, *n*) [Function File]

Convert audio data from mu-law to linear. Mu-law values are 8-bit unsigned integers. Linear values use *n*-bit signed integers or floating point values in the range $-1 \le y \le 1$ if *n* is 0.

If *n* is not specified it defaults to 0.

See also: [lin2mu], page 669, [loadaudio], page 669, [saveaudio], page 669.

`loadaudio` (*name*, *ext*, *bps*) [Function File]

Load audio data from the file '*name.ext*' into the vector *x*.

The extension *ext* determines how the data in the audio file is interpreted; the extensions '`lin`' (default) and '`raw`' correspond to linear, the extensions '`au`', '`mu`', or '`snd`' to mu-law encoding.

The argument *bps* can be either 8 (default) or 16, and specifies the number of bits per sample used in the audio file.

See also: [lin2mu], page 669, [mu2lin], page 669, [saveaudio], page 669, [playaudio], page 670, [setaudio], page 670, [record], page 670.

`saveaudio` (*name*, *x*, *ext*, *bps*) [Function File]

Save a vector *x* of audio data to the file '*name.ext*'. The optional parameters *ext* and *bps* determine the encoding and the number of bits per sample used in the audio file (see `loadaudio`); defaults are '`lin`' and 8, respectively.

See also: [lin2mu], page 669, [mu2lin], page 669, [loadaudio], page 669, [playaudio], page 670, [setaudio], page 670, [record], page 670.

The following functions for audio I/O require special A/D hardware and operating system support. It is assumed that audio data in linear encoding can be played and recorded by reading from and writing to '/dev/dsp', and that similarly '/dev/audio' is used for mu-law encoding. These file names are system-dependent. Improvements so that these functions will work without modification on a wide variety of hardware are welcome.

`playaudio (name, ext)` [Function File]
`playaudio (x)` [Function File]

Play the audio file '`name.ext`' or the audio data stored in the vector *x*.

See also: [lin2mu], page 669, [mu2lin], page 669, [loadaudio], page 669, [saveaudio], page 669, [setaudio], page 670, [record], page 670.

`record (sec, sampling_rate)` [Function File]

Record *sec* seconds of audio input into the vector *x*. The default value for *sampling_rate* is 8000 samples per second, or 8kHz. The program waits until the user types RET and then immediately starts to record.

See also: [lin2mu], page 669, [mu2lin], page 669, [loadaudio], page 669, [saveaudio], page 669, [playaudio], page 670, [setaudio], page 670.

`setaudio ()` [Function File]
`setaudio (w_type)` [Function File]
`setaudio (w_type, value)` [Function File]

Execute the shell command '`mixer`', possibly with optional arguments *w_type* and *value*.

`y = wavread (filename)` [Function File]
`[y, Fs, bps] = wavread (filename)` [Function File]
`[...] = wavread (filename, n)` [Function File]
`[...] = wavread (filename, [n1 n2])` [Function File]
`[samples, channels] = wavread (filename, "size")` [Function File]

Load the RIFF/WAVE sound file *filename*, and return the samples in vector *y*. If the file contains multichannel data, then *y* is a matrix with the channels represented as columns.

`[y, Fs, bps] = wavread (filename)`

Additionally return the sample rate (*fs*) in Hz and the number of bits per sample (*bps*).

`[...] = wavread (filename, n)`

Read only the first *n* samples from each channel.

`wavread (filename, [n1 n2])`

Read only samples *n1* through *n2* from each channel.

`[samples, channels] = wavread (filename, "size")`

Return the number of samples (*n*) and number of channels (*ch*) instead of the audio data.

See also: [wavwrite], page 671.

`wavwrite` (*y*, *filename*) [Function File]
`wavwrite` (*y*, *Fs*, *filename*) [Function File]
`wavwrite` (*y*, *Fs*, *bps*, *filename*) [Function File]

Write *y* to the canonical RIFF/WAVE sound file *filename* with sample rate *Fs* and bits per sample *bps*. The default sample rate is 8000 Hz with 16-bits per sample. Each column of the data represents a separate channel. If *y* is either a row vector or a column vector, it is written as a single channel.

See also: [wavread], page 670.

34 Object Oriented Programming

Octave includes the capability to include user classes, including the features of operator and function overloading. Equally a user class can be used to encapsulate certain properties of the class so that they cannot be altered accidentally and can be set up to address the issue of class precedence in mixed class operations.

This chapter discussions the means of constructing a user class with the example of a polynomial class, how to query and set the properties of this class, together with the means to overload operators and functions.

34.1 Creating a Class

We use in the following text a polynomial class to demonstrate the use of object oriented programming within Octave. This class was chosen as it is simple, and so doesn't distract unnecessarily from the discussion of the programming features of Octave. However, even still a small understand of the polynomial class itself is necessary to fully grasp the techniques described.

The polynomial class is used to represent polynomials of the form

$$a_0 + a_1 x + a_2 x^2 + \dots a_n x^n$$

where a_0, a_1, etc. are elements of $\Re$. Thus the polynomial can be represented by a vector

```
a = [a0, a1, a2, ..., an];
```

We therefore now have sufficient information about the requirements of the class constructor for our polynomial class to write it. All object oriented classes in Octave, must be contained with a directory taking the name of the class, prepended with the @ symbol. For example, with our polynomial class, we would place the methods defining the class in the @polynomial directory.

The constructor of the class, must have the name of the class itself and so in our example the constructor with have the name '@polynomial/polynomial.m'. Also ideally when the constructor is called with no arguments to should return a value object. So for example our polynomial might look like

```
## -*- texinfo -*-
## @deftypefn  {Function File} {} polynomial ()
## @deftypefnx {Function File} {} polynomial (@var{a})
## Create a polynomial object representing the polynomial
##
## @example
## a0 + a1 * x + a2 * x^2 + @dots{} + an * x^n
## @end example
##
## @noindent
## from a vector of coefficients [a0 a1 a2 @dots{} an].
## @end deftypefn

function p = polynomial (a)
```

```
  if (nargin == 0)
    p.poly = [0];
    p = class (p, "polynomial");
  elseif (nargin == 1)
    if (strcmp (class (a), "polynomial"))
      p = a;
    elseif (isvector (a) && isreal (a))
      p.poly = a(:).';
      p = class (p, "polynomial");
    else
      error ("polynomial: expecting real vector");
    endif
  else
    print_usage ();
  endif
endfunction
```

Note that the return value of the constructor must be the output of the **class** function called with the first argument being a structure and the second argument being the class name. An example of the call to this constructor function is then

```
p = polynomial ([1, 0, 1]);
```

Note that methods of a class can be documented. The help for the constructor itself can be obtained with the constructor name, that is for the polynomial constructor **help polynomial** will return the help string. Also the help can be obtained by restricting the search for the help to a particular class, for example **help @polynomial/polynomial**. This second method is the only means of getting help for the overloaded methods and functions of the class.

The same is true for other Octave functions that take a function name as an argument. For example **type @polynomial/display** will print the code of the display method of the polynomial class to the screen, and **dbstop @polynomial/display** will set a breakpoint at the first executable line of the display method of the polynomial class.

To check where a variable is a user class, the **isobject** and **isa** functions can be used. For example:

```
p = polynomial ([1, 0, 1]);
isobject (p)
  ⇒ 1
isa (p, "polynomial")
  ⇒ 1
```

`isobject (`*x*`)` [Built-in Function]

Return true if *x* is a class object.

See also: [class], page 39, [typeinfo], page 39, [isa], page 39, [ismethod], page 675.

The available methods of a class can be displayed with the **methods** function.

`methods (`*obj*`)` [Function File]
`methods ("`*classname*`")` [Function File]

mtds = methods (...) [Function File]

Return a cell array containing the names of the methods for the object *obj* or the named class *classname*. *obj* may be an Octave class object or a Java object.

See also: [fieldnames], page 105.

To inquire whether a particular method is available to a user class, the **ismethod** function can be used.

ismethod (*x*, *method*) [Built-in Function]

Return true if *x* is a class object and the string *method* is a method of this class.

See also: [isprop], page 336, [isobject], page 674.

For example:

```
p = polynomial ([1, 0, 1]);
ismethod (p, "roots")
  ⇒ 1
```

34.2 Manipulating Classes

There are a number of basic classes methods that can be defined to allow the contents of the classes to be queried and set. The most basic of these is the `display` method. The `display` method is used by Octave when displaying a class on the screen, due to an expression that is not terminated with a semicolon. If this method is not defined, then Octave will printed nothing when displaying the contents of a class.

display (a) [Function File]

Display the contents of an object. If *a* is an object of the class "myclass", then `display` is called in a case like

```
myclass (...)
```

where Octave is required to display the contents of a variable of the type "myclass".

See also: [class], page 39, [subsref], page 678, [subsasgn], page 680.

An example of a display method for the polynomial class might be

```
function display (p)
  a = p.poly;
  first = true;
  fprintf ("%s =", inputname (1));
  for i = 1 : length (a);
    if (a(i) != 0)
      if (first)
        first = false;
      elseif (a(i) > 0)
        fprintf (" +");
      endif
      if (a(i) < 0)
        fprintf (" -");
      endif
```

```
        if (i == 1)
          fprintf (" %g", abs (a(i)));
        elseif (abs(a(i)) != 1)
          fprintf (" %g *", abs (a(i)));
        endif
        if (i > 1)
          fprintf (" X");
        endif
        if (i > 2)
          fprintf (" ^ %d", i - 1);
        endif
      endif
    endfor
    if (first)
      fprintf (" 0");
    endif
    fprintf ("\n");
  endfunction
```

Note that in the display method, it makes sense to start the method with the line `fprintf ("%s =", inputname (1))` to be consistent with the rest of Octave and print the variable name to be displayed when displaying the class.

To be consistent with the Octave graphic handle classes, a class should also define the `get` and `set` methods. The `get` method should accept one or two arguments, and given one argument of the appropriate class it should return a structure with all of the properties of the class. For example:

```
function s = get (p, f)
  if (nargin == 1)
    s.poly = p.poly;
  elseif (nargin == 2)
    if (ischar (f))
      switch (f)
        case "poly"
          s = p.poly;
        otherwise
          error ("get: invalid property %s", f);
      endswitch
    else
      error ("get: expecting the property to be a string");
    endif
  else
    print_usage ();
  endif
endfunction
```

Similarly, the `set` method should taken as its first argument an object to modify, and then take property/value pairs to be modified.

```
function s = set (p, varargin)
  s = p;
  if (length (varargin) < 2 || rem (length (varargin), 2) != 0)
    error ("set: expecting property/value pairs");
  endif
  while (length (varargin) > 1)
    prop = varargin{1};
    val = varargin{2};
    varargin(1:2) = [];
    if (ischar (prop) && strcmp (prop, "poly"))
      if (isvector (val) && isreal (val))
        s.poly = val(:).';
      else
        error ("set: expecting the value to be a real vector");
      endif
    else
      error ("set: invalid property of polynomial class");
    endif
  endwhile
endfunction
```

Note that as Octave does not implement pass by reference, than the modified object is the return value of the `set` method and it must be called like

```
p = set (p, "a", [1, 0, 0, 0, 1]);
```

Also the `set` method makes use of the `subsasgn` method of the class, and this method must be defined. The `subsasgn` method is discussed in the next section.

Finally, user classes can be considered as a special type of a structure, and so they can be saved to a file in the same manner as a structure. For example:

```
p = polynomial ([1, 0, 1]);
save userclass.mat p
clear p
load userclass.mat
```

All of the file formats supported by `save` and `load` are supported. In certain circumstances, a user class might either contain a field that it makes no sense to save or a field that needs to be initialized before it is saved. This can be done with the `saveobj` method of the class

b = `saveobj (a)` [Function File]

Method of a class to manipulate an object prior to saving it to a file. The function `saveobj` is called when the object *a* is saved using the `save` function. An example of the use of `saveobj` might be to remove fields of the object that don't make sense to be saved or it might be used to ensure that certain fields of the object are initialized before the object is saved. For example:

```
function b = saveobj (a)
  b = a;
  if (isempty (b.field))
     b.field = initfield (b);
  endif
endfunction
```

See also: [loadobj], page 678, [class], page 39.

`saveobj` is called just prior to saving the class to a file. Likely, the `loadobj` method is called just after a class is loaded from a file, and can be used to ensure that any removed fields are reinserted into the user object.

b = `loadobj (a)` [Function File]

Method of a class to manipulate an object after loading it from a file. The function `loadobj` is called when the object *a* is loaded using the `load` function. An example of the use of `saveobj` might be to add fields to an object that don't make sense to be saved. For example:

```
function b = loadobj (a)
  b = a;
  b.addmissingfield = addfield (b);
endfunction
```

See also: [saveobj], page 677, [class], page 39.

34.3 Indexing Objects

34.3.1 Defining Indexing And Indexed Assignment

Objects can be indexed with parentheses, either like `a (idx)` or like `a {idx}`, or even like `a (idx).field`. However, it is up to the user to decide what this indexing actually means. In the case of our polynomial class `p (n)` might mean either the coefficient of the *n*-th power of the polynomial, or it might be the evaluation of the polynomial at *n*. The meaning of this subscripted referencing is determined by the `subsref` method.

`subsref (val, idx)` [Built-in Function]

Perform the subscripted element selection operation according to the subscript specified by *idx*.

The subscript *idx* is expected to be a structure array with fields '`type`' and '`subs`'. Valid values for '`type`' are '`"()"`', '`"{}"`', and '`"."`'. The '`subs`' field may be either '`":"`' or a cell array of index values.

The following example shows how to extract the first two columns of a matrix

```
val = magic (3)
    ⇒ val = [ 8   1   6
              3   5   7
              4   9   2 ]
idx.type = "()";
idx.subs = {":", 1:2};
subsref (val, idx)
     ⇒ [ 8   1
         3   5
         4   9 ]
```

Note that this is the same as writing `val(:,1:2)`.

If *idx* is an empty structure array with fields '`type`' and '`subs`', return *val*.

See also: [subsasgn], page 680, [substruct], page 108.

For example we might decide that indexing with `"()"` evaluates the polynomial and indexing with `"{}"` returns the *n*-th coefficient (of *n*-th power). In this case the `subsref` method of our polynomial class might look like

```
function b = subsref (a, s)
  if (isempty (s))
    error ("polynomial: missing index");
  endif
  switch (s(1).type)
    case "()"
      ind = s(1).subs;
      if (numel (ind) != 1)
        error ("polynomial: need exactly one index");
      else
        b = polyval (fliplr (a.poly), ind{1});
      endif
    case "{}"
      ind = s(1).subs;
      if (numel (ind) != 1)
        error ("polynomial: need exactly one index");
      else
        if (isnumeric (ind{1}))
          b = a.poly(ind{1}+1);
        else
          b = a.poly(ind{1});
        endif
      endif
    case "."
      fld = s.subs;
      if (strcmp (fld, "poly"))
        b = a.poly;
      else
        error ("@polynomial/subsref: invalid property \"%s\"", fld);
```

```
        endif
      otherwise
        error ("invalid subscript type");
    endswitch
    if (numel (s) > 1)
      b = subsref (b, s(2:end));
    endif
  endfunction
```

The equivalent functionality for subscripted assignments uses the `subsasgn` method.

`subsasgn (val, idx, rhs)` [Built-in Function]

Perform the subscripted assignment operation according to the subscript specified by *idx*.

The subscript *idx* is expected to be a structure array with fields '`type`' and '`subs`'. Valid values for '`type`' are '`"()"`', '`"{}"`', and '`"."`'. The '`subs`' field may be either '`":"`' or a cell array of index values.

The following example shows how to set the two first columns of a 3-by-3 matrix to zero.

```
val = magic (3);
idx.type = "()";
idx.subs = {":", 1:2};
subsasgn (val, idx, 0)
     ⇒  [ 0   0   6
           0   0   7
           0   0   2 ]
```

Note that this is the same as writing `val(:,1:2) = 0`.

If *idx* is an empty structure array with fields '`type`' and '`subs`', return *rhs*.

See also: [subsref], page 678, [substruct], page 108.

`val = optimize_subsasgn_calls ()` [Built-in Function]
`old_val = optimize_subsasgn_calls (new_val)` [Built-in Function]
`optimize_subsasgn_calls (new_val, "local")` [Built-in Function]

Query or set the internal flag for subsasgn method call optimizations.

If true, Octave will attempt to eliminate the redundant copying when calling the subsasgn method of a user-defined class.

When called from inside a function with the `"local"` option, the variable is changed locally for the function and any subroutines it calls. The original variable value is restored when exiting the function.

Note that the `subsref` and `subsasgn` methods always receive the whole index chain, while they usually handle only the first element. It is the responsibility of these methods to handle the rest of the chain (if needed), usually by forwarding it again to `subsref` or `subsasgn`.

If you wish to use the end keyword in subscripted expressions of an object, then the user needs to define the end method for the class. For example, the end method for our polynomial class might look like

```
function r = end (obj, index_pos, num_indices)

  if (num_indices != 1)
    error ("polynomial object may only have one index")
  endif

  r = length (obj.poly) - 1;

endfunction
```

which is a fairly generic end method that has a behavior similar to the end keyword for Octave Array classes. It can then be used as follows:

```
p = polynomial ([1,2,3,4]);
p(end-1)
  ⇒ 3
```

Objects can also be used as the index in a subscripted expression themselves and this is controlled with the subsindex function.

idx = subsindex (a) [Function File]

Convert an object to an index vector. When *a* is a class object defined with a class constructor, then subsindex is the overloading method that allows the conversion of this class object to a valid indexing vector. It is important to note that subsindex must return a zero-based real integer vector of the class "double". For example, if the class constructor

```
function b = myclass (a)
  b = class (struct ("a", a), "myclass");
endfunction
```

then the subsindex function

```
function idx = subsindex (a)
  idx = double (a.a) - 1.0;
endfunction
```

can then be used as follows

```
a = myclass (1:4);
b = 1:10;
b(a)
⇒ 1  2  3  4
```

See also: [class], page 39, [subsref], page 678, [subsasgn], page 680.

Finally, objects can equally be used like ranges, using the colon method

r = colon (*a*, *b*) [Function File]

r = colon (*a*, *b*, *c*) [Function File]

Method of a class to construct a range with the : operator. For example:

```
a = myclass (...);
b = myclass (...);
c = a : b
```

See also: [class], page 39, [subsref], page 678, [subsasgn], page 680.

34.3.2 Indexed Assignment Optimization

Octave's ubiquitous lazily-copied pass-by-value semantics implies a problem for performance of user-defined subsasgn methods. Imagine a call to subsasgn:

```
ss = substruct ("()",{1});
x = subsasgn (x, ss, 1);
```

and the corresponding method looking like this:

```
function x = subsasgn (x, ss, val)
  ...
  x.myfield (ss.subs{1}) = val;
endfunction
```

The problem is that on entry to the subsasgn method, `x` is still referenced from the caller's scope, which means that the method will first need to unshare (copy) `x` and `x.myfield` before performing the assignment. Upon completing the call, unless an error occurs, the result is immediately assigned to `x` in the caller's scope, so that the previous value of `x.myfield` is forgotten. Hence, the Octave language implies a copy of N elements (N being the size of `x.myfield`), where modifying just a single element would actually suffice, i.e., degrades a constant-time operation to linear-time one. This may be a real problem for user classes that intrinsically store large arrays.

To partially solve the problem, Octave uses a special optimization for user-defined subsasgn methods coded as m-files. When the method gets called as a result of the built-in assignment syntax (not direct subsasgn call as shown above), i.e.

```
x(1) = 1;
```

AND if the subsasgn method is declared with identical input and output argument, like in the example above, then Octave will ignore the copy of `x` inside the caller's scope; therefore, any changes made to `x` during the method execution will directly affect the caller's copy as well. This allows, for instance, defining a polynomial class where modifying a single element takes constant time.

It is important to understand the implications that this optimization brings. Since no extra copy of `x` in the caller's scope will exist, it is *solely* the callee's responsibility to not leave `x` in an invalid state if an error occurs throughout the execution. Also, if the method partially changes `x` and then errors out, the changes *will* affect `x` in the caller's scope. Deleting or completely replacing `x` inside subsasgn will not do anything, however, only indexed assignments matter.

Since this optimization may change the way code works (especially if badly written), a built-in variable `optimize_subsasgn_calls` is provided to control it. It is on by default. Another option to avoid the effect is to declare subsasgn methods with different output and input arguments, like this:

```
function y = subsasgn (x, ss, val)
  ...
endfunction
```

34.4 Overloading Objects

34.4.1 Function Overloading

Any Octave function can be overloaded, and allows an object specific version of this function to be called as needed. A pertinent example for our polynomial class might be to overload the `polyval` function like

```
function [y, dy] = polyval (p, varargin)
  if (nargout == 2)
    [y, dy] = polyval (fliplr (p.poly), varargin{:});
  else
    y = polyval (fliplr (p.poly), varargin{:});
  endif
endfunction
```

This function just hands off the work to the normal Octave `polyval` function. Another interesting example for an overloaded function for our polynomial class is the `plot` function.

```
function h = plot (p, varargin)
  n = 128;
  rmax = max (abs (roots (p.poly)));
  x = [0 : (n - 1)] / (n - 1) * 2.2 * rmax - 1.1 * rmax;
  if (nargout > 0)
    h = plot (x, p(x), varargin{:});
  else
    plot (x, p(x), varargin{:});
  endif
endfunction
```

which allows polynomials to be plotted in the domain near the region of the roots of the polynomial.

Functions that are of particular interest to be overloaded are the class conversion functions such as `double`. Overloading these functions allows the `cast` function to work with the user class and can aid in the use of methods of other classes with the user class. An example `double` function for our polynomial class might look like.

```
function b = double (a)
  b = a.poly;
endfunction
```

34.4.2 Operator Overloading

The following table shows, for each built-in numerical operation, the corresponding function name to use when providing an overloaded method for a user class.

Operation	Method	Description
$a + b$	plus (a, b)	Binary addition operator
$a - b$	minus (a, b)	Binary subtraction operator
$+a$	uplus (a)	Unary addition operator
$-a$	uminus (a)	Unary subtraction operator
$a. * b$	times (a, b)	Element-wise multiplication operator
$a * b$	mtimes (a, b)	Matrix multiplication operator
$a./b$	rdivide (a, b)	Element-wise right division operator
a/b	mrdivide (a, b)	Matrix right division operator
$a.\backslash b$	ldivide (a, b)	Element-wise left division operator
$a\backslash b$	mldivide (a, b)	Matrix left division operator
$a.\hat{}b$	power (a, b)	Element-wise power operator
$a\hat{}b$	mpower (a, b)	Matrix power operator
$a < b$	lt (a, b)	Less than operator
$a <= b$	le (a, b)	Less than or equal to operator
$a > b$	gt (a, b)	Greater than operator
$a >= b$	ge (a, b)	Greater than or equal to operator
$a == b$	eq (a, b)	Equal to operator
$a! = b$	ne (a, b)	Not equal to operator
$a\&b$	and (a, b)	Logical and operator
$a \vert b$	or (a, b)	Logical or operator
$!b$	not (a)	Logical not operator
a'	ctranspose (a)	Complex conjugate transpose operator
$a.'$	transpose (a)	Transpose operator
$a : b$	colon (a, b)	Two element range operator
$a : b : c$	colon (a, b, c)	Three element range operator
$[a, b]$	horzcat (a, b)	Horizontal concatenation operator
$[a; b]$	vertcat (a, b)	Vertical concatenation operator
$a(s_1, \ldots, s_n)$	subsref (a, s)	Subscripted reference
$a(s_1, \ldots, s_n) = b$	subsasgn (a, s, b)	Subscripted assignment
$b(a)$	subsindex (a)	Convert to zero-based index
display	display (a)	Commandline display function

Table 34.1: Available overloaded operators and their corresponding class method

An example `mtimes` method for our polynomial class might look like

```
function y = mtimes (a, b)
  y = polynomial (conv (double (a), double (b)));
endfunction
```

34.4.3 Precedence of Objects

Many functions and operators take two or more arguments and so the case can easily arise that these functions are called with objects of different classes. It is therefore necessary to determine the precedence of which method of which class to call when there are mixed objects given to a function or operator. To do this the `superiorto` and `inferiorto` functions can be used

superiorto (*class_name*, ...) [Built-in Function]

When called from a class constructor, mark the object currently constructed as having a higher precedence than *class_name*. More that one such class can be specified in a single call. This function may only be called from a class constructor.

See also: [inferiorto], page 685.

inferiorto (*class_name*, ...) [Built-in Function]

When called from a class constructor, mark the object currently constructed as having a lower precedence than *class_name*. More that one such class can be specified in a single call. This function may only be called from a class constructor.

See also: [superiorto], page 685.

For example with our polynomial class consider the case

```
2 * polynomial ([1, 0, 1]);
```

That mixes an object of the class `"double"` with an object of the class `"polynomial"`. In this case we like to ensure that the return type of the above is of the type `"polynomial"` and so we use the `superiorto` function in the class constructor. In particular our polynomial class constructor would be modified to be

```
## -*- texinfo -*-
## @deftypefn  {Function File} {} polynomial ()
## @deftypefnx {Function File} {} polynomial (@var{a})
## Create a polynomial object representing the polynomial
##
## @example
## a0 + a1 * x + a2 * x^2 + @dots{} + an * x^n
## @end example
##
## @noindent
## from a vector of coefficients [a0 a1 a2 @dots{} an].
## @end deftypefn

function p = polynomial (a)
  if (nargin == 0)
    p.poly = [0];
    p = class (p, "polynomial");
  elseif (nargin == 1)
    if (strcmp (class (a), "polynomial"))
      p = a;
    elseif (isvector (a) && isreal (a))
      p.poly = a(:).';
      p = class (p, "polynomial");
    else
```

```
        error ("polynomial: expecting real vector");
      endif
    else
      print_usage ();
    endif
    superiorto ("double");
  endfunction
```

Note that user classes always have higher precedence than built-in Octave types. So in fact marking our polynomial class higher than the `"double"` class is in fact not necessary.

When faced with two objects that have the same precedence, Octave will use the method of the object that appears first on the list of arguments.

34.5 Inheritance and Aggregation

Using classes to build new classes is supported by octave through the use of both inheritance and aggregation.

Class inheritance is provided by octave using the `class` function in the class constructor. As in the case of the polynomial class, the octave programmer will create a struct that contains the data fields required by the class, and then call the class function to indicate that an object is to be created from the struct. Creating a child of an existing object is done by creating an object of the parent class and providing that object as the third argument of the class function.

This is easily demonstrated by example. Suppose the programmer needs an FIR filter, i.e., a filter with a numerator polynomial but a unity denominator polynomial. In traditional octave programming, this would be performed as follows.

```
octave:1> x = [some data vector];
octave:2> n = [some coefficient vector];
octave:3> y = filter (n, 1, x);
```

The equivalent class could be implemented in a class directory @FIRfilter that is on the octave path. The constructor is a file FIRfilter.m in the class directory.

```
## -*- texinfo -*-
## @deftypefn  {Function File} {} FIRfilter ()
## @deftypefnx {Function File} {} FIRfilter (@var{p})
## Create a FIR filter with polynomial @var{p} as coefficient vector.
## @end deftypefn

function f = FIRfilter (p)

  f.polynomial = [];
  if (nargin == 0)
    p = @polynomial ([1]);
  elseif (nargin == 1)
    if (!isa (p, "polynomial"))
      error ("FIRfilter: expecting polynomial as input argument");
    endif
```

```
    else
      print_usage ();
    endif
    f = class (f, "FIRfilter", p);
  endfunction
```

As before, the leading comments provide command-line documentation for the class constructor. This constructor is very similar to the polynomial class constructor, except that we pass a polynomial object as the third argument to the class function, telling octave that the FIRfilter class will be derived from the polynomial class. Our FIR filter does not have any data fields, but we must provide a struct to the `class` function. The `class` function will add an element named polynomial to the object struct, so we simply add a dummy element named polynomial as the first line of the constructor. This dummy element will be overwritten by the class function.

Note further that all our examples provide for the case in which no arguments are supplied. This is important since octave will call the constructor with no arguments when loading objects from save files to determine the inheritance structure.

A class may be a child of more than one class (see the documentation for the `class` function), and inheritance may be nested. There is no limitation to the number of parents or the level of nesting other than memory or other physical issues.

As before, we need a `display` method. A simple example might be

```
function display (f)

  display (f.polynomial);

endfunction
```

Note that we have used the polynomial field of the struct to display the filter coefficients.

Once we have the class constructor and display method, we may create an object by calling the class constructor. We may also check the class type and examine the underlying structure.

```
octave:1> f = FIRfilter (polynomial ([1 1 1]/3))
f.polynomial = 0.333333 + 0.333333 * X + 0.333333 * X ^ 2
octave:2> class (f)
ans = FIRfilter
octave:3> isa (f,"FIRfilter")
ans =  1
octave:4> isa (f,"polynomial")
ans =  1
octave:5> struct (f)
ans =
{
polynomial = 0.333333 + 0.333333 * X + 0.333333 * X ^ 2
}
```

We only need to define a method to actually process data with our filter and our class is usable. It is also useful to provide a means of changing the data stored in the class. Since the fields in the underlying struct are private by default, we could provide a mechanism to access the fields. The `subsref` method may be used for both.

```
function out = subsref (f, x)
  switch (x.type)
    case "()"
      n = f.polynomial;
      out = filter (n.poly, 1, x.subs{1});
    case "."
      fld = x.subs;
      if (strcmp (fld, "polynomial"))
        out = f.polynomial;
      else
        error ("@FIRfilter/subsref: invalid property \"%s\"", fld);
      endif
    otherwise
      error ("@FIRfilter/subsref: invalid subscript type for FIR filter");
  endswitch
endfunction
```

The `"()"` case allows us to filter data using the polynomial provided to the constructor.

```
octave:2> f = FIRfilter (polynomial ([1 1 1]/3));
octave:3> x = ones (5,1);
octave:4> y = f(x)
y =

   0.33333
   0.66667
   1.00000
   1.00000
   1.00000
```

The `"."` case allows us to view the contents of the polynomial field.

```
octave:1> f = FIRfilter (polynomial ([1 1 1]/3));
octave:2> f.polynomial
ans = 0.333333 + 0.333333 * X + 0.333333 * X ^ 2
```

In order to change the contents of the object, we need to define a `subsasgn` method. For example, we may make the polynomial field publicly writable.

```
function out = subsasgn (f, index, val)
  switch (index.type)
    case "."
      fld = index.subs;
      if (strcmp (fld, "polynomial"))
        out = f;
        out.polynomial = val;
      else
        error ("@FIRfilter/subsref: invalid property \"%s\"", fld);
      endif
    otherwise
      error ("FIRfilter/subsagn: Invalid index type")
  endswitch
endfunction
```

So that

```
octave:6> f = FIRfilter ();
octave:7> f.polynomial = polynomial ([1 2 3]);
f.polynomial = 1 + 2 * X + 3 * X ^ 2
```

Defining the FIRfilter class as a child of the polynomial class implies that and FIRfilter object may be used any place that a polynomial may be used. This is not a normal use of a filter, so that aggregation may be a more sensible design approach. In this case, the polynomial is simply a field in the class structure. A class constructor for this case might be

```
## -*- texinfo -*-
## @deftypefn  {Function File} {} FIRfilter ()
## @deftypefnx {Function File} {} FIRfilter (@var{p})
## Create a FIR filter with polynomial @var{p} as coefficient vector.
## @end deftypefn

function f = FIRfilter (p)

  if (nargin == 0)
    f.polynomial = @polynomial ([1]);
  elseif (nargin == 1)
    if (isa (p, "polynomial"))
      f.polynomial = p;
    else
      error ("FIRfilter: expecting polynomial as input argument");
    endif
  else
    print_usage ();
  endif
  f = class (f, "FIRfilter");
endfunction
```

For our example, the remaining class methods remain unchanged.

35 GUI Development

Octave is principally a batch or command-line language. However, it does offer some limited features for constructing graphical interfaces for interacting with users.

The GUI elements available are I/O dialogs and a progress bar. For example, rather than hardcoding a filename for output results a script can open a dialog box and allow the user to choose a file. Similarly, if a calculation is expected to take a long time a script can display a progress bar.

Several utility functions make it possible to store private data for use with a GUI which will not pollute the user's variable space.

Finally, a program written in Octave might want to have long term storage of preferences or state variables. This can be done with user-defined preferences.

35.1 I/O Dialogs

Simple dialog menus are available for choosing directories or files. They return a string variable which can then be used with any command requiring a file name.

dirname = **uigetdir** () [Function File]
dirname = **uigetdir** (*init_path*) [Function File]
dirname = **uigetdir** (*init_path*, *dialog_name*) [Function File]

Open a GUI dialog for selecting a directory. If *init_path* is not given the current working directory is used. *dialog_name* may be used to customize the dialog title.

See also: [uigetfile], page 691, [uiputfile], page 692.

[*fname*, *fpath*, *fltidx*] = **uigetfile** () [Function File]
[...] = **uigetfile** (*flt*) [Function File]
[...] = **uigetfile** (*flt*, *dialog_name*) [Function File]
[...] = **uigetfile** (*flt*, *dialog_name*, *default_file*) [Function File]
[...] = **uigetfile** (..., "*Position*", [*px py*]) [Function File]
[...] = **uigetfile** (..., "*MultiSelect*", *mode*) [Function File]

Open a GUI dialog for selecting a file and return the filename *fname*, the path to this file *fpath*, and the filter index *fltidx*. *flt* contains a (list of) file filter string(s) in one of the following formats:

`"/path/to/filename.ext"`
: If a filename is given then the file extension is extracted and used as filter. In addition, the path is selected as current path and the filename is selected as default file. Example: `uigetfile ("myfun.m")`

A single file extension `"*.ext"`
: Example: `uigetfile ("*.ext")`

A 2-column cell array
: containing a file extension in the first column and a brief description in the second column. Example: `uigetfile ({"*.ext", "My Description";"*.xyz", "XYZ-Format"})`

The filter string can also contain a semicolon separated list of filter extensions. Example: `uigetfile ({"*.gif;*.png;*.jpg", "Supported Picture Formats"})`

dialog_name can be used to customize the dialog title. If *default_file* is given then it will be selected in the GUI dialog. If, in addition, a path is given it is also used as current path.

The screen position of the GUI dialog can be set using the `"Position"` key and a 2-element vector containing the pixel coordinates. Two or more files can be selected when setting the `"MultiSelect"` key to `"on"`. In that case *fname* is a cell array containing the files.

See also: [uiputfile], page 692, [uigetdir], page 691.

`[fname, fpath, fltidx] = uiputfile ()` [Function File]
`[fname, fpath, fltidx] = uiputfile (flt)` [Function File]
`[fname, fpath, fltidx] = uiputfile (flt, dialog_name)` [Function File]
`[fname, fpath, fltidx] = uiputfile (flt, dialog_name, default_file)` [Function File]

Open a GUI dialog for selecting a file. *flt* contains a (list of) file filter string(s) in one of the following formats:

`"/path/to/filename.ext"`
: If a filename is given the file extension is extracted and used as filter. In addition the path is selected as current path and the filename is selected as default file. Example: `uiputfile ("myfun.m")`

`"*.ext"`
: A single file extension. Example: `uiputfile ("*.ext")`

`{"*.ext", "My Description"}`
: A 2-column cell array containing the file extension in the 1st column and a brief description in the 2nd column. Example: `uiputfile ({"*.ext","My Description";"*.xyz", "XYZ-Format"})`

The filter string can also contain a semicolon separated list of filter extensions. Example: `uiputfile ({"*.gif;*.png;*.jpg", "Supported Picture Formats"})`

dialog_name can be used to customize the dialog title. If *default_file* is given it is preselected in the GUI dialog. If, in addition, a path is given it is also used as current path.

See also: [uigetfile], page 691, [uigetdir], page 691.

35.2 Progress Bar

`h = waitbar (frac)` [Function File]
`h = waitbar (frac, msg)` [Function File]
`h = waitbar (..., "FigureProperty", "Value", ...)` [Function File]
`waitbar (frac)` [Function File]
`waitbar (frac, hwbar)` [Function File]
`waitbar (frac, hwbar, msg)` [Function File]

Return a handle *h* to a new waitbar object.

The waitbar is filled to fraction *frac* which must be in the range [0, 1]. The optional message *msg* is centered and displayed above the waitbar. The appearance of the waitbar figure window can be configured by passing property/value pairs to the function.

When called with a single input the current waitbar, if it exists, is updated to the new value *frac*. If there are multiple outstanding waitbars they can be updated individually by passing the handle *hwbar* of the specific waitbar to modify.

35.3 GUI Utility Functions

These functions do not implement a GUI element but are useful when developing programs that do. **Warning:** The functions `uiwait`, `uiresume`, and `waitfor` are only available for the FLTK tooolkit.

used = *desktop* ("*-inuse*") [Function File]

Return true if the desktop (GUI) is currently in use.

See also: [isguirunning], page 693.

data = `guidata` (*h*) [Function File]
`guidata` (*h*, *data*) [Function File]

Query or set user-custom GUI data.

The GUI data is stored in the figure handle *h*. If *h* is not a figure handle then it's parent figure will be used for storage.

data must be a single object which means it is usually preferable for it to be a data container such as a cell array or struct so that additional data items can be added easily.

See also: [getappdata], page 363, [setappdata], page 363, [get], page 341, [set], page 342, [getpref], page 695, [setpref], page 695.

hdata = `guihandles` (*h*) [Function File]
hdata = `guihandles` [Function File]

Return a structure of object handles for the figure associated with handle *h*.

If no handle is specified the current figure returned by `gcf` is used.

The fieldname for each entry of *hdata* is taken from the `"tag"` property of the graphic object. If the tag is empty then the handle is not returned. If there are multiple graphic objects with the same tag then the entry in *hdata* will be a vector of handles. `guihandles` includes all possible handles, including those for which `"HandleVisibility"` is `"off"`.

See also: [guidata], page 693, [findobj], page 359, [findall], page 359, [allchild], page 342.

`isguirunning` () [Built-in Function]

Return true if Octave is running in GUI mode and false otherwise.

`uiwait` [Function File]
`uiwait` (*h*) [Function File]

`uiwait (h, timeout)` [Function File]

Suspend program execution until the figure with handle *h* is deleted or `uiresume` is called. When no figure handle is specified, this function uses the current figure.

If the figure handle is invalid or there is no current figure, this functions returns immediately.

When specified, *timeout* defines the number of seconds to wait for the figure deletion or the `uiresume` call. The timeout value must be at least 1. If a smaller value is specified, a warning is issued and a timeout value of 1 is used instead. If a non-integer value is specified, it is truncated towards 0. If *timeout* is not specified, the program execution is suspended indefinitely.

See also: [uiresume], page 694, [waitfor], page 694.

`uiresume (h)` [Function File]

Resume program execution suspended with `uiwait`. The handle *h* must be the same as the on specified in `uiwait`. If the handle is invalid or there is no `uiwait` call pending for the figure with handle *h*, this function does nothing.

See also: [uiwait], page 693.

`waitfor (h)` [Built-in Function]
`waitfor (h, prop)` [Built-in Function]
`waitfor (h, prop, value)` [Built-in Function]
`waitfor (..., "timeout", timeout)` [Built-in Function]

Suspend the execution of the current program until a condition is satisfied on the graphics handle *h*.

While the program is suspended graphics events are still being processed normally, allowing callbacks to modify the state of graphics objects. This function is reentrant and can be called from a callback, while another `waitfor` call is pending at the top-level.

In the first form, program execution is suspended until the graphics object *h* is destroyed. If the graphics handle is invalid, the function returns immediately.

In the second form, execution is suspended until the graphics object is destroyed or the property named *prop* is modified. If the graphics handle is invalid or the property does not exist, the function returns immediately.

In the third form, execution is suspended until the graphics object is destroyed or the property named *prop* is set to *value*. The function `isequal` is used to compare property values. If the graphics handle is invalid, the property does not exist or the property is already set to *value*, the function returns immediately.

An optional timeout can be specified using the property `timeout`. This timeout value is the number of seconds to wait for the condition to be true. *timeout* must be at least 1. If a smaller value is specified, a warning is issued and a value of 1 is used instead. If the timeout value is not an integer, it is truncated towards 0.

To define a condition on a property named `timeout`, use the string `\timeout` instead.

In all cases, typing CTRL-C stops program execution immediately.

See also: [waitforbuttonpress], page 332, [isequal], page 141.

35.4 User-Defined Preferences

`getpref` (*group*, *pref*, *default*) [Function File]

Return the preference value corresponding to the named preference *pref* in the preference group *group*.

The named preference group must be a character string.

If *pref* does not exist in *group* and *default* is specified, return *default*.

The preference *pref* may be a character string or a cell array of character strings. The corresponding default value *default* may be any value, or, if *pref* is a cell array of strings, *default* must be a cell array of values with the same size as *pref*.

If neither *pref* nor *default* are specified, return a structure of preferences for the preference group *group*.

If no arguments are specified, return a structure containing all groups of preferences and their values.

See also: [addpref], page 695, [setpref], page 695, [ispref], page 696, [rmpref], page 695.

`setpref` (*group*, *pref*, *val*) [Function File]

Set a preference *pref* to the given *val* in the named preference group *group*.

The named preference group must be a character string.

The preference *pref* may be a character string or a cell array of character strings. The corresponding value *val* may be any value, or, if *pref* is a cell array of strings, *val* must be a cell array of values with the same size as *pref*.

If the named preference or group does not exist, it is added.

See also: [addpref], page 695, [getpref], page 695, [ispref], page 696, [rmpref], page 695.

`addpref` (*group*, *pref*, *val*) [Function File]

Add a preference *pref* and associated value *val* to the named preference group *group*.

The named preference group must be a character string.

The preference *pref* may be a character string or a cell array of character strings. The corresponding value *val* may be any value, or, if *pref* is a cell array of strings, *val* must be a cell array of values with the same size as *pref*.

See also: [setpref], page 695, [getpref], page 695, [ispref], page 696, [rmpref], page 695.

`rmpref` (*group*) [Function File]
`rmpref` (*group*, *pref*) [Function File]

Remove the named preference *pref* from the preference group *group*.

The named preference group must be a character string.

The preference *pref* may be a character string or cell array of strings.

If *pref* is not specified, remove the preference group *group*.

It is an error to remove a nonexistent preference or group.

See also: [addpref], page 695, [ispref], page 696, [setpref], page 695, [getpref], page 695.

`ispref (group, pref)` [Function File]

Return true if the named preference *pref* exists in the preference group *group*.

The named preference group must be a character string.

The preference *pref* may be a character string or a cell array of character strings.

If *pref* is not specified, return true if the preference group *group* exists.

See also: [getpref], page 695, [addpref], page 695, [setpref], page 695, [rmpref], page 695.

`prefdir` [Command]
`dir = prefdir` [Command]

Return the directory that contains the preferences for Octave.

Examples:

Display the preferences directory

```
prefdir
```

Change to the preferences folder

```
cd (prefdir)
```

See also: [getpref], page 695, [setpref], page 695, [addpref], page 695, [rmpref], page 695, [ispref], page 696.

`preferences` [Command]

Display the GUI preferences dialog window for Octave.

36 System Utilities

This chapter describes the functions that are available to allow you to get information about what is happening outside of Octave, while it is still running, and use this information in your program. For example, you can get information about environment variables, the current time, and even start other programs from the Octave prompt.

36.1 Timing Utilities

Octave's core set of functions for manipulating time values are patterned after the corresponding functions from the standard C library. Several of these functions use a data structure for time that includes the following elements:

`usec` Microseconds after the second (0-999999).

`sec` Seconds after the minute (0-60). This number can be 60 to account for leap seconds.

`min` Minutes after the hour (0-59).

`hour` Hours since midnight (0-23).

`mday` Day of the month (1-31).

`mon` Months since January (0-11).

`year` Years since 1900.

`wday` Days since Sunday (0-6).

`yday` Days since January 1 (0-365).

`isdst` Daylight Savings Time flag.

`zone` Time zone.

In the descriptions of the following functions, this structure is referred to as a *tm_struct*.

seconds = `time ()` [Built-in Function]

Return the current time as the number of seconds since the epoch. The epoch is referenced to 00:00:00 CUT (Coordinated Universal Time) 1 Jan 1970. For example, on Monday February 17, 1997 at 07:15:06 CUT, the value returned by `time` was 856163706.

See also: [strftime], page 699, [strptime], page 701, [localtime], page 698, [gmtime], page 698, [mktime], page 699, [now], page 697, [date], page 701, [clock], page 701, [datenum], page 703, [datestr], page 704, [datevec], page 705, [calendar], page 706, [weekday], page 706.

`t = now ()` [Function File]

Return the current local date/time as a serial day number (see `datenum`).

The integral part, `floor (now)` corresponds to the number of days between today and Jan 1, 0000.

The fractional part, `rem (now, 1)` corresponds to the current time.

See also: [clock], page 701, [date], page 701, [datenum], page 703.

`ctime (t)` [Function File]

Convert a value returned from `time` (or any other non-negative integer), to the local time and return a string of the same form as `asctime`. The function `ctime (time)` is equivalent to `asctime (localtime (time))`. For example:

```
ctime (time ())
    ⇒ "Mon Feb 17 01:15:06 1997"
```

See also: [asctime], page 699, [time], page 697, [localtime], page 698.

tm_struct = `gmtime (t)` [Built-in Function]

Given a value returned from `time`, or any non-negative integer, return a time structure corresponding to CUT (Coordinated Universal Time). For example:

```
gmtime (time ())
     ⇒ {
           usec = 0
           sec = 6
           min = 15
           hour = 7
           mday = 17
           mon = 1
           year = 97
           wday = 1
           yday = 47
           isdst = 0
           zone = CST
        }
```

See also: [strftime], page 699, [strptime], page 701, [localtime], page 698, [mktime], page 699, [time], page 697, [now], page 697, [date], page 701, [clock], page 701, [datenum], page 703, [datestr], page 704, [datevec], page 705, [calendar], page 706, [weekday], page 706.

tm_struct = `localtime (t)` [Built-in Function]

Given a value returned from `time`, or any non-negative integer, return a time structure corresponding to the local time zone.

```
localtime (time ())
     ⇒ {
           usec = 0
           sec = 6
           min = 15
           hour = 1
           mday = 17
           mon = 1
           year = 97
           wday = 1
           yday = 47
           isdst = 0
           zone = CST
        }
```

See also: [strftime], page 699, [strptime], page 701, [gmtime], page 698, [mktime], page 699, [time], page 697, [now], page 697, [date], page 701, [clock], page 701, [datenum], page 703, [datestr], page 704, [datevec], page 705, [calendar], page 706, [weekday], page 706.

seconds = mktime (*tm_struct*) [Built-in Function]

Convert a time structure corresponding to the local time to the number of seconds since the epoch. For example:

```
mktime (localtime (time ()))
     ⇒ 856163706
```

See also: [strftime], page 699, [strptime], page 701, [localtime], page 698, [gmtime], page 698, [time], page 697, [now], page 697, [date], page 701, [clock], page 701, [datenum], page 703, [datestr], page 704, [datevec], page 705, [calendar], page 706, [weekday], page 706.

asctime (*tm_struct*) [Function File]

Convert a time structure to a string using the following format: "ddd mmm mm HH:MM:SS yyyy". For example:

```
asctime (localtime (time ()))
     ⇒ "Mon Feb 17 01:15:06 1997"
```

This is equivalent to `ctime (time ())`.

See also: [ctime], page 698, [localtime], page 698, [time], page 697.

strftime (*fmt*, *tm_struct*) [Built-in Function]

Format the time structure *tm_struct* in a flexible way using the format string *fmt* that contains '%' substitutions similar to those in `printf`. Except where noted, substituted fields have a fixed size; numeric fields are padded if necessary. Padding is with zeros by default; for fields that display a single number, padding can be changed or inhibited by following the '%' with one of the modifiers described below. Unknown field specifiers are copied as normal characters. All other characters are copied to the output without change. For example:

```
strftime ("%r (%Z) %A %e %B %Y", localtime (time ()))
      ⇒ "01:15:06 AM (CST) Monday 17 February 1997"
```

Octave's `strftime` function supports a superset of the ANSI C field specifiers.

Literal character fields:

`%%` % character.

`%n` Newline character.

`%t` Tab character.

Numeric modifiers (a nonstandard extension):

`- (dash)` Do not pad the field.

`_ (underscore)`
Pad the field with spaces.

Time fields:

`%H`	Hour (00-23).
`%I`	Hour (01-12).
`%k`	Hour (0-23).
`%l`	Hour (1-12).
`%M`	Minute (00-59).
`%p`	Locale's AM or PM.
`%r`	Time, 12-hour (hh:mm:ss [AP]M).
`%R`	Time, 24-hour (hh:mm).
`%s`	Time in seconds since 00:00:00, Jan 1, 1970 (a nonstandard extension).
`%S`	Second (00-61).
`%T`	Time, 24-hour (hh:mm:ss).
`%X`	Locale's time representation (%H:%M:%S).
`%Z`	Time zone (EDT), or nothing if no time zone is determinable.

Date fields:

`%a`	Locale's abbreviated weekday name (Sun-Sat).
`%A`	Locale's full weekday name, variable length (Sunday-Saturday).
`%b`	Locale's abbreviated month name (Jan-Dec).
`%B`	Locale's full month name, variable length (January-December).
`%c`	Locale's date and time (Sat Nov 04 12:02:33 EST 1989).
`%C`	Century (00-99).
`%d`	Day of month (01-31).
`%e`	Day of month (1-31).
`%D`	Date (mm/dd/yy).
`%h`	Same as %b.
`%j`	Day of year (001-366).
`%m`	Month (01-12).
`%U`	Week number of year with Sunday as first day of week (00-53).
`%w`	Day of week (0-6).
`%W`	Week number of year with Monday as first day of week (00-53).
`%x`	Locale's date representation (mm/dd/yy).
`%y`	Last two digits of year (00-99).
`%Y`	Year (1970-).

See also: [strptime], page 701, [localtime], page 698, [gmtime], page 698, [mktime], page 699, [time], page 697, [now], page 697, [date], page 701, [clock], page 701, [datenum], page 703, [datestr], page 704, [datevec], page 705, [calendar], page 706, [weekday], page 706.

`[tm_struct, nchars] = strptime (str, fmt)` [Built-in Function]

Convert the string *str* to the time structure *tm_struct* under the control of the format string *fmt*.

If *fmt* fails to match, *nchars* is 0; otherwise, it is set to the position of last matched character plus 1. Always check for this unless you're absolutely sure the date string will be parsed correctly.

See also: [strftime], page 699, [localtime], page 698, [gmtime], page 698, [mktime], page 699, [time], page 697, [now], page 697, [date], page 701, [clock], page 701, [datenum], page 703, [datestr], page 704, [datevec], page 705, [calendar], page 706, [weekday], page 706.

Most of the remaining functions described in this section are not patterned after the standard C library. Some are available for compatibility with MATLAB and others are provided because they are useful.

`clock ()` [Function File]

Return the current local date and time as a date vector. The date vector contains the following fields: current year, month (1-12), day (1-31), hour (0-23), minute (0-59), and second (0-61). The seconds field has a fractional part after the decimal point for extended accuracy.

For example:

```
fix (clock ())
     ⇒ [ 1993, 8, 20, 4, 56, 1 ]
```

The function clock is more accurate on systems that have the `gettimeofday` function.

See also: [now], page 697, [date], page 701, [datevec], page 705.

`date ()` [Function File]

Return the current date as a character string in the form DD-MMM-YYYY.

For example:

```
date ()
  ⇒ "20-Aug-1993"
```

See also: [now], page 697, [clock], page 701, [datestr], page 704, [localtime], page 698.

`etime (t2, t1)` [Function File]

Return the difference in seconds between two time values returned from `clock` ($t2 - t1$). For example:

```
t0 = clock ();
# many computations later...
elapsed_time = etime (clock (), t0);
```

will set the variable `elapsed_time` to the number of seconds since the variable `t0` was set.

See also: [tic], page 702, [toc], page 702, [clock], page 701, [cputime], page 702, [addtodate], page 706.

`[total, user, system] = cputime ();` [Built-in Function]

Return the CPU time used by your Octave session. The first output is the total time spent executing your process and is equal to the sum of second and third outputs, which are the number of CPU seconds spent executing in user mode and the number of CPU seconds spent executing in system mode, respectively. If your system does not have a way to report CPU time usage, `cputime` returns 0 for each of its output values. Note that because Octave used some CPU time to start, it is reasonable to check to see if `cputime` works by checking to see if the total CPU time used is nonzero.

See also: [tic], page 702, [toc], page 702.

`is_leap_year ()` [Function File]
`is_leap_year (year)` [Function File]

Return true if *year* is a leap year and false otherwise. If no year is specified, `is_leap_year` uses the current year. For example:

```
is_leap_year (2000)
   ⇒ 1
```

See also: [weekday], page 706, [eomday], page 707, [calendar], page 706.

`tic ()` [Built-in Function]
`id = tic ()` [Built-in Function]
`toc ()` [Built-in Function]
`toc (id)` [Built-in Function]
`val = toc (...)` [Built-in Function]

Set or check a wall-clock timer. Calling `tic` without an output argument sets the internal timer state. Subsequent calls to `toc` return the number of seconds since the timer was set. For example,

```
tic ();
# many computations later...
elapsed_time = toc ();
```

will set the variable `elapsed_time` to the number of seconds since the most recent call to the function `tic`.

If called with one output argument, `tic` returns a scalar of type `uint64` that may be later passed to `toc`.

```
id = tic; sleep (5); toc (id)
      ⇒ 5.0010
```

Calling `tic` and `toc` this way allows nested timing calls.

If you are more interested in the CPU time that your process used, you should use the `cputime` function instead. The `tic` and `toc` functions report the actual wall clock time that elapsed between the calls. This may include time spent processing other jobs or doing nothing at all.

See also: [toc], page 702, [cputime], page 702.

`pause ()` [Built-in Function]
`pause (n)` [Built-in Function]

Suspend the execution of the program for *n* seconds.

n is a positive real value and may be a fraction of a second. If invoked without an input arguments then the program is suspended until a character is typed.

The following example prints a message and then waits 5 seconds before clearing the screen.

```
fprintf (stderr, "wait please...\n");
pause (5);
clc;
```

See also: [kbhit], page 229, [sleep], page 703.

`sleep (`*`seconds`*`)` [Built-in Function]

Suspend the execution of the program for the given number of seconds.

See also: [usleep], page 703, [pause], page 702.

`usleep (`*`microseconds`*`)` [Built-in Function]

Suspend the execution of the program for the given number of microseconds. On systems where it is not possible to sleep for periods of time less than one second, `usleep` will pause the execution for `round (`*`microseconds`*` / 1e6)` seconds.

See also: [sleep], page 703, [pause], page 702.

days = `datenum (`*datevec*`)` [Function File]
days = `datenum (`*year*, *month*, *day*`)` [Function File]
days = `datenum (`*year*, *month*, *day*, *hour*`)` [Function File]
days = `datenum (`*year*, *month*, *day*, *hour*, *minute*`)` [Function File]
days = `datenum (`*year*, *month*, *day*, *hour*, *minute*, *second*`)` [Function File]
days = `datenum ("`*datestr*`")` [Function File]
days = `datenum ("`*datestr*`",` *p*`)` [Function File]
[*days*, *secs*] = `datenum (...)` [Function File]

Return the date/time input as a serial day number, with Jan 1, 0000 defined as day 1.

The integer part, `floor (`*`days`*`)` counts the number of complete days in the date input.

The fractional part, `rem (`*`days`*`, 1)` corresponds to the time on the given day.

The input may be a date vector (see `datevec`), datestr (see `datestr`), or directly specified as input.

When processing input datestrings, *p* is the year at the start of the century to which two-digit years will be referenced. If not specified, it defaults to the current year minus 50.

The optional output *secs* holds the time on the specified day with greater precision than *days*.

Notes:

- Years can be negative and/or fractional.
- Months below 1 are considered to be January.
- Days of the month start at 1.
- Days beyond the end of the month go into subsequent months.

- Days before the beginning of the month go to the previous month.
- Days can be fractional.

Caution: this function does not attempt to handle Julian calendars so dates before October 15, 1582 are wrong by as much as eleven days. Also, be aware that only Roman Catholic countries adopted the calendar in 1582. It took until 1924 for it to be adopted everywhere. See the Wikipedia entry on the Gregorian calendar for more details.

Warning: leap seconds are ignored. A table of leap seconds is available on the Wikipedia entry for leap seconds.

See also: [datestr], page 704, [datevec], page 705, [now], page 697, [clock], page 701, [date], page 701.

str = `datestr` (*date*) [Function File]
str = `datestr` (*date*, *f*) [Function File]
str = `datestr` (*date*, *f*, *p*) [Function File]

Format the given date/time according to the format `f` and return the result in *str*. *date* is a serial date number (see `datenum`) or a date vector (see `datevec`). The value of *date* may also be a string or cell array of strings.

f can be an integer which corresponds to one of the codes in the table below, or a date format string.

p is the year at the start of the century in which two-digit years are to be interpreted in. If not specified, it defaults to the current year minus 50.

For example, the date 730736.65149 (2000-09-07 15:38:09.0934) would be formatted as follows:

Code	**Format**	**Example**
0	dd-mmm-yyyy HH:MM:SS	07-Sep-2000 15:38:09
1	dd-mmm-yyyy	07-Sep-2000
2	mm/dd/yy	09/07/00
3	mmm	Sep
4	m	S
5	mm	09
6	mm/dd	09/07
7	dd	07
8	ddd	Thu
9	d	T
10	yyyy	2000
11	yy	00
12	mmmyy	Sep00
13	HH:MM:SS	15:38:09
14	HH:MM:SS PM	03:38:09 PM
15	HH:MM	15:38
16	HH:MM PM	03:38 PM
17	QQ-YY	Q3-00
18	QQ	Q3

19	dd/mm	07/09
20	dd/mm/yy	07/09/00
21	mmm.dd,yyyy HH:MM:SS	Sep.07,2000 15:38:08
22	mmm.dd,yyyy	Sep.07,2000
23	mm/dd/yyyy	09/07/2000
24	dd/mm/yyyy	07/09/2000
25	yy/mm/dd	00/09/07
26	yyyy/mm/dd	2000/09/07
27	QQ-YYYY	Q3-2000
28	mmmyyyy	Sep2000
29	yyyy-mm-dd	2000-09-07
30	yyyymmddTHHMMSS	20000907T153808
31	yyyy-mm-dd HH:MM:SS	2000-09-07 15:38:08

If *f* is a format string, the following symbols are recognized:

Symbol	**Meaning**	**Example**
yyyy	Full year	2005
yy	Two-digit year	05
mmmm	Full month name	December
mmm	Abbreviated month name	Dec
mm	Numeric month number (padded with zeros)	01, 08, 12
m	First letter of month name (capitalized)	D
dddd	Full weekday name	Sunday
ddd	Abbreviated weekday name	Sun
dd	Numeric day of month (padded with zeros)	11
d	First letter of weekday name (capitalized)	S
HH	Hour of day, padded with zeros if PM is set	09:00
	and not padded with zeros otherwise	9:00 AM
MM	Minute of hour (padded with zeros)	10:05
SS	Second of minute (padded with zeros)	10:05:03
FFF	Milliseconds of second (padded with zeros)	10:05:03.012
AM	Use 12-hour time format	11:30 AM
PM	Use 12-hour time format	11:30 PM

If *f* is not specified or is `-1`, then use 0, 1 or 16, depending on whether the date portion or the time portion of *date* is empty.

If *p* is nor specified, it defaults to the current year minus 50.

If a matrix or cell array of dates is given, a column vector of date strings is returned.

See also: [datenum], page 703, [datevec], page 705, [date], page 701, [now], page 697, [clock], page 701.

v = `datevec` (*date*) [Function File]
v = `datevec` (*date*, *f*) [Function File]
v = `datevec` (*date*, *p*) [Function File]
v = `datevec` (*date*, *f*, *p*) [Function File]

`[y, m, d, h, mi, s] = datevec (...)` [Function File]

Convert a serial date number (see `datenum`) or date string (see `datestr`) into a date vector.

A date vector is a row vector with six members, representing the year, month, day, hour, minute, and seconds respectively.

f is the format string used to interpret date strings (see `datestr`). If *date* is a string, but no format is specified, then a relatively slow search is performed through various formats. It is always preferable to specify the format string *f* if it is known. Formats which do not specify a particular time component will have the value set to zero. Formats which do not specify a date will default to January 1st of the current year.

p is the year at the start of the century to which two-digit years will be referenced. If not specified, it defaults to the current year minus 50.

See also: [datenum], page 703, [datestr], page 704, [clock], page 701, [now], page 697, [date], page 701.

`d = addtodate (d, q, f)` [Function File]

Add *q* amount of time (with units *f*) to the serial datenum, *d*.

f must be one of `"year"`, `"month"`, `"day"`, `"hour"`, `"minute"`, `"second"`, or `"millisecond"`.

See also: [datenum], page 703, [datevec], page 705, [etime], page 701.

`c = calendar ()` [Function File]
`c = calendar (d)` [Function File]
`c = calendar (y, m)` [Function File]
`calendar (...)` [Function File]

Return the current monthly calendar in a 6x7 matrix.

If *d* is specified, return the calendar for the month containing the date *d*, which must be a serial date number or a date string.

If *y* and *m* are specified, return the calendar for year *y* and month *m*.

If no output arguments are specified, print the calendar on the screen instead of returning a matrix.

See also: [datenum], page 703, [datestr], page 704.

`[n, s] = weekday (d)` [Function File]
`[n, s] = weekday (d, format)` [Function File]

Return the day of the week as a number in *n* and as a string in *s*. The days of the week are numbered 1–7 with the first day being Sunday.

d is a serial date number or a date string.

If the string *format* is not present or is equal to `"short"` then *s* will contain the abbreviated name of the weekday. If *format* is `"long"` then *s* will contain the full name.

Table of return values based on *format*:

n `"short"` `"long"`

1	Sun	Sunday
2	Mon	Monday
3	Tue	Tuesday
4	Wed	Wednesday
5	Thu	Thursday
6	Fri	Friday
7	Sat	Saturday

See also: [eomday], page 707, [is_leap_year], page 702, [calendar], page 706, [datenum], page 703, [datevec], page 705.

`e = eomday (y, m)` [Function File]

Return the last day of the month *m* for the year *y*.

See also: [weekday], page 706, [datenum], page 703, [datevec], page 705, [is_leap_year], page 702, [calendar], page 706.

`datetick ()` [Function File]
`datetick (form)` [Function File]
`datetick (axis, form)` [Function File]
`datetick (..., "keeplimits")` [Function File]
`datetick (..., "keepticks")` [Function File]
`datetick (hax, ...)` [Function File]

Add date formatted tick labels to an axis. The axis to apply the ticks to is determined by *axis* which can take the values `"x"`, `"y"`, or `"z"`. The default value is `"x"`. The formatting of the labels is determined by the variable *form*, which can either be a string or positive integer that `datestr` accepts.

See also: [datenum], page 703, [datestr], page 704.

36.2 Filesystem Utilities

Octave includes many utility functions for copying, moving, renaming, and deleting files; for creating, reading, and deleting directories; for retrieving status information on files; and for manipulating file and path names.

`movefile (f1)` [Function File]
`movefile (f1, f2)` [Function File]
`movefile (f1, f2, 'f')` [Function File]
`[status, msg, msgid] = movefile (...)` [Function File]

Move the file *f1* to the destination *f2*.

The name *f1* may contain globbing patterns. If *f1* expands to multiple file names, *f2* must be a directory. If no destination *f2* is specified then the destination is the present working directory. If *f2* is a file name then *f1* is renamed to *f2*. When the force flag `'f'` is given any existing files will be overwritten without prompting.

If successful, *status* is 1, and *msg*, *msgid* are empty character strings (`""`). Otherwise, *status* is 0, *msg* contains a system-dependent error message, and *msgid* contains a unique message identifier. Note that the status code is exactly opposite that of the `system` command.

See also: [rename], page 708, [copyfile], page 708, [unlink], page 708, [delete], page 323, [glob], page 713.

`rename old new` [Built-in Function]
`[err, msg] = rename (old, new)` [Built-in Function]

Change the name of file *old* to *new*.

If successful, *err* is 0 and *msg* is an empty string. Otherwise, *err* is nonzero and *msg* contains a system-dependent error message.

See also: [movefile], page 707, [copyfile], page 708, [ls], page 729, [dir], page 729.

`[status, msg, msgid] = copyfile (f1, f2)` [Function File]
`[status, msg, msgid] = copyfile (f1, f2, 'f')` [Function File]

Copy the file *f1* to the destination *f2*.

The name *f1* may contain globbing patterns. If *f1* expands to multiple file names, *f2* must be a directory. when the force flag `'f'` is given any existing files will be overwritten without prompting.

If successful, *status* is 1, and *msg*, *msgid* are empty character strings (`""`). Otherwise, *status* is 0, *msg* contains a system-dependent error message, and *msgid* contains a unique message identifier. Note that the status code is exactly opposite that of the `system` command.

See also: [movefile], page 707, [rename], page 708, [unlink], page 708, [delete], page 323, [glob], page 713.

`[err, msg] = unlink (file)` [Built-in Function]

Delete the file named *file*.

If successful, *err* is 0 and *msg* is an empty string. Otherwise, *err* is nonzero and *msg* contains a system-dependent error message.

`link old new` [Built-in Function]
`[err, msg] = link (old, new)` [Built-in Function]

Create a new link (also known as a hard link) to an existing file.

If successful, *err* is 0 and *msg* is an empty string. Otherwise, *err* is nonzero and *msg* contains a system-dependent error message.

See also: [symlink], page 708, [unlink], page 708, [readlink], page 708, [lstat], page 710.

`symlink old new` [Built-in Function]
`[err, msg] = symlink (old, new)` [Built-in Function]

Create a symbolic link *new* which contains the string *old*.

If successful, *err* is 0 and *msg* is an empty string. Otherwise, *err* is nonzero and *msg* contains a system-dependent error message.

See also: [link], page 708, [unlink], page 708, [readlink], page 708, [lstat], page 710.

`readlink symlink` [Built-in Function]
`[result, err, msg] = readlink (symlink)` [Built-in Function]

Read the value of the symbolic link *symlink*.

If successful, *result* contains the contents of the symbolic link *symlink*, *err* is 0, and *msg* is an empty string. Otherwise, *err* is nonzero and *msg* contains a system-dependent error message.

See also: [lstat], page 710, [symlink], page 708, [link], page 708, [unlink], page 708, [delete], page 323.

`mkdir` *dir* [Built-in Function]
`mkdir (`*parent*, *dir*`)` [Built-in Function]
`[`*status*, *msg*, *msgid*`] = mkdir (...)` [Built-in Function]

Create a directory named *dir* in the directory *parent*.

If no *parent* directory is specified the present working directory is used.

If successful, *status* is 1, and *msg*, *msgid* are empty character strings (""). Otherwise, *status* is 0, *msg* contains a system-dependent error message, and *msgid* contains a unique message identifier.

See also: [rmdir], page 709, [pwd], page 730, [cd], page 729.

`rmdir` *dir* [Built-in Function]
`rmdir (`*dir*, "s"`)` [Built-in Function]
`[`*status*, *msg*, *msgid*`] = rmdir (...)` [Built-in Function]

Remove the directory named *dir*.

If successful, *status* is 1, and *msg*, *msgid* are empty character strings (""). Otherwise, *status* is 0, *msg* contains a system-dependent error message, and *msgid* contains a unique message identifier.

If the optional second parameter is supplied with value `"s"`, recursively remove all subdirectories as well.

See also: [mkdir], page 709, [confirm_recursive_rmdir], page 709, [pwd], page 730.

val `= confirm_recursive_rmdir ()` [Built-in Function]
old_val `= confirm_recursive_rmdir (`*new_val*`)` [Built-in Function]
`confirm_recursive_rmdir (`*new_val*, "*local*"`)` [Built-in Function]

Query or set the internal variable that controls whether Octave will ask for confirmation before recursively removing a directory tree.

When called from inside a function with the `"local"` option, the variable is changed locally for the function and any subroutines it calls. The original variable value is restored when exiting the function.

See also: [rmdir], page 709.

`mkfifo (`*name*, *mode*`)` [Built-in Function]
`[`*err*, *msg*`] = mkfifo (`*name*, *mode*`)` [Built-in Function]

Create a FIFO special file named *name* with file mode *mode*

If successful, *err* is 0 and *msg* is an empty string. Otherwise, *err* is nonzero and *msg* contains a system-dependent error message.

See also: [pipe], page 724.

`umask (mask)` [Built-in Function]

Set the permission mask for file creation. The parameter *mask* is an integer, interpreted as an octal number. If successful, returns the previous value of the mask (as an integer to be interpreted as an octal number); otherwise an error message is printed.

`[info, err, msg] = stat (file)` [Built-in Function]
`[info, err, msg] = stat (fid)` [Built-in Function]
`[info, err, msg] = lstat (file)` [Built-in Function]
`[info, err, msg] = lstat (fid)` [Built-in Function]

Return a structure *info* containing the following information about *file* or file identifier *fid*.

`dev` ID of device containing a directory entry for this file.

`ino` File number of the file.

`mode` File mode, as an integer. Use the functions `S_ISREG`, `S_ISDIR`, `S_ISCHR`, `S_ISBLK`, `S_ISFIFO`, `S_ISLNK`, or `S_ISSOCK` to extract information from this value.

`modestr` File mode, as a string of ten letters or dashes as would be returned by `ls -l`.

`nlink` Number of links.

`uid` User ID of file's owner.

`gid` Group ID of file's group.

`rdev` ID of device for block or character special files.

`size` Size in bytes.

`atime` Time of last access in the same form as time values returned from `time`. See Section 36.1 [Timing Utilities], page 697.

`mtime` Time of last modification in the same form as time values returned from `time`. See Section 36.1 [Timing Utilities], page 697.

`ctime` Time of last file status change in the same form as time values returned from `time`. See Section 36.1 [Timing Utilities], page 697.

`blksize` Size of blocks in the file.

`blocks` Number of blocks allocated for file.

If the call is successful *err* is 0 and *msg* is an empty string. If the file does not exist, or some other error occurs, *info* is an empty matrix, *err* is −1, and *msg* contains the corresponding system error message.

If *file* is a symbolic link, `stat` will return information about the actual file that is referenced by the link. Use `lstat` if you want information about the symbolic link itself.

For example:

```
[info, err, msg] = stat ("/vmlinuz")
  ⇒ info =
     {
       atime = 855399756
       rdev = 0
       ctime = 847219094
       uid = 0
       size = 389218
       blksize = 4096
       mtime = 847219094
       gid = 6
       nlink = 1
       blocks = 768
       mode = -rw-r--r--
       modestr = -rw-r--r--
       ino = 9316
       dev = 2049
     }
  ⇒ err = 0
  ⇒ msg =
```

See also: [lstat], page 710, [ls], page 729, [dir], page 729.

S_ISBLK (*mode*) [Built-in Function]

Return true if *mode* corresponds to a block device.

The value of *mode* is assumed to be returned from a call to `stat`.

See also: [stat], page 710, [lstat], page 710.

S_ISCHR (*mode*) [Built-in Function]

Return true if *mode* corresponds to a character device.

The value of *mode* is assumed to be returned from a call to `stat`.

See also: [stat], page 710, [lstat], page 710.

S_ISDIR (*mode*) [Built-in Function]

Return true if *mode* corresponds to a directory.

The value of *mode* is assumed to be returned from a call to `stat`.

See also: [stat], page 710, [lstat], page 710.

S_ISFIFO (*mode*) [Built-in Function]

Return true if *mode* corresponds to a fifo.

The value of *mode* is assumed to be returned from a call to `stat`.

See also: [stat], page 710, [lstat], page 710.

S_ISLNK (*mode*) [Built-in Function]

Return true if *mode* corresponds to a symbolic link.

The value of *mode* is assumed to be returned from a call to `stat`.

See also: [stat], page 710, [lstat], page 710.

`S_ISREG (mode)` [Built-in Function]

Return true if *mode* corresponds to a regular file.

The value of *mode* is assumed to be returned from a call to `stat`.

See also: [stat], page 710, [lstat], page 710.

`S_ISSOCK (mode)` [Built-in Function]

Return true if *mode* corresponds to a socket.

The value of *mode* is assumed to be returned from a call to `stat`.

See also: [stat], page 710, [lstat], page 710.

`[status, result, msgid] = fileattrib (file)` [Function File]

Return information about *file*.

If successful, *status* is 1, with *result* containing a structure with the following fields:

`Name` Full name of *file*.

`archive` True if *file* is an archive (Windows).

`system` True if *file* is a system file (Windows).

`hidden` True if *file* is a hidden file (Windows).

`directory`
True if *file* is a directory.

`UserRead`
`GroupRead`
`OtherRead`
True if the user (group; other users) has read permission for *file*.

`UserWrite`
`GroupWrite`
`OtherWrite`
True if the user (group; other users) has write permission for *file*.

`UserExecute`
`GroupExecute`
`OtherExecute`
True if the user (group; other users) has execute permission for *file*.

If an attribute does not apply (i.e., archive on a Unix system) then the field is set to NaN.

With no input arguments, return information about the current directory.

If *file* contains globbing characters, return information about all the matching files.

See also: [glob], page 713.

`isdir (f)` [Function File]

Return true if *f* is a directory.

See also: [exist], page 128, [stat], page 710, [is_absolute_filename], page 715, [is_rooted_relative_filename], page 715.

files = readdir (*dir*) [Built-in Function]
[*files*, *err*, *msg*] = readdir (*dir*) [Built-in Function]

Return the names of files in the directory *dir* as a cell array of strings.

If an error occurs, return an empty cell array in *files*. If successful, *err* is 0 and *msg* is an empty string. Otherwise, *err* is nonzero and *msg* contains a system-dependent error message.

See also: [ls], page 729, [dir], page 729, [glob], page 713, [what], page 130.

glob (*pattern*) [Built-in Function]

Given an array of pattern strings (as a char array or a cell array) in *pattern*, return a cell array of file names that match any of them, or an empty cell array if no patterns match. The pattern strings are interpreted as filename globbing patterns (as they are used by Unix shells). Within a pattern

`*` matches any string, including the null string,

`?` matches any single character, and

`[...]` matches any of the enclosed characters.

Tilde expansion is performed on each of the patterns before looking for matching file names. For example:

```
ls
   ⇒
      file1  file2  file3  myfile1 myfile1b
glob ("*file1")
   ⇒
      {
        [1,1] = file1
        [2,1] = myfile1
      }
glob ("myfile?")
   ⇒
      {
        [1,1] = myfile1
      }
glob ("file[12]")
   ⇒
      {
        [1,1] = file1
        [2,1] = file2
      }
```

See also: [ls], page 729, [dir], page 729, [readdir], page 713, [what], page 130, [fnmatch], page 713.

fnmatch (*pattern*, *string*) [Built-in Function]

Return true or false for each element of *string* that matches any of the elements of the string array *pattern*, using the rules of filename pattern matching. For example:

```
fnmatch ("a*b", {"ab"; "axyzb"; "xyzab"})
     ⇒ [ 1; 1; 0 ]
```

See also: [glob], page 713, [regexp], page 84.

`file_in_path (`*path*, *file*`)` [Built-in Function]
`file_in_path (`*path*, *file*, "*all*"`)` [Built-in Function]

Return the absolute name of *file* if it can be found in *path*. The value of *path* should be a colon-separated list of directories in the format described for `path`. If no file is found, return an empty character string. For example:

```
file_in_path (EXEC_PATH, "sh")
     ⇒ "/bin/sh"
```

If the second argument is a cell array of strings, search each directory of the path for element of the cell array and return the first that matches.

If the third optional argument `"all"` is supplied, return a cell array containing the list of all files that have the same name in the path. If no files are found, return an empty cell array.

See also: [file_in_loadpath], page 182, [find_dir_in_path], page 182, [path], page 181.

`filesep ()` [Built-in Function]
`filesep ("`*all*`")` [Built-in Function]

Return the system-dependent character used to separate directory names.

If `"all"` is given, the function returns all valid file separators in the form of a string. The list of file separators is system-dependent. It is '/' (forward slash) under UNIX or Mac OS X, '/' and '\' (forward and backward slashes) under Windows.

See also: [pathsep], page 182.

val `= filemarker ()` [Built-in Function]
`filemarker (`*new_val*`)` [Built-in Function]
`filemarker (`*new_val*, "*local*"`)` [Built-in Function]

Query or set the character used to separate filename from the the subfunction names contained within the file. This can be used in a generic manner to interact with subfunctions. For example,

```
help (["myfunc", filemarker, "mysubfunc"])
```

returns the help string associated with the subfunction `mysubfunc` of the function `myfunc`. Another use of `filemarker` is when debugging it allows easier placement of breakpoints within subfunctions. For example,

```
dbstop (["myfunc", filemarker, "mysubfunc"])
```

will set a breakpoint at the first line of the subfunction `mysubfunc`.

When called from inside a function with the `"local"` option, the variable is changed locally for the function and any subroutines it calls. The original variable value is restored when exiting the function.

`[`*dir*, *name*, *ext*, *ver*`] = fileparts (`*filename*`)` [Function File]

Return the directory, name, extension, and version components of *filename*.

See also: [fullfile], page 715.

filename = fullfile (*dir1*, *dir2*, ..., *file*) [Function File]

Return a complete filename constructed from the given components.

See also: [fileparts], page 714.

tilde_expand (*string*) [Built-in Function]

Perform tilde expansion on *string*. If *string* begins with a tilde character, ('~'), all of the characters preceding the first slash (or all characters, if there is no slash) are treated as a possible user name, and the tilde and the following characters up to the slash are replaced by the home directory of the named user. If the tilde is followed immediately by a slash, the tilde is replaced by the home directory of the user running Octave. For example:

```
tilde_expand ("~joeuser/bin")
     ⇒ "/home/joeuser/bin"
tilde_expand ("~/bin")
     ⇒ "/home/jwe/bin"
```

[*cname*, *status*, *msg*] = canonicalize_file_name (*fname*) [Built-in Function]

Return the canonical name of file *fname*. If the file does not exist the empty string ("") is returned.

See also: [make_absolute_filename], page 715, [is_absolute_filename], page 715, [is_rooted_relative_filename], page 715.

make_absolute_filename (*file*) [Built-in Function]

Return the full name of *file* beginning from the root of the file system. No check is done for the existence of *file*.

See also: [canonicalize_file_name], page 715, [is_absolute_filename], page 715, [is_rooted_relative_filename], page 715, [isdir], page 712.

is_absolute_filename (*file*) [Built-in Function]

Return true if *file* is an absolute filename.

See also: [is_rooted_relative_filename], page 715, [make_absolute_filename], page 715, [isdir], page 712.

is_rooted_relative_filename (*file*) [Built-in Function]

Return true if *file* is a rooted-relative filename.

See also: [is_absolute_filename], page 715, [make_absolute_filename], page 715, [isdir], page 712.

P_tmpdir () [Built-in Function]

Return the default name of the directory for temporary files on this system. The name of this directory is system dependent.

dir = tempdir () [Function File]

Return the name of the system's directory for temporary files.

`tempname ()` [Function File]
`tempname (dir)` [Function File]
`tempname (dir, prefix)` [Function File]

This function is an alias for `tmpnam`.

See also: [tmpnam], page 255.

`current_state = recycle ()` [Function File]
`old_state = recycle (new_state)` [Function File]

Query or set the preference for recycling deleted files.

Recycling files, instead of permanently deleting them, is not currently implemented in Octave. To help avoid accidental data loss an error will be raised if an attempt is made to enable file recycling.

See also: [delete], page 323.

36.3 File Archiving Utilities

`bunzip2 (bzfile)` [Function File]
`bunzip2 (bzfile, dir)` [Function File]

Unpack the bzip2 archive *bzfile* to the directory *dir*. If *dir* is not specified, it defaults to the current directory.

See also: [bzip2], page 717, [unpack], page 717, [gunzip], page 716, [unzip], page 717, [untar], page 717.

`entries = gzip (files)` [Function File]
`entries = gzip (files, outdir)` [Function File]

Compress the list of files and/or directories specified in *files*. Each file is compressed separately and a new file with a '".gz"' extension is created. The original files are not modified. Existing compressed files are silently overwritten. If *outdir* is defined the compressed files are placed in this directory.

See also: [gunzip], page 716, [bzip2], page 717, [zip], page 717, [tar], page 716.

`gunzip (gzfile, dir)` [Function File]

Unpack the gzip archive *gzfile* to the directory *dir*. If *dir* is not specified, it defaults to the current directory. If *gzfile* is a directory, all gzfiles in the directory will be recursively gunzipped.

See also: [gzip], page 716, [unpack], page 717, [bunzip2], page 716, [unzip], page 717, [untar], page 717.

`entries = tar (tarfile, files)` [Function File]
`entries = tar (tarfile, files, root)` [Function File]

Pack *files files* into the TAR archive *tarfile*. The list of files must be a string or a cell array of strings.

The optional argument *root* changes the relative path of *files* from the current directory.

If an output argument is requested the entries in the archive are returned in a cell array.

See also: [untar], page 717, [bzip2], page 717, [gzip], page 716, [zip], page 717.

`untar (tarfile)` [Function File]
`untar (tarfile, dir)` [Function File]

Unpack the TAR archive *tarfile* to the directory *dir*. If *dir* is not specified, it defaults to the current directory.

See also: [tar], page 716, [unpack], page 717, [bunzip2], page 716, [gunzip], page 716, [unzip], page 717.

`entries = zip (zipfile, files)` [Function File]
`entries = zip (zipfile, files, rootdir)` [Function File]

Compress the list of files and/or directories specified in *files* into the archive *zipfile* in the same directory. If *rootdir* is defined the *files* are located relative to *rootdir* rather than the current directory.

See also: [unzip], page 717, [bzip2], page 717, [gzip], page 716, [tar], page 716.

`unzip (zipfile)` [Function File]
`unzip (zipfile, dir)` [Function File]

Unpack the ZIP archive *zipfile* to the directory *dir*. If *dir* is not specified, it defaults to the current directory.

See also: [zip], page 717, [unpack], page 717, [bunzip2], page 716, [gunzip], page 716, [untar], page 717.

`files = unpack (file)` [Function File]
`files = unpack (file, dir)` [Function File]
`files = unpack (file, dir, filetype)` [Function File]

Unpack the archive *file* based on its extension to the directory *dir*. If *file* is a list of strings, then each file is unpacked individually. If *dir* is not specified, it defaults to the current directory. If a directory is in the file list, then the *filetype* must also be specified.

The optional return value is a list of *files* unpacked.

See also: [bzip2], page 717, [gzip], page 716, [zip], page 717, [tar], page 716.

`entries = bzip2 (files)` [Function File]
`entries = bzip2 (files, outdir)` [Function File]

Compress the list of files specified in *files*. Each file is compressed separately and a new file with a '`".bz2"`' extension is created. The original files are not modified. Existing compressed files are silently overwritten. If *outdir* is defined the compressed files are placed in this directory.

See also: [bunzip2], page 716, [gzip], page 716, [zip], page 717, [tar], page 716.

36.4 Networking Utilities

`gethostname ()` [Built-in Function]

Return the hostname of the system where Octave is running.

36.4.1 FTP Objects

Octave supports the FTP protocol through an object-oriented interface. Use the function `ftp` to create an FTP object which represents the connection. All FTP functions take an FTP object as the first argument.

f = ftp (*host*) [Function File]
f = ftp (*host*, *username*, *password*) [Function File]

Connect to the FTP server *host* with *username* and *password*. If *username* and *password* are not specified, user `"anonymous"` with no password is used. The returned FTP object *f* represents the established FTP connection.

The list of actions for an FTP object are shown below. All functions require an FTP object as the first argument.

Method	**Description**
ascii	Set transfer type to ascii
binary	Set transfer type to binary
cd	Change remote working directory
close	Close FTP connection
delete	Delete remote file
dir	List remote directory contents
mget	Download remote files
mkdir	Create remote directory
mput	Upload local files
rename	Rename remote file or directory
rmdir	Remove remote directory

close (*f*) [Function File]

Close the FTP connection represented by the FTP object *f*.

f is an FTP object returned by the `ftp` function.

mget (*f*, *file*) [Function File]
mget (*f*, *dir*) [Function File]
mget (*f*, *remote_name*, *target*) [Function File]

Download a remote file *file* or directory *dir* to the local directory on the FTP connection *f*. *f* is an FTP object returned by the `ftp` function.

The arguments *file* and *dir* can include wildcards and any files or directories on the remote server that match will be downloaded.

If a third argument *target* is given, then a single file or directory will be downloaded to the local directory and the local name will be changed to *target*.

mput (*f*, *file*) [Function File]

Upload the local file *file* into the current remote directory on the FTP connection *f*. *f* is an FTP object returned by the ftp function.

The argument *file* is passed through the `glob` function and any files that match the wildcards in *file* will be uploaded.

`cd (f)` [Function File]
`cd (f, path)` [Function File]

Get or set the remote directory on the FTP connection *f*.

f is an FTP object returned by the `ftp` function.

If *path* is not specified, return the remote current working directory. Otherwise, set the remote directory to *path* and return the new remote working directory.

If the directory does not exist, an error message is printed and the working directory is not changed.

`lst = dir (f)` [Function File]

List the current directory in verbose form for the FTP connection *f*.

f is an FTP object returned by the `ftp` function.

`ascii (f)` [Function File]

Set the FTP connection *f* to use ASCII mode for transfers. ASCII mode is only appropriate for text files as it will convert the remote host's newline representation to the local host's newline representation.

f is an FTP object returned by the `ftp` function.

`binary (f)` [Function File]

Set the FTP connection *f* to use binary mode for transfers. In binary mode there is no conversion of newlines from the remote representation to the local representation.

f is an FTP object returned by the `ftp` function.

`delete (f, file)` [Function File]

Delete the remote file *file* over the FTP connection *f*.

f is an FTP object returned by the `ftp` function.

`rename (f, oldname, newname)` [Function File]

Rename or move the remote file or directory *oldname* to *newname*, over the FTP connection *f*.

f is an FTP object returned by the ftp function.

`mkdir (f, path)` [Function File]

Create the remote directory *path*, over the FTP connection *f*.

f is an FTP object returned by the `ftp` function.

`rmdir (f, path)` [Function File]

Remove the remote directory *path*, over the FTP connection *f*.

f is an FTP object returned by the `ftp` function.

36.4.2 URL Manipulation

`s = urlread (url)` [Loadable Function]
`[s, success] = urlread (url)` [Loadable Function]
`[s, success, message] = urlread (url)` [Loadable Function]
`[...] = urlread (url, method, param)` [Loadable Function]

Download a remote file specified by its *url* and return its content in string *s*. For example:

```
s = urlread ("ftp://ftp.octave.org/pub/octave/README");
```

The variable *success* is 1 if the download was successful, otherwise it is 0 in which case *message* contains an error message. If no output argument is specified and an error occurs, then the error is signaled through Octave's error handling mechanism.

This function uses libcurl. Curl supports, among others, the HTTP, FTP and FILE protocols. Username and password may be specified in the URL. For example:

```
s = urlread ("http://user:password@example.com/file.txt");
```

GET and POST requests can be specified by *method* and *param*. The parameter *method* is either 'get' or 'post' and *param* is a cell array of parameter and value pairs. For example:

```
s = urlread ("http://www.google.com/search", "get",
            {"query", "octave"});
```

See also: [urlwrite], page 720.

urlwrite (*url*, *localfile*) [Loadable Function]
f = urlwrite (*url*, *localfile*) [Loadable Function]
[*f*, *success*] = urlwrite (*url*, *localfile*) [Loadable Function]
[*f*, *success*, *message*] = urlwrite (*url*, *localfile*) [Loadable Function]

Download a remote file specified by its *url* and save it as *localfile*. For example:

```
urlwrite ("ftp://ftp.octave.org/pub/octave/README",
          "README.txt");
```

The full path of the downloaded file is returned in *f*. The variable *success* is 1 if the download was successful, otherwise it is 0 in which case *message* contains an error message. If no output argument is specified and an error occurs, then the error is signaled through Octave's error handling mechanism.

This function uses libcurl. Curl supports, among others, the HTTP, FTP and FILE protocols. Username and password may be specified in the URL, for example:

```
urlwrite ("http://username:password@example.com/file.txt",
          "file.txt");
```

GET and POST requests can be specified by *method* and *param*. The parameter *method* is either 'get' or 'post' and *param* is a cell array of parameter and value pairs. For example:

```
urlwrite ("http://www.google.com/search", "search.html",
          "get", {"query", "octave"});
```

See also: [urlread], page 719.

36.4.3 Base64 and Binary Data Transmission

Some transmission channels can not accept binary data. It is customary to encode binary data in Base64 for transmission and to decode the data upon reception.

s = base64_encode (*x*) [Built-in Function]

Encode a double matrix or array *x* into the base64 format string *s*.

See also: [base64_decode], page 721.

x = base64_decode (*s*) [Built-in Function]
x = base64_decode (*s*, *dims*) [Built-in Function]

Decode the double matrix or array *x* from the base64 encoded string *s*. The optional input parameter *dims* should be a vector containing the dimensions of the decoded array.

See also: [base64_encode], page 720.

36.5 Controlling Subprocesses

Octave includes some high-level commands like `system` and `popen` for starting subprocesses. If you want to run another program to perform some task and then look at its output, you will probably want to use these functions.

Octave also provides several very low-level Unix-like functions which can also be used for starting subprocesses, but you should probably only use them if you can't find any way to do what you need with the higher-level functions.

system ("*string*") [Built-in Function]
system ("*string*", *return_output*) [Built-in Function]
system ("*string*", *return_output*, *type*) [Built-in Function]
[*status*, *output*] = system (...) [Built-in Function]

Execute a shell command specified by *string*. If the optional argument *type* is `"async"`, the process is started in the background and the process ID of the child process is returned immediately. Otherwise, the child process is started and Octave waits until it exits. If the *type* argument is omitted, it defaults to the value `"sync"`.

If *system* is called with one or more output arguments, or if the optional argument *return_output* is true and the subprocess is started synchronously, then the output from the command is returned as a variable. Otherwise, if the subprocess is executed synchronously, its output is sent to the standard output. To send the output of a command executed with `system` through the pager, use a command like

```
[output, text] = system ("cmd");
disp (text);
```

or

```
printf ("%s\n", nthargout (2, "system", "cmd"));
```

The `system` function can return two values. The first is the exit status of the command and the second is any output from the command that was written to the standard output stream. For example,

```
[status, output] = system ("echo foo; exit 2");
```

will set the variable `output` to the string '`foo`', and the variable `status` to the integer '`2`'.

For commands run asynchronously, *status* is the process id of the command shell that is started to run the command.

See also: [unix], page 721, [dos], page 722.

unix ("*command*") [Function File]
status = unix ("*command*") [Function File]

`[`*status*`,` *text*`] = unix ("`*command*`")` [Function File]
`[...] = unix ("`*command*`", "`*-echo*`")` [Function File]

Execute a system command if running under a Unix-like operating system, otherwise do nothing. Return the exit status of the program in *status* and any output from the command in *text*. When called with no output argument, or the `"-echo"` argument is given, then *text* is also sent to standard output.

See also: [dos], page 722, [system], page 721, [isunix], page 733, [ispc], page 732.

`dos ("`*command*`")` [Function File]
status `= dos ("`*command*`")` [Function File]
`[`*status*`,` *text*`] = dos ("`*command*`")` [Function File]
`[...] = dos ("`*command*`", "`*-echo*`")` [Function File]

Execute a system command if running under a Windows-like operating system, otherwise do nothing. Return the exit status of the program in *status* and any output from the command in *text*. When called with no output argument, or the `"-echo"` argument is given, then *text* is also sent to standard output.

See also: [unix], page 721, [system], page 721, [isunix], page 733, [ispc], page 732.

output `= perl (`*scriptfile*`)` [Function File]
output `= perl (`*scriptfile*, *argument1*, *argument2*, ...`)` [Function File]
`[`*output*`,` *status*`] = perl (...)` [Function File]

Invoke Perl script *scriptfile*, possibly with a list of command line arguments. Return output in *output* and optional status in *status*. If *scriptfile* is not an absolute file name it is is searched for in the current directory and then in the Octave loadpath.

See also: [system], page 721, [python], page 722.

output `= python (`*scriptfile*`)` [Function File]
output `= python (`*scriptfile*, *argument1*, *argument2*, ...`)` [Function File]
`[`*output*`,` *status*`] = python (...)` [Function File]

Invoke Python script *scriptfile*, possibly with a list of command line arguments. Return output in *output* and optional status in *status*. If *scriptfile* is not an absolute file name it is is searched for in the current directory and then in the Octave loadpath.

See also: [system], page 721, [perl], page 722.

fid `= popen (`*command*, *mode*`)` [Built-in Function]

Start a process and create a pipe. The name of the command to run is given by *command*. The file identifier corresponding to the input or output stream of the process is returned in *fid*. The argument *mode* may be

`"r"` The pipe will be connected to the standard output of the process, and open for reading.

`"w"` The pipe will be connected to the standard input of the process, and open for writing.

For example:

```
fid = popen ("ls -ltr / | tail -3", "r");
while (ischar (s = fgets (fid)))
  fputs (stdout, s);
endwhile

   ⊣ drwxr-xr-x  33 root  root  3072 Feb 15 13:28 etc
   ⊣ drwxr-xr-x   3 root  root  1024 Feb 15 13:28 lib
   ⊣ drwxrwxrwt  15 root  root  2048 Feb 17 14:53 tmp
```

pclose (*fid*) [Built-in Function]

Close a file identifier that was opened by popen. You may also use fclose for the same purpose.

[*in*, *out*, *pid*] = popen2 (*command*, *args*) [Built-in Function]

Start a subprocess with two-way communication. The name of the process is given by *command*, and *args* is an array of strings containing options for the command. The file identifiers for the input and output streams of the subprocess are returned in *in* and *out*. If execution of the command is successful, *pid* contains the process ID of the subprocess. Otherwise, *pid* is -1.

For example:

```
[in, out, pid] = popen2 ("sort", "-r");
fputs (in, "these\nare\nsome\nstrings\n");
fclose (in);
EAGAIN = errno ("EAGAIN");
done = false;
do
  s = fgets (out);
  if (ischar (s))
    fputs (stdout, s);
  elseif (errno () == EAGAIN)
    sleep (0.1);
    fclear (out);
  else
    done = true;
  endif
until (done)
fclose (out);
waitpid (pid);

   ⊣ these
   ⊣ strings
   ⊣ some
   ⊣ are
```

Note that popen2, unlike popen, will not "reap" the child process. If you don't use waitpid to check the child's exit status, it will linger until Octave exits.

See also: [popen], page 722, [waitpid], page 725.

val = EXEC_PATH () [Built-in Function]
old_val = EXEC_PATH (*new_val*) [Built-in Function]
EXEC_PATH (*new_val*, "*local*") [Built-in Function]

Query or set the internal variable that specifies a colon separated list of directories to append to the shell PATH when executing external programs. The initial value of is taken from the environment variable `OCTAVE_EXEC_PATH`, but that value can be overridden by the command line argument '`--exec-path PATH`'.

When called from inside a function with the `"local"` option, the variable is changed locally for the function and any subroutines it calls. The original variable value is restored when exiting the function.

See also: [IMAGE_PATH], page 653, [OCTAVE_HOME], page 733.

In most cases, the following functions simply decode their arguments and make the corresponding Unix system calls. For a complete example of how they can be used, look at the definition of the function `popen2`.

[*pid*, *msg*] = fork () [Built-in Function]

Create a copy of the current process.

Fork can return one of the following values:

> 0 — You are in the parent process. The value returned from `fork` is the process id of the child process. You should probably arrange to wait for any child processes to exit.

0 — You are in the child process. You can call `exec` to start another process. If that fails, you should probably call `exit`.

< 0 — The call to `fork` failed for some reason. You must take evasive action. A system dependent error message will be waiting in *msg*.

[*err*, *msg*] = exec (*file*, *args*) [Built-in Function]

Replace current process with a new process. Calling `exec` without first calling `fork` will terminate your current Octave process and replace it with the program named by *file*. For example,

```
exec ("ls" "-l")
```

will run `ls` and return you to your shell prompt.

If successful, `exec` does not return. If `exec` does return, *err* will be nonzero, and *msg* will contain a system-dependent error message.

[*read_fd*, *write_fd*, *err*, *msg*] = pipe () [Built-in Function]

Create a pipe and return the reading and writing ends of the pipe into *read_fd* and *write_fd* respectively.

If successful, *err* is 0 and *msg* is an empty string. Otherwise, *err* is nonzero and *msg* contains a system-dependent error message.

See also: [mkfifo], page 709.

[*fid*, *msg*] = dup2 (*old*, *new*) [Built-in Function]

Duplicate a file descriptor.

If successful, *fid* is greater than zero and contains the new file ID. Otherwise, *fid* is negative and *msg* contains a system-dependent error message.

`[pid, status, msg] = waitpid (pid, options)` [Built-in Function]

Wait for process *pid* to terminate. The *pid* argument can be:

−1 Wait for any child process.

0 Wait for any child process whose process group ID is equal to that of the Octave interpreter process.

> 0 Wait for termination of the child process with ID *pid*.

The *options* argument can be a bitwise OR of zero or more of the following constants:

`0` Wait until signal is received or a child process exits (this is the default if the *options* argument is missing).

`WNOHANG` Do not hang if status is not immediately available.

`WUNTRACED`
Report the status of any child processes that are stopped, and whose status has not yet been reported since they stopped.

`WCONTINUE`
Return if a stopped child has been resumed by delivery of `SIGCONT`. This value may not be meaningful on all systems.

If the returned value of *pid* is greater than 0, it is the process ID of the child process that exited. If an error occurs, *pid* will be less than zero and *msg* will contain a system-dependent error message. The value of *status* contains additional system-dependent information about the subprocess that exited.

See also: [WCONTINUE], page 725, [WCOREDUMP], page 725, [WEXITSTATUS], page 725, [WIFCONTINUED], page 726, [WIFSIGNALED], page 726, [WIFSTOPPED], page 726, [WNOHANG], page 726, [WSTOPSIG], page 726, [WTERMSIG], page 727, [WUNTRACED], page 727.

`WCONTINUE ()` [Built-in Function]

Return the numerical value of the option argument that may be passed to `waitpid` to indicate that it should also return if a stopped child has been resumed by delivery of a `SIGCONT` signal.

See also: [waitpid], page 725, [WNOHANG], page 726, [WUNTRACED], page 727.

`WCOREDUMP (status)` [Built-in Function]

Given *status* from a call to `waitpid`, return true if the child produced a core dump. This function should only be employed if `WIFSIGNALED` returned true. The macro used to implement this function is not specified in POSIX.1-2001 and is not available on some Unix implementations (e.g., AIX, SunOS).

See also: [waitpid], page 725, [WIFEXITED], page 726, [WEXITSTATUS], page 725, [WIFSIGNALED], page 726, [WTERMSIG], page 727, [WIFSTOPPED], page 726, [WSTOPSIG], page 726, [WIFCONTINUED], page 726.

`WEXITSTATUS (status)` [Built-in Function]

Given *status* from a call to `waitpid`, return the exit status of the child. This function should only be employed if `WIFEXITED` returned true.

See also: [waitpid], page 725, [WIFEXITED], page 726, [WIFSIGNALED], page 726, [WTERMSIG], page 727, [WCOREDUMP], page 725, [WIFSTOPPED], page 726, [WSTOPSIG], page 726, [WIFCONTINUED], page 726.

`WIFCONTINUED` (*status*) [Built-in Function]

Given *status* from a call to `waitpid`, return true if the child process was resumed by delivery of `SIGCONT`.

See also: [waitpid], page 725, [WIFEXITED], page 726, [WEXITSTATUS], page 725, [WIFSIGNALED], page 726, [WTERMSIG], page 727, [WCOREDUMP], page 725, [WIFSTOPPED], page 726, [WSTOPSIG], page 726.

`WIFSIGNALED` (*status*) [Built-in Function]

Given *status* from a call to `waitpid`, return true if the child process was terminated by a signal.

See also: [waitpid], page 725, [WIFEXITED], page 726, [WEXITSTATUS], page 725, [WTERMSIG], page 727, [WCOREDUMP], page 725, [WIFSTOPPED], page 726, [WSTOPSIG], page 726, [WIFCONTINUED], page 726.

`WIFSTOPPED` (*status*) [Built-in Function]

Given *status* from a call to `waitpid`, return true if the child process was stopped by delivery of a signal; this is only possible if the call was done using `WUNTRACED` or when the child is being traced (see ptrace(2)).

See also: [waitpid], page 725, [WIFEXITED], page 726, [WEXITSTATUS], page 725, [WIFSIGNALED], page 726, [WTERMSIG], page 727, [WCOREDUMP], page 725, [WSTOPSIG], page 726, [WIFCONTINUED], page 726.

`WIFEXITED` (*status*) [Built-in Function]

Given *status* from a call to `waitpid`, return true if the child terminated normally.

See also: [waitpid], page 725, [WEXITSTATUS], page 725, [WIFSIGNALED], page 726, [WTERMSIG], page 727, [WCOREDUMP], page 725, [WIFSTOPPED], page 726, [WSTOPSIG], page 726, [WIFCONTINUED], page 726.

`WNOHANG` () [Built-in Function]

Return the numerical value of the option argument that may be passed to `waitpid` to indicate that it should return its status immediately instead of waiting for a process to exit.

See also: [waitpid], page 725, [WUNTRACED], page 727, [WCONTINUE], page 725.

`WSTOPSIG` (*status*) [Built-in Function]

Given *status* from a call to `waitpid`, return the number of the signal which caused the child to stop. This function should only be employed if `WIFSTOPPED` returned true.

See also: [waitpid], page 725, [WIFEXITED], page 726, [WEXITSTATUS], page 725, [WIFSIGNALED], page 726, [WTERMSIG], page 727, [WCOREDUMP], page 725, [WIFSTOPPED], page 726, [WIFCONTINUED], page 726.

`WTERMSIG (status)` [Built-in Function]

Given *status* from a call to `waitpid`, return the number of the signal that caused the child process to terminate. This function should only be employed if `WIFSIGNALED` returned true.

See also: [waitpid], page 725, [WIFEXITED], page 726, [WEXITSTATUS], page 725, [WIFSIGNALED], page 726, [WCOREDUMP], page 725, [WIFSTOPPED], page 726, [WSTOPSIG], page 726, [WIFCONTINUED], page 726.

`WUNTRACED ()` [Built-in Function]

Return the numerical value of the option argument that may be passed to `waitpid` to indicate that it should also return if the child process has stopped but is not traced via the `ptrace` system call

See also: [waitpid], page 725, [WNOHANG], page 726, [WCONTINUE], page 725.

`[err, msg] = fcntl (fid, request, arg)` [Built-in Function]

Change the properties of the open file *fid*. The following values may be passed as *request*:

`F_DUPFD` Return a duplicate file descriptor.

`F_GETFD` Return the file descriptor flags for *fid*.

`F_SETFD` Set the file descriptor flags for *fid*.

`F_GETFL` Return the file status flags for *fid*. The following codes may be returned (some of the flags may be undefined on some systems).

- `O_RDONLY` Open for reading only.
- `O_WRONLY` Open for writing only.
- `O_RDWR` Open for reading and writing.
- `O_APPEND` Append on each write.
- `O_CREAT` Create the file if it does not exist.
- `O_NONBLOCK` Non-blocking mode.
- `O_SYNC` Wait for writes to complete.
- `O_ASYNC` Asynchronous I/O.

`F_SETFL` Set the file status flags for *fid* to the value specified by *arg*. The only flags that can be changed are `O_APPEND` and `O_NONBLOCK`.

If successful, *err* is 0 and *msg* is an empty string. Otherwise, *err* is nonzero and *msg* contains a system-dependent error message.

`[err, msg] = kill (pid, sig)` [Built-in Function]

Send signal *sig* to process *pid*.

If *pid* is positive, then signal *sig* is sent to *pid*.

If *pid* is 0, then signal *sig* is sent to every process in the process group of the current process.

If *pid* is -1, then signal *sig* is sent to every process except process 1.

If *pid* is less than -1, then signal *sig* is sent to every process in the process group -*pid*.

If *sig* is 0, then no signal is sent, but error checking is still performed.

Return 0 if successful, otherwise return -1.

`SIG ()` [Built-in Function]

Return a structure containing Unix signal names and their defined values.

36.6 Process, Group, and User IDs

`pgid = getpgrp ()` [Built-in Function]

Return the process group id of the current process.

`pid = getpid ()` [Built-in Function]

Return the process id of the current process.

`pid = getppid ()` [Built-in Function]

Return the process id of the parent process.

`euid = geteuid ()` [Built-in Function]

Return the effective user id of the current process.

`uid = getuid ()` [Built-in Function]

Return the real user id of the current process.

`egid = getegid ()` [Built-in Function]

Return the effective group id of the current process.

`gid = getgid ()` [Built-in Function]

Return the real group id of the current process.

36.7 Environment Variables

`getenv (var)` [Built-in Function]

Return the value of the environment variable *var*. For example,

```
getenv ("PATH")
```

returns a string containing the value of your path.

`putenv (var, value)` [Built-in Function]
`setenv (var, value)` [Built-in Function]

Set the value of the environment variable *var* to *value*.

36.8 Current Working Directory

`cd` *dir* [Command]
`cd` [Command]
old_dir = `cd` *dir* [Built-in Function]
`chdir ...` [Command]

Change the current working directory to *dir*.

If *dir* is omitted, the current directory is changed to the user's home directory ("~").

For example,

```
cd ~/octave
```

changes the current working directory to '~/octave'. If the directory does not exist, an error message is printed and the working directory is not changed.

`chdir` is an alias for `cd` and can be used in all of the same calling formats.

Compatibility Note: When called with no arguments, MATLAB prints the present working directory rather than changing to the user's home directory.

See also: [pwd], page 730, [mkdir], page 709, [rmdir], page 709, [dir], page 729, [ls], page 729.

`ls` [Command]
`ls` *filenames* [Command]
`ls` *options* [Command]
`ls` *options filenames* [Command]

List directory contents. For example:

```
ls -l
     ⊣ total 12
     ⊣ -rw-r--r--   1 jwe  users  4488 Aug 19 04:02 foo.m
     ⊣ -rw-r--r--   1 jwe  users  1315 Aug 17 23:14 bar.m
```

The `dir` and `ls` commands are implemented by calling your system's directory listing command, so the available options will vary from system to system.

Filenames are subject to shell expansion if they contain any wildcard characters '*', '?', '[]'. If you want to find a literal example of a wildcard character you must escape it using the backslash operator '\'.

See also: [dir], page 729, [readdir], page 713, [glob], page 713, [what], page 130, [stat], page 710, [filesep], page 714, [ls_command], page 729.

val = `ls_command ()` [Function File]
old_val = `ls_command (`*new_val*`)` [Function File]

Query or set the shell command used by Octave's `ls` command.

See also: [ls], page 729.

`dir` [Function File]
`dir (`*directory*`)` [Function File]
`[`*list*`] = dir (`*directory*`)` [Function File]

Display file listing for directory *directory*.

If *directory* is not specified then list the present working directory.

If a return value is requested, return a structure array with the fields

name — File or directory name.

date — Timestamp of file modification (string value).

bytes — File size in bytes.

isdir — True if name is a directory.

datenum — Timestamp of file modification as serial date number (double).

statinfo — Information structure returned from `stat`.

If *directory* is a filename, rather than a directory, then return information about the named file. *directory* may also be a list rather than a single directory or file.

directory is subject to shell expansion if it contains any wildcard characters '*', '?', '[]'. If you want to find a literal example of a wildcard character you must escape it using the backslash operator '\'.

Note that for symbolic links, `dir` returns information about the file that the symbolic link points to rather than the link itself. However, if the link points to a nonexistent file, `dir` returns information about the link.

See also: [ls], page 729, [readdir], page 713, [glob], page 713, [what], page 130, [stat], page 710.

`pwd ()` [Built-in Function]
dir `= pwd ()` [Built-in Function]

Return the current working directory.

See also: [cd], page 729, [dir], page 729, [ls], page 729, [mkdir], page 709, [rmdir], page 709.

36.9 Password Database Functions

Octave's password database functions return information in a structure with the following fields.

`name` — The user name.

`passwd` — The encrypted password, if available.

`uid` — The numeric user id.

`gid` — The numeric group id.

`gecos` — The GECOS field.

`dir` — The home directory.

`shell` — The initial shell.

In the descriptions of the following functions, this data structure is referred to as a *pw_struct*.

pw_struct `= getpwent ()` [Built-in Function]

Return a structure containing an entry from the password database, opening it if necessary. Once the end of the data has been reached, `getpwent` returns 0.

pw_struct = **getpwuid** (*uid*). [Built-in Function]
Return a structure containing the first entry from the password database with the user ID *uid*. If the user ID does not exist in the database, **getpwuid** returns 0.

pw_struct = **getpwnam** (*name*) [Built-in Function]
Return a structure containing the first entry from the password database with the user name *name*. If the user name does not exist in the database, **getpwname** returns 0.

setpwent () [Built-in Function]
Return the internal pointer to the beginning of the password database.

endpwent () [Built-in Function]
Close the password database.

36.10 Group Database Functions

Octave's group database functions return information in a structure with the following fields.

`name` The user name.

`passwd` The encrypted password, if available.

`gid` The numeric group id.

`mem` The members of the group.

In the descriptions of the following functions, this data structure is referred to as a *grp_struct*.

grp_struct = **getgrent** () [Built-in Function]
Return an entry from the group database, opening it if necessary. Once the end of data has been reached, **getgrent** returns 0.

grp_struct = **getgrgid** (*gid*). [Built-in Function]
Return the first entry from the group database with the group ID *gid*. If the group ID does not exist in the database, **getgrgid** returns 0.

grp_struct = **getgrnam** (*name*) [Built-in Function]
Return the first entry from the group database with the group name *name*. If the group name does not exist in the database, **getgrnam** returns 0.

setgrent () [Built-in Function]
Return the internal pointer to the beginning of the group database.

endgrent () [Built-in Function]
Close the group database.

36.11 System Information

`[c, maxsize, endian] = computer ()` [Function File]
`arch = computer ("arch")` [Function File]

Print or return a string of the form *cpu-vendor-os* that identifies the kind of computer Octave is running on. If invoked with an output argument, the value is returned instead of printed. For example:

```
computer ()
    -| i586-pc-linux-gnu

x = computer ()
    ⇒ x = "i586-pc-linux-gnu"
```

If two output arguments are requested, also return the maximum number of elements for an array.

If three output arguments are requested, also return the byte order of the current system as a character ("B" for big-endian or "L" for little-endian).

If the argument "arch" is specified, return a string indicating the architecture of the computer on which Octave is running.

`[uts, err, msg] = uname ()` [Built-in Function]

Return system information in the structure. For example:

```
uname ()
    ⇒ {
          sysname = x86_64
          nodename = segfault
          release = 2.6.15-1-amd64-k8-smp
          version = Linux
          machine = #2 SMP Thu Feb 23 04:57:49 UTC 2006
       }
```

If successful, *err* is 0 and *msg* is an empty string. Otherwise, *err* is nonzero and *msg* contains a system-dependent error message.

`nproc ()` [Built-in Function]
`nproc (query)` [Built-in Function]

Return the current number of available processors.

If called with the optional argument *query*, modify how processors are counted as follows:

`all` total number of processors.

`current` processors available to the current process.

`overridable`
likewise, but overridable through the `OMP_NUM_THREADS` environment variable.

`ispc ()` [Function File]

Return true if Octave is running on a Windows system and false otherwise.

See also: [isunix], page 733, [ismac], page 733.

`isunix ()` [Function File]

Return true if Octave is running on a Unix-like system and false otherwise.

See also: [ismac], page 733, [ispc], page 732.

`ismac ()` [Function File]

Return true if Octave is running on a Mac OS X system and false otherwise.

See also: [isunix], page 733, [ispc], page 732.

`isieee ()` [Built-in Function]

Return true if your computer *claims* to conform to the IEEE standard for floating point calculations. No actual tests are performed.

`isdeployed ()` [Function File]

Return true if the current program has been compiled and is running separately from the Octave interpreter and false if it is running in the Octave interpreter. Currently, this function always returns false in Octave.

`OCTAVE_HOME ()` [Built-in Function]

Return the name of the top-level Octave installation directory.

See also: [EXEC_PATH], page 724, [IMAGE_PATH], page 653.

`matlabroot ()` [Function File]

Return the name of the top-level Octave installation directory.

This is an alias for the function `OCTAVE_HOME` provided for compatibility.

See also: [OCTAVE_HOME], page 733.

`OCTAVE_VERSION ()` [Built-in Function]

Return the version number of Octave, as a string.

`version ()` [Function File]

Return the version number of Octave, as a string.

This is an alias for the function `OCTAVE_VERSION` provided for compatibility.

See also: [OCTAVE_VERSION], page 733.

`ver ()` [Function File]
`v = ver ()` [Function File]
`v = ver ("`*Octave*`")` [Function File]
`v = ver (`*package*`)` [Function File]

Display a header containing the current Octave version number, license string, and operating system followed by a list of installed packages, versions, and installation directories.

`v = ver ()`

Return a vector of structures describing Octave and each installed package. The structure includes the following fields.

`Name` Package name.

`Version` Version of the package.

`Revision` Revision of the package.

`Date` Date of the version/revision.

`v = ver ("Octave")`

Return version information for Octave only.

`v = ver (`*`package`*`)`

Return version information for *package*.

See also: [version], page 733, [octave_config_info], page 735.

`compare_versions (`*`v1, v2, operator`*`)` [Function File]

Compare two version strings using the given *operator*.

This function assumes that versions *v1* and *v2* are arbitrarily long strings made of numeric and period characters possibly followed by an arbitrary string (e.g., `"1.2.3"`, `"0.3"`, `"0.1.2+"`, or `"1.2.3.4-test1"`).

The version is first split into numeric and character portions and then the parts are padded to be the same length (i.e., `"1.1"` would be padded to be `"1.1.0"` when being compared with `"1.1.1"`, and separately, the character parts of the strings are padded with nulls).

The operator can be any logical operator from the set

- `"=="` equal
- `"<"` less than
- `"<="` less than or equal to
- `">"` greater than
- `">="` greater than or equal to
- `"!="` not equal
- `"~="` not equal

Note that version `"1.1-test2"` will compare as greater than `"1.1-test10"`. Also, since the numeric part is compared first, `"a"` compares less than `"1a"` because the second string starts with a numeric part even though `double ("a")` is greater than `double ("1")`.

`license` [Command]
`license ("`*`inuse`*`")` [Function File]
`retval`` = license ("`*`inuse`*`")` [Function File]
`retval`` = license ("`*`test`*`", `*`feature`*`)` [Function File]
`license ("`*`test`*`", `*`feature, toggle`*`)` [Function File]
`retval`` = license ("`*`checkout`*`", `*`feature`*`)` [Function File]

Display the license of Octave.

`license ("inuse")`

Display a list of packages currently being used.

`retval`` = license ("inuse")`

Return a structure containing the fields `feature` and `user`.

`retval`` = license ("test", `*`feature`*`)`

Return 1 if a license exists for the product identified by the string *feature* and 0 otherwise. The argument *feature* is case insensitive and only the first 27 characters are checked.

`license ("test", feature, toggle)`

Enable or disable license testing for *feature*, depending on *toggle*, which may be one of:

`"enable"` Future tests for the specified license of *feature* are conducted as usual.

`"disable"`
Future tests for the specified license of *feature* return 0.

`retval = license ("checkout", feature)`

Check out a license for *feature*, returning 1 on success and 0 on failure.

This function is provided for compatibility with MATLAB.

See also: [ver], page 733, [version], page 733.

`octave_config_info ()` [Built-in Function]
`octave_config_info (option)` [Built-in Function]

Return a structure containing configuration and installation information for Octave.

If *option* is a string, return the configuration information for the specified option.

`getrusage ()` [Built-in Function]

Return a structure containing a number of statistics about the current Octave process. Not all fields are available on all systems. If it is not possible to get CPU time statistics, the CPU time slots are set to zero. Other missing data are replaced by NaN. The list of possible fields is:

`idrss` Unshared data size.

`inblock` Number of block input operations.

`isrss` Unshared stack size.

`ixrss` Shared memory size.

`majflt` Number of major page faults.

`maxrss` Maximum data size.

`minflt` Number of minor page faults.

`msgrcv` Number of messages received.

`msgsnd` Number of messages sent.

`nivcsw` Number of involuntary context switches.

`nsignals` Number of signals received.

`nswap` Number of swaps.

`nvcsw` Number of voluntary context switches.

`oublock` Number of block output operations.

`stime` A structure containing the system CPU time used. The structure has the elements `sec` (seconds) `usec` (microseconds).

`utime` A structure containing the user CPU time used. The structure has the elements `sec` (seconds) `usec` (microseconds).

36.12 Hashing Functions

It is often necessary to find if two strings or files are identical. This might be done by comparing them character by character and looking for differences. However, this can be slow, and so comparing a hash of the string or file can be a rapid way of finding if the files differ.

Another use of the hashing function is to check for file integrity. The user can check the hash of the file against a known value and find if the file they have is the same as the one that the original hash was produced with.

Octave supplies the `md5sum` function to perform MD5 hashes on strings and files. An example of the use of `md5sum` function might be

```
if exist (file, "file")
  hash = md5sum (file);
else
  # Treat the variable "file" as a string
  hash = md5sum (file, true);
endif
```

md5sum (*file*) [Built-in Function]
md5sum (*str*, *opt*) [Built-in Function]

Calculate the MD5 sum of the file *file*. If the second parameter *opt* exists and is true, then calculate the MD5 sum of the string *str*.

37 Java Interface

The Java Interface is designed for calling Java functions from within Octave. If you want to do the reverse, and call Octave from within Java, try a library like `javaOctave` (`http://kenai.com/projects/javaOctave`) or `joPas` (`http://jopas.sourceforge.net`).

37.1 Java Interface Functions

The following functions are the core of the Java Interface. They provide a way to create a Java object, get and set its data fields, and call Java methods which return results to Octave.

jobj = `javaObject` (*classname*) [Built-in Function]
jobj = `javaObject` (*classname*, *arg1*, ...) [Built-in Function]

Create a Java object of class *classsname*, by calling the class constructor with the arguments *arg1*, ...

The first example below creates an uninitialized object, while the second example supplies an initial argument to the constructor.

```
x = javaObject ("java.lang.StringBuffer")
x = javaObject ("java.lang.StringBuffer", "Initial string")
```

See also: [javaMethod], page 739, [javaArray], page 737.

jary = `javaArray` (*classname*, *sz*) [Function File]
jary = `javaArray` (*classname*, *m*, *n*, ...) [Function File]

Create a Java array of size *sz* with elements of class *classname*. *classname* may be a Java object representing a class or a string containing the fully qualified class name. The size of the object may also be specified with individual integer arguments *m*, *n*, etc.

The generated array is uninitialized. All elements are set to null if *classname* is a reference type, or to a default value (usually 0) if *classname* is a primitive type.

Sample code:

```
jary = javaArray ("java.lang.String", 2, 2);
jary(1,1) = "Hello";
```

See also: [javaObject], page 737.

There are many different variable types in Octave but only ones created through `javaObject` can use Java functions. Before using Java with an unknown object the type can be checked with `isjava`.

`isjava` (*x*) [Built-in Function]

Return true if *x* is a Java object.

See also: [class], page 39, [typeinfo], page 39, [isa], page 39, [javaObject], page 737.

Once an object has been created it is natural to find out what fields the object has and to read (get) and write (set) them.

In Octave the `fieldnames` function for structures has been overloaded to return the fields of a Java object. For example:

```
dobj = javaObject ("java.lang.Double", pi);
fieldnames (dobj)
⇒
{
  [1,1] = public static final double java.lang.Double.POSITIVE_INFINITY
  [1,2] = public static final double java.lang.Double.NEGATIVE_INFINITY
  [1,3] = public static final double java.lang.Double.NaN
  [1,4] = public static final double java.lang.Double.MAX_VALUE
  [1,5] = public static final double java.lang.Double.MIN_NORMAL
  [1,6] = public static final double java.lang.Double.MIN_VALUE
  [1,7] = public static final int java.lang.Double.MAX_EXPONENT
  [1,8] = public static final int java.lang.Double.MIN_EXPONENT
  [1,9] = public static final int java.lang.Double.SIZE
  [1,10] = public static final java.lang.Class java.lang.Double.TYPE
}
```

The analogy of objects with structures is carried over into reading and writing object fields. To read a field the object is indexed with the '.' operator from structures. This is the preferred method for reading fields, but Octave also provides a function interface to read fields with `java_get`. An example of both styles is shown below.

```
dobj = javaObject ("java.lang.Double", pi);
dobj.MAX_VALUE
⇒  1.7977e+308
java_get ("java.lang.Float", "MAX_VALUE")
⇒  3.4028e+38
```

`val = java_get (obj, name)` [Function File]

Get the value of the field *name* of the Java object *obj*. For static fields, *obj* can be a string representing the fully qualified name of the corresponding class.

When *obj* is a regular Java object, structure-like indexing can be used as a shortcut syntax. For instance, the two following statements are equivalent

```
java_get (x, "field1")
x.field1
```

See also: [java_set], page 738, [javaMethod], page 739, [javaObject], page 737.

`obj = java_set (obj, name, val)` [Function File]

Set the value of the field *name* of the Java object *obj* to *val*. For static fields, *obj* can be a string representing the fully qualified named of the corresponding Java class.

When *obj* is a regular Java object, structure-like indexing can be used as a shortcut syntax. For instance, the two following statements are equivalent

```
java_set (x, "field1", val)
x.field1 = val
```

See also: [java_get], page 738, [javaMethod], page 739, [javaObject], page 737.

To see what functions can be called with an object use `methods`. For example, using the previously created *dobj*:

```
methods (dobj)
⇒
Methods for class java.lang.Double:
boolean equals(java.lang.Object)
java.lang.String toString(double)
java.lang.String toString()
...
```

To call a method of an object the same structure indexing operator '.' is used. Octave also provides a functional interface to calling the methods of an object through `javaMethod`. An example showing both styles is shown below.

```
dobj = javaObject ("java.lang.Double", pi);
dobj.equals (3)
⇒ 0
javaMethod ("equals", dobj, pi)
⇒ 1
```

ret = javaMethod (*methodname*, *obj*) [Built-in Function]
ret = javaMethod (*methodname*, *obj*, *arg1*, ...) [Built-in Function]

Invoke the method *methodname* on the Java object *obj* with the arguments *arg1*, ... For static methods, *obj* can be a string representing the fully qualified name of the corresponding class. The function returns the result of the method invocation.

When *obj* is a regular Java object, structure-like indexing can be used as a shortcut syntax. For instance, the two following statements are equivalent

```
ret = javaMethod ("method1", x, 1.0, "a string")
ret = x.method1 (1.0, "a string")
```

See also: [methods], page 674, [javaObject], page 737.

The following three functions are used to display and modify the class path used by the Java Virtual Machine. This is entirely separate from Octave's PATH variable and is used by the JVM to find the correct code to execute.

javaclasspath () [Function File]
dpath = javaclasspath () [Function File]
[*dpath*, *spath*] = javaclasspath () [Function File]
clspath = javaclasspath (*what*) [Function File]

Return the class path of the Java virtual machine in the form of a cell array of strings.

If called with no inputs:

- If no output is requested, the dynamic and static classpaths are printed to the standard output.
- If one output value *dpath* is requested, the result is the dynamic classpath.
- If two output values*dpath* and *spath* are requested, the first variable will contain the dynamic classpath and the second will be contain the static classpath.

If called with a single input parameter *what*:

`"-dynamic"`
Return the dynamic classpath.

`"-static"`
: Return the static classpath.

`"-all"` Return both the static and dynamic classpath in a single cellstr.

See also: [javaaddpath], page 740, [javarmpath], page 740.

`javaaddpath (`*`clspath`*`)` [Function File]
`javaaddpath (`*`clspath1`*`, ...)` [Function File]

Add *clspath* to the dynamic class path of the Java virtual machine. *clspath* may either be a directory where '`.class`' files are found, or a '`.jar`' file containing Java classes. Multiple paths may be added at once by specifying additional arguments.

See also: [javarmpath], page 740, [javaclasspath], page 739.

`javarmpath (`*`clspath`*`)` [Function File]
`javarmpath (`*`clspath1`*`, ...)` [Function File]

Remove *clspath* from the dynamic class path of the Java virtual machine. *clspath* may either be a directory where '`.class`' files are found, or a '`.jar`' file containing Java classes. Multiple paths may be removed at once by specifying additional arguments.

See also: [javaaddpath], page 740, [javaclasspath], page 739.

The following four functions provide information and control over the interface between Octave and the Java Virtual Machine.

`usejava (`*`feature`*`)` [Function File]

Return true if the Java element *feature* is available.

Possible features are:

`"awt"` Abstract Window Toolkit for GUIs.

`"desktop"`
: Interactive desktop is running.

`"jvm"` Java Virtual Machine.

`"swing"` Swing components for lightweight GUIs.

`usejava` determines if specific Java features are available in an Octave session. This function is provided for scripts which may alter their behavior based on the availability of Java. The feature `"desktop"` always returns `false` as Octave has no Java-based desktop. Other features may be available if Octave was compiled with the Java Interface and Java is installed.

`javamem ()` [Function File]
`jmem` `= javamem ()` [Function File]

Show the current memory usage of the Java virtual machine (JVM) and run the garbage collector.

When no return argument is given the info is printed to the screen. Otherwise, the output cell array *jmem* contains Maximum, Total, and Free memory (in bytes).

All Java-based routines are run in the JVM's shared memory pool, a dedicated and separate part of memory claimed by the JVM from your computer's total memory (which comprises physical RAM and virtual memory / swap space on hard disk).

The maximum allowable memory usage can be configured using the file '`java.opts`'. The directory where this file resides is determined by the environment variable `OCTAVE_JAVA_DIR`. If unset, the directory where '`javaaddpath.m`' resides is used instead (typically '`OCTAVE_HOME/share/octave/OCTAVE_VERSION/m/java/`'

'`java.opts`' is a plain text file with one option per line. The default initial memory size and default maximum memory size (which are both system dependent) can be overridden like so:

-Xms64m

-Xmx512m

(in megabytes in this example). You can adapt these values to your own requirements if your system has limited available physical memory or if you get Java memory errors.

"Total memory" is what the operating system has currently assigned to the JVM and depends on actual and active memory usage. "Free memory" is self-explanatory. During operation of Java-based Octave functions the amount of Total and Free memory will vary, due to Java's own cleaning up and your operating system's memory management.

val = `java_matrix_autoconversion ()` [Built-in Function]
old_val = `java_matrix_autoconversion (`*new_val*`)` [Built-in Function]
`java_matrix_autoconversion (`*new_val*, "*local*"`)` [Built-in Function]

Query or set the internal variable that controls whether Java arrays are automatically converted to Octave matrices. The default value is false.

When called from inside a function with the `"local"` option, the variable is changed locally for the function and any subroutines it calls. The original variable value is restored when exiting the function.

See also: [java_unsigned_autoconversion], page 741, [debug_java], page 741.

val = `java_unsigned_autoconversion ()` [Built-in Function]
old_val = `java_unsigned_autoconversion (`*new_val*`)` [Built-in Function]
`java_unsigned_autoconversion (`*new_val*, "*local*"`)` [Built-in Function]

Query or set the internal variable that controls how integer classes are converted when `java_matrix_autoconversion` is enabled. When enabled, Java arrays of class Byte or Integer are converted to matrices of class uint8 or uint32 respectively. The default value is true.

When called from inside a function with the `"local"` option, the variable is changed locally for the function and any subroutines it calls. The original variable value is restored when exiting the function.

See also: [java_matrix_autoconversion], page 741, [debug_java], page 741.

val = `debug_java ()` [Built-in Function]
old_val = `debug_java (`*new_val*`)` [Built-in Function]
`debug_java (`*new_val*, "*local*"`)` [Built-in Function]

Query or set the internal variable that determines whether extra debugging information regarding the initialization of the JVM and any Java exceptions is printed.

When called from inside a function with the `"local"` option, the variable is changed locally for the function and any subroutines it calls. The original variable value is restored when exiting the function.

See also: [java_matrix_autoconversion], page 741, [java_unsigned_autoconversion], page 741.

37.2 Dialog Box Functions

The following functions all use the Java Interface to provide some form of dialog box.

h = `msgbox` (*msg*) [Function File]
h = `msgbox` (*msg*, *title*) [Function File]
h = `msgbox` (*msg*, *title*, *icon*) [Function File]

Display *msg* using a message dialog box.

The message may have multiple lines separated by newline characters (`"\n"`), or it may be a cellstr array with one element for each line. The optional input *title* (character string) can be used to decorate the dialog caption.

The optional argument *icon* selects a dialog icon. It can be one of `"none"` (default), `"error"`, `"help"`, or `"warn"`.

The return value is always 1.

See also: [errordlg], page 742, [helpdlg], page 742, [inputdlg], page 743, [listdlg], page 743, [questdlg], page 744, [warndlg], page 744.

h = `errordlg` (*msg*) [Function File]
h = `errordlg` (*msg*, *title*) [Function File]

Display *msg* using an error dialog box.

The message may have multiple lines separated by newline characters (`"\n"`), or it may be a cellstr array with one element for each line. The optional input *title* (character string) can be used to set the dialog caption. The default title is `"Error Dialog"`.

The return value is always 1.

See also: [helpdlg], page 742, [inputdlg], page 743, [listdlg], page 743, [msgbox], page 742, [questdlg], page 744, [warndlg], page 744.

h = `helpdlg` (*msg*) [Function File]
h = `helpdlg` (*msg*, *title*) [Function File]

Display *msg* in a help dialog box.

The message may have multiple lines separated by newline characters (`"\n"`), or it may be a cellstr array with one element for each line. The optional input *title* (character string) can be used to set the dialog caption. The default title is `"Help Dialog"`.

The return value is always 1.

See also: [errordlg], page 742, [inputdlg], page 743, [listdlg], page 743, [msgbox], page 742, [questdlg], page 744, [warndlg], page 744.

`cstr = inputdlg (prompt)` [Function File]
`cstr = inputdlg (prompt, title)` [Function File]
`cstr = inputdlg (prompt, title, rowscols)` [Function File]
`cstr = inputdlg (prompt, title, rowscols, defaults)` [Function File]

Return user input from a multi-textfield dialog box in a cell array of strings, or an empty cell array if the dialog is closed by the Cancel button.

Inputs:

prompt A cell array with strings labeling each text field. This input is required.

title String to use for the caption of the dialog. The default is `"Input Dialog"`.

rowscols Specifies the size of the text fields and can take three forms:

1. a scalar value which defines the number of rows used for each text field.
2. a vector which defines the individual number of rows used for each text field.
3. a matrix which defines the individual number of rows and columns used for each text field. In the matrix each row describes a single text field. The first column specifies the number of input rows to use and the second column specifies the text field width.

defaults A list of default values to place in each text fields. It must be a cell array of strings with the same size as *prompt*.

See also: [errordlg], page 742, [helpdlg], page 742, [listdlg], page 743, [msgbox], page 742, [questdlg], page 744, [warndlg], page 744.

`[sel, ok] = listdlg (key, value, ...)` [Function File]

Return user inputs from a list dialog box in a vector of selection indices *sel* and a flag *ok* indicating how the user closed the dialog box. The value of *ok* is 1 if the user closed the box with the OK button, otherwise it is 0 and *sel* is empty.

The indices in *sel* are 1-based.

The arguments are specified in form of *key*, *value* pairs. The `"ListString"` argument pair must be specified.

Valid *key* and *value* pairs are:

`"ListString"`
a cell array of strings comprising the content of the list.

`"SelectionMode"`
can be either `"Single"` or `"Multiple"` (default).

`"ListSize"`
a vector with two elements *width* and *height* defining the size of the list field in pixels. Default is [160 300].

`"InitialValue"`
a vector containing 1-based indices of preselected elements. Default is 1 (first item).

`"Name"` a string to be used as the dialog caption. Default is `""`.

`"PromptString"`
: a cell array of strings to be displayed above the list field. Default is `{}`.

`"OKString"`
: a string used to label the OK button. Default is `"OK"`.

`"CancelString"`
: a string used to label the Cancel button. Default is `"Cancel"`.

Example:

```
[sel, ok] = listdlg ("ListString", {"An item", "another", "yet another"},
                     "SelectionMode", "Multiple");
if (ok == 1)
  for i = 1:numel (sel)
    disp (sel(i));
  endfor
endif
```

See also: [menu], page 229, [errordlg], page 742, [helpdlg], page 742, [inputdlg], page 743, [msgbox], page 742, [questdlg], page 744, [warndlg], page 744.

btn = **questdlg** (*msg*) [Function File]
btn = **questdlg** (*msg*, *title*) [Function File]
btn = **questdlg** (*msg*, *title*, *default*) [Function File]
btn = **questdlg** (*msg*, *title*, *btn1*, *btn2*, *default*) [Function File]
btn = **questdlg** (*msg*, *title*, *btn1*, *btn2*, *btn3*, *default*) [Function File]

Display *msg* using a question dialog box and return the caption of the activated button.

The dialog may contain two or three buttons which will all close the dialog.

The message may have multiple lines separated by newline characters ("\n"), or it may be a cellstr array with one element for each line. The optional *title* (character string) can be used to decorate the dialog caption.

The string *default* identifies the default button, which is activated by pressing the ENTER key. It must match one of the strings given in *btn1*, *btn2*, or *btn3*.

If only *msg* and *title* are specified, three buttons with the default captions `"Yes"`, `"No"`, and `"Cancel"` are used.

If only two button captions, *btn1* and *btn2*, are specified the dialog will have only these two buttons.

See also: [errordlg], page 742, [helpdlg], page 742, [inputdlg], page 743, [listdlg], page 743, [warndlg], page 744.

h = **warndlg** (*msg*) [Function File]
h = **warndlg** (*msg*, *title*) [Function File]

Display *msg* using a warning dialog box.

The message may have multiple lines separated by newline characters ("\n"), or it may be a cellstr array with one element for each line. The optional input *title* (character string) can be used to set the dialog caption. The default title is "Warning Dialog".

See also: [helpdlg], page 742, [inputdlg], page 743, [listdlg], page 743, [questdlg], page 744.

37.3 FAQ - Frequently asked Questions

37.3.1 How to distinguish between Octave and Matlab?

Octave and MATLAB are very similar, but handle Java slightly different. Therefore it may be necessary to detect the environment and use the appropriate functions. The following function can be used to detect the environment. Due to the persistent variable it can be called repeatedly without a heavy performance hit.

Example:

```
%%
%% Return: true if the environment is Octave.
%%
function retval = isOctave
  persistent cacheval;  % speeds up repeated calls

  if isempty (cacheval)
    cacheval = (exist ("OCTAVE_VERSION", "builtin") > 0);
  end

  retval = cacheval;
end
```

37.3.2 How to make Java classes available to Octave?

Java finds classes by searching a *classpath*. This is a list of Java archive files and/or directories containing class files. In Octave the *classpath* is composed of two parts:

- the *static classpath* is initialized once at startup of the JVM, and
- the *dynamic classpath* which can be modified at runtime.

Octave searches the *static classpath* first, then the *dynamic classpath*. Classes appearing in the *static* as well as in the *dynamic classpath* will therefore be found in the *static classpath* and loaded from this location. Classes which will be used frequently or must be available to all users should be added to the *static classpath*. The *static classpath* is populated once from the contents of a plain text file named '`javaclasspath.txt`' (or '`classpath.txt`' historically) when the Java Virtual Machine starts. This file contains one line for each individual classpath to be added to the *static classpath*. These lines can identify single class files, directories containing class files, or Java archives with complete class file hierarchies. Comment lines starting with a '`#`' or a '`%`' character are ignored.

The search rules for the file '`javaclasspath.txt`' (or '`classpath.txt`') are:

- First, Octave tries to locate it in the current directory (where Octave was started from). If such a file is found, it is read and defines the initial *static classpath*. Thus, it is possible to define a static classpath on a 'per Octave invocation' basis.

- Next, Octave searches in the user's home directory. If a file '`javaclasspath.txt`' exists here, its contents are appended to the static classpath (if any). Thus, it is possible to build an initial static classpath on a 'per user' basis.
- Finally, Octave looks for a next occurrence of file '`javaclasspath.txt`' in the m-files directory where Octave Java functions live. This is where '`javaclasspath.m`' resides, usually something like '`OCTAVE_HOME/share/octave/OCTAVE_VERSION/m/java/`'. You can find this directory by executing the command

  ```
  which javaclasspath
  ```

  If this file exists here, its contents are also appended to the static classpath. Note that the archives and class directories defined in this last step will affect all users.

Classes which are used only by a specific script should be placed in the *dynamic classpath*. This portion of the classpath can be modified at runtime using the `javaaddpath` and `javarmpath` functions.

Example:

```
octave> base_path = "C:/Octave/java_files";

octave> % add two JARchives to the dynamic classpath
octave> javaaddpath ([base_path, "/someclasses.jar"]);
octave> javaaddpath ([base_path, "/moreclasses.jar"]);

octave> % check the dynamic classpath
octave> p = javaclasspath;
octave> disp (p{1});
C:/Octave/java_files/someclasses.jar
octave> disp (p{2});
C:/Octave/java_files/moreclasses.jar

octave> % remove the first element from the classpath
octave> javarmpath ([base_path, "/someclasses.jar"]);
octave> p = javaclasspath;
octave> disp (p{1});
C:/Octave/java_files/moreclasses.jar

octave> % provoke an error
octave> disp (p{2});
error: A(I): Index exceeds matrix dimension.
```

Another way to add files to the *dynamic classpath* exclusively for your user account is to use the file '`.octaverc`' which is stored in your home directory. All Octave commands in this file are executed each time you start a new instance of Octave. The following example adds the directory '`octave`' to Octave's search path and the archive '`myclasses.jar`' in this directory to the Java search path.

```
% contents of .octaverc:
addpath ("~/octave");
javaaddpath ("~/octave/myclasses.jar");
```

37.3.3 How to create an instance of a Java class?

The function `javaObject` can be used to create Java objects..

Example:

```
Passenger = javaObject ("package.FirstClass", row, seat);
```

37.3.4 How can I handle memory limitations?

In order to execute Java code Octave creates a Java Virtual Machine (JVM). Such a JVM allocates a fixed amount of initial memory and may expand this pool up to a fixed maximum memory limit. The default values depend on the Java version (see [javamem], page 740). The memory pool is shared by all Java objects running in the JVM. This strict memory limit is intended mainly to avoid that runaway applications inside web browsers or in enterprise servers can consume all memory and crash the system. When the maximum memory limit is hit, Java code will throw exceptions so that applications will fail or behave unexpectedly.

You can specify options for the creation of the JVM inside a file named '`java.opts`'. This is a text file where you can enter lines containing '`-X`' and '`-D`' options handed to the JVM during initialization.

The directory where the Java options file is located is specified by the environment variable `OCTAVE_JAVA_DIR`. If unset the directory where '`javaclasspath.m`' resides is used instead (typically '`OCTAVE_HOME/share/octave/OCTAVE_VERSION/m/java/`'). You can find this directory by executing

```
which javaclasspath
```

The '`-X`' options allow you to increase the maximum amount of memory available to the JVM. The following example allows up to 256 Megabytes to be used by adding the following line to the '`java.opts`' file:

```
-Xmx256m
```

The maximum possible amount of memory depends on your system. On a Windows system with 2 Gigabytes main memory you should be able to set this maximum to about 1 Gigabyte.

If your application requires a large amount of memory from the beginning, you can also specify the initial amount of memory allocated to the JVM. Adding the following line to the '`java.opts`' file starts the JVM with 64 Megabytes of initial memory:

```
-Xms64m
```

For more details on the available '`-X`' options of your Java Virtual Machine issue the command '`java -X`' at the operating system command prompt and consult the Java documentation.

The '`-D`' options can be used to define system properties which can then be used by Java classes inside Octave. System properties can be retrieved by using the `getProperty()` methods of the `java.lang.System` class. The following example line defines the property *MyProperty* and assigns it the string `12.34`.

```
-DMyProperty=12.34
```

The value of this property can then be retrieved as a string by a Java object or in Octave:

```
octave> javaMethod ("getProperty", "java.lang.System", "MyProperty");
ans = 12.34
```

See also: javamem.

37.3.5 Which TEX symbols are implemented in dialog functions?

The dialog functions contain a translation table for TEX like symbol codes. Thus messages and labels can be tailored to show some common mathematical symbols or Greek characters. No further TEX formatting codes are supported. The characters are translated to their Unicode equivalent. However, not all characters may be displayable on your system. This depends on the font used by the Java system on your computer.

Each TEX symbol code must be terminated by a space character to make it distinguishable from the surrounding text. Therefore the string '`\alpha =12.0`' will produce the desired result, whereas '`\alpha=12.0`' would produce the literal text *'\alpha=12.0'*.

See also: errordlg, helpdlg, inputdlg, listdlg, msgbox, questdlg, warndlg.

The table below shows each TEX character code and the corresponding Unicode character:

TEX code	Symbol	TEX code	Symbol	TEX code	Symbol
\alpha	'α'	\beta	'β'	\gamma	'γ'
\delta	'δ'	\epsilon	'ϵ'	\zeta	'ζ'
\eta	'η'	\theta	'θ'	\vartheta	'ϑ'
\iota	'ι'	\kappa	'κ'	\lambda	'λ'
\mu	'μ'	\nu	'ν'	\xi	'ξ'
\pi	'π'	\rho	'ρ'	\sigma	'σ'
\varsigma	'ς'	\tau	'τ'	\phi	'ϕ'
\chi	'χ'	\psi	'ψ'	\omega	'ω'
\upsilon	'υ'	\Gamma	'Γ'	\Delta	'Δ'
\Theta	'Θ'	\Lambda	'Λ'	\Pi	'Π'
\Xi	'Ξ'	\Sigma	'Σ'	\Upsilon	'Υ'
\Phi	'Φ'	\Psi	'Ψ'	\Omega	'Ω'
\Im	'$\Im$'	\Re	'$\Re$'	\leq	'$\leq$'
\geq	'$\geq$'	\neq	'$\neq$'	\pm	'$\pm$'
\infty	'∞'	\partial	'∂'	\approx	'$\approx$'
\circ	'$\circ$'	\bullet	'$\bullet$'	\times	'$\times$'
\sim	'$\sim$'	\nabla	'∇'	\ldots	'$\ldots$'
\exists	'$\exists$'	\neg	'$\neg$'	\aleph	'$\aleph$'
\forall	'$\forall$'	\cong	'$\cong$'	\wp	'$\wp$'
\propto	'$\propto$'	\otimes	'$\otimes$'	\oplus	'$\oplus$'
\oslash	'$\oslash$'	\cap	'$\cap$'	\cup	'$\cup$'
\ni	'$\ni$'	\in	'$\in$'	\div	'$\div$'
\equiv	'$\equiv$'	\int	'$\int$'	\perp	'$\perp$'
\wedge	'$\wedge$'	\vee	'$\vee$'	\supseteq	'$\supseteq$'
\supset	'$\supset$'	\subseteq	'$\subseteq$'	\subset	'$\subset$'
\clubsuit	'$\clubsuit$'	\spadesuit	'$\spadesuit$'	\heartsuit	'$\heartsuit$'
\diamondsuit	'$\diamondsuit$'	\copyright	'$\copyright$'	\leftarrow	'$\leftarrow$'
\uparrow	'$\uparrow$'	\rightarrow	'$\rightarrow$'	\downarrow	'$\downarrow$'
\leftrightarrow	'$\leftrightarrow$'	\updownarrow	'$\updownarrow$'		

Table: TEX character codes and the resulting symbols.

38 Packages

Since Octave is Free Software users are encouraged to share their programs amongst each other. To aid this sharing Octave supports the installation of extra packages. The 'Octave-Forge' project is a community-maintained set of packages that can be downloaded and installed in Octave. At the time of writing the 'Octave-Forge' project can be found online at `http://octave.sourceforge.net`, but since the Internet is an ever-changing place this may not be true at the time of reading. Therefore it is recommended to see the Octave website for an updated reference.

38.1 Installing and Removing Packages

Assuming a package is available in the file '`image-1.0.0.tar.gz`' it can be installed from the Octave prompt with the command

```
pkg install image-1.0.0.tar.gz
```

If the package is installed successfully nothing will be printed on the prompt, but if an error occurred during installation it will be reported. It is possible to install several packages at once by writing several package files after the `pkg install` command. If a different version of the package is already installed it will be removed prior to installing the new package. This makes it easy to upgrade and downgrade the version of a package, but makes it impossible to have several versions of the same package installed at once.

To see which packages are installed type

```
pkg list
⊣ Package Name  | Version | Installation directory
⊣ --------------+---------+-----------------------
⊣        image *|   1.0.0 | /home/jwe/octave/image-1.0.0
```

In this case only version 1.0.0 of the `image` package is installed. The `'*'` character next to the package name shows that the image package is loaded and ready for use.

It is possible to remove a package from the system using the `pkg uninstall` command like this

```
pkg uninstall image
```

If the package is removed successfully nothing will be printed in the prompt, but if an error occurred it will be reported. It should be noted that the package file used for installation is not needed for removal, and that only the package name as reported by `pkg list` should be used when removing a package. It is possible to remove several packages at once by writing several package names after the `pkg uninstall` command.

To minimize the amount of code duplication between packages it is possible that one package depends on another one. If a package depends on another, it will check if that package is installed during installation. If it is not, an error will be reported and the package will not be installed. This behavior can be disabled by passing the '`-nodeps`' flag to the `pkg install` command

```
pkg install -nodeps my_package_with_dependencies.tar.gz
```

Since the installed package expects its dependencies to be installed it may not function correctly. Because of this it is not recommended to disable dependency checking.

pkg *command pkg_name* [Command]
pkg *command option pkg_name* [Command]

Manage packages (groups of add-on functions) for Octave. Different actions are available depending on the value of *command*.

Available commands:

'install' Install named packages. For example,

```
pkg install image-1.0.0.tar.gz
```

installs the package found in the file 'image-1.0.0.tar.gz'.

The *option* variable can contain options that affect the manner in which a package is installed. These options can be one or more of

-nodeps The package manager will disable dependency checking. With this option it is possible to install a package even when it depends on another package which is not installed on the system. **Use this option with care.**

-noauto The package manager will not automatically load the installed package when starting Octave. This overrides any setting within the package.

-auto The package manager will automatically load the installed package when starting Octave. This overrides any setting within the package.

-local A local installation (package available only to current user) is forced, even if the user has system privileges.

-global A global installation (package available to all users) is forced, even if the user doesn't normally have system privileges.

-forge Install a package directly from the Octave-Forge repository. This requires an internet connection and the cURL library.

-verbose The package manager will print the output of all commands as they are performed.

'update' Check installed Octave-Forge packages against repository and update any outdated items. This requires an internet connection and the cURL library. Usage:

```
pkg update
```

'uninstall'

Uninstall named packages. For example,

```
pkg uninstall image
```

removes the image package from the system. If another installed package depends on the image package an error will be issued. The package can be uninstalled anyway by using the '-nodeps' option.

'load' Add named packages to the path. After loading a package it is possible to use the functions provided by the package. For example,

```
pkg load image
```

adds the **image** package to the path. It is possible to load all installed packages at once with the keyword '**all**'. Usage:

```
pkg load all
```

'**unload**' Remove named packages from the path. After unloading a package it is no longer possible to use the functions provided by the package. It is possible to unload all installed packages at once with the keyword '**all**'. Usage:

```
pkg unload all
```

'**list**' Show the list of currently installed packages. For example,

```
installed_packages = pkg ("list")
```

returns a cell array containing a structure for each installed package.

If two output arguments are requested **pkg** splits the list of installed packages into those which were installed by the current user, and those which were installed by the system administrator.

```
[user_packages, system_packages] = pkg ("list")
```

The option "-forge" lists packages available at the Octave-Forge repository. This requires an internet connection and the cURL library. For example:

```
oct_forge_pkgs = pkg ("list", "-forge")
```

'**describe**'

Show a short description of the named installed packages, with the option "-verbose" also list functions provided by the package. For example,

```
pkg describe -verbose all
```

will describe all installed packages and the functions they provide. If one output is requested a cell of structure containing the description and list of functions of each package is returned as output rather than printed on screen:

```
desc = pkg ("describe", "secs1d", "image")
```

If any of the requested packages is not installed, pkg returns an error, unless a second output is requested:

```
[desc, flag] = pkg ("describe", "secs1d", "image")
```

flag will take one of the values "Not installed", "Loaded", or "Not loaded" for each of the named packages.

'**prefix**' Set the installation prefix directory. For example,

```
pkg prefix ~/my_octave_packages
```

sets the installation prefix to '~/my_octave_packages'. Packages will be installed in this directory.

It is possible to get the current installation prefix by requesting an output argument. For example:

```
pfx = pkg ("prefix")
```

The location in which to install the architecture dependent files can be independently specified with an addition argument. For example:

```
pkg prefix ~/my_octave_packages ~/my_arch_dep_pkgs
```

'local_list'
: Set the file in which to look for information on locally installed packages. Locally installed packages are those that are available only to the current user. For example:

```
pkg local_list ~/.octave_packages
```

It is possible to get the current value of local_list with the following

```
pkg local_list
```

'global_list'
: Set the file in which to look for information on globally installed packages. Globally installed packages are those that are available to all users. For example:

```
pkg global_list /usr/share/octave/octave_packages
```

It is possible to get the current value of global_list with the following

```
pkg global_list
```

'build'
: Build a binary form of a package or packages. The binary file produced will itself be an Octave package that can be installed normally with pkg. The form of the command to build a binary package is

```
pkg build builddir image-1.0.0.tar.gz ...
```

where builddir is the name of a directory where the temporary installation will be produced and the binary packages will be found. The options '-verbose' and '-nodeps' are respected, while all other options are ignored.

'rebuild'
: Rebuild the package database from the installed directories. This can be used in cases where the package database has been corrupted. It can also take the '-auto' and '-noauto' options to allow the autoloading state of a package to be changed. For example,

```
pkg rebuild -noauto image
```

will remove the autoloading status of the image package.

38.2 Using Packages

By default installed packages are not available from the Octave prompt, but it is possible to control this using the pkg load and pkg unload commands. The functions from a package can be added to the Octave path by typing

```
pkg load package_name
```

where package_name is the name of the package to be added to the path.

In much the same way a package can be removed from the Octave path by typing

```
pkg unload package_name
```

38.3 Administrating Packages

On UNIX-like systems it is possible to make both per-user and system-wide installations of a package. If the user performing the installation is `root` the packages will be installed in a system-wide directory that defaults to '`OCTAVE_HOME/share/octave/packages/`'. If the user is not `root` the default installation directory is '`~/octave/`'. Packages will be installed in a subdirectory of the installation directory that will be named after the package. It is possible to change the installation directory by using the `pkg prefix` command

```
pkg prefix new_installation_directory
```

The current installation directory can be retrieved by typing

```
current_installation_directory = pkg prefix
```

To function properly the package manager needs to keep some information about the installed packages. For per-user packages this information is by default stored in the file '`~/.octave_packages`' and for system-wide installations it is stored in '`OCTAVE_HOME/share/octave/octave_packages`'. The path to the per-user file can be changed with the `pkg local_list` command

```
pkg local_list /path/to/new_file
```

For system-wide installations this can be changed in the same way using the `pkg global_list` command. If these commands are called without a new path, the current path will be returned.

38.4 Creating Packages

Internally a package is simply a gzipped tar file that contains a top level directory of any given name. This directory will in the following be referred to as `package` and may contain the following files:

`package/CITATION`
: This is am optional file describing instructions on how to cite the package for publication. It will be displayed verbatim by the function `citation`.

`package/COPYING`
: This is a required file containing the license of the package. No restrictions is made on the license in general. If however the package contains dynamically linked functions the license must be compatible with the GNU General Public License.

`package/DESCRIPTION`
: This is a required file containing information about the package. See Section 38.4.1 [The DESCRIPTION File], page 755, for details on this file.

`package/ChangeLog`
: This is an optional file describing all the changes made to the package source files.

`package/INDEX`
: This is an optional file describing the functions provided by the package. If this file is not given then one with be created automatically from the functions in the package and the `Categories` keyword in the '`DESCRIPTION`' file. See Section 38.4.2 [The INDEX File], page 757, for details on this file.

package/NEWS
: This is an optional file describing all user-visible changes worth mentioning. As this file increases on size, old entries can be moved into 'package/ONEWS'.

package/ONEWS
: This is an optional file describing old entries from the 'NEWS' file.

package/PKG_ADD
: An optional file that includes commands that are run when the package is added to the users path. Note that PKG_ADD directives in the source code of the package will also be added to this file by the Octave package manager. Note that symbolic links are to be avoided in packages, as symbolic links do not exist on some file systems, and so a typical use for this file is the replacement of the symbolic link

```
ln -s foo.oct bar.oct
```

with an autoload directive like

```
autoload ('bar', which ('foo'));
```

See Section 38.4.3 [PKG_ADD and PKG_DEL Directives], page 758, for details on PKG_ADD directives.

package/PKG_DEL
: An optional file that includes commands that are run when the package is removed from the users path. Note that PKG_DEL directives in the source code of the package will also be added to this file by the Octave package manager. See Section 38.4.3 [PKG_ADD and PKG_DEL Directives], page 758, for details on PKG_DEL directives.

package/pre_install.m
: This is an optional function that is run prior to the installation of a package. This function is called with a single argument, a struct with fields names after the data in the 'DESCRIPTION', and the paths where the package functions will be installed.

package/post_install.m
: This is an optional function that is run after the installation of a package. This function is called with a single argument, a struct with fields names after the data in the 'DESCRIPTION', and the paths where the package functions were installed.

package/on_uninstall.m
: This is an optional function that is run prior to the removal of a package. This function is called with a single argument, a struct with fields names after the data in the 'DESCRIPTION', the paths where the package functions are installed, and whether the package is currently loaded.

Besides the above mentioned files, a package can also contain one or more of the following directories:

package/inst
: An optional directory containing any files that are directly installed by the package. Typically this will include any m-files.

`package/src`
: An optional directory containing code that must be built prior to the packages installation. The Octave package manager will execute '`./configure`' in this directory if this script exists, and will then call `make` if a file '`Makefile`' exists in this directory. `make install` will however not be called. The environment variables `MKOCTFILE`, `OCTAVE_CONFIG`, and `OCTAVE` will be set to the full paths of the programs `mkoctfile`, `octave-config`, and `octave`, respectively, of the correct version when `configure` and `make` are called. If a file called `FILES` exists all files listed there will be copied to the `inst` directory, so they also will be installed. If the `FILES` file doesn't exist, '`src/*.m`' and '`src/*.oct`' will be copied to the `inst` directory.

`package/doc`
: An optional directory containing documentation for the package. The files in this directory will be directly installed in a sub-directory of the installed package for future reference.

`package/bin`
: An optional directory containing files that will be added to the Octave `EXEC_PATH` when the package is loaded. This might contain external scripts, etc., called by functions within the package.

38.4.1 The DESCRIPTION File

The '`DESCRIPTION`' file contains various information about the package, such as its name, author, and version. This file has a very simple format

- Lines starting with '`#`' are comments.
- Lines starting with a blank character are continuations from the previous line.
- Everything else is of the form `NameOfOption: ValueOfOption`.

The following is a simple example of a '`DESCRIPTION`' file

```
Name: The name of my package
Version: 1.0.0
Date: 2007-18-04
Author: The name (and possibly email) of the package author.
Maintainer: The name (and possibly email) of the current
 package maintainer.
Title: The title of the package
Description: A short description of the package.  If this
 description gets too long for one line it can continue
 on the next by adding a space to the beginning of the
 following lines.
License: GPLv3+
```

The package manager currently recognizes the following keywords

`Name`
: Name of the package.

`Version`
: Version of the package. A package version must be 3 numbers separated by dots.

`Date` Date of last update.

`Author` Original author of the package.

`Maintainer`
Maintainer of the package.

`Title` A one line description of the package.

`Description`
A one paragraph description of the package.

`Categories`
Optional keyword describing the package (if no 'INDEX' file is given this is mandatory).

`Problems` Optional list of known problems.

`Url` Optional list of homepages related to the package.

`Autoload` Optional field that sets the default loading behavior for the package. If set to `yes`, `true` or `on`, then Octave will automatically load the package when starting. Otherwise the package must be manually loaded with the pkg load command. This default behavior can be overridden when the package is installed.

`Depends` A list of other Octave packages that this package depends on. This can include dependencies on particular versions, with a format

```
Depends: package (>= 1.0.0)
```

Possible operators are `<`, `<=`, `==`, `>=` or `>`. If the part of the dependency in `()` is missing, any version of the package is acceptable. Multiple dependencies can be defined either as a comma separated list or on separate `Depends` lines.

`License` An optional short description of the used license (e.g., GPL version 3 or newer). This is optional since the file 'COPYING' is mandatory.

`SystemRequirements`
These are the external install dependencies of the package and are not checked by the package manager. This is here as a hint to the distribution packager. They follow the same conventions as the `Depends` keyword.

`BuildRequires`
These are the external build dependencies of the package and are not checked by the package manager. This is here as a hint to the distribution packager. They follow the same conventions as the `Depends` keyword. Note that in general, packaging systems such as `rpm` or `deb` and autoprobe the install dependencies from the build dependencies, and therefore the often a `BuildRequires` dependency removes the need for a `SystemRequirements` dependency.

The developer is free to add additional arguments to the 'DESCRIPTION' file for their own purposes. One further detail to aid the packager is that the `SystemRequirements` and `BuildRequires` keywords can have a distribution dependent section, and the automatic build process will use these. An example of the format of this is

```
BuildRequires: libtermcap-devel [Mandriva] libtermcap2-devel
```

where the first package name will be used as a default and if the RPMs are built on a Mandriva distribution, then the second package name will be used instead.

38.4.2 The INDEX File

The optional 'INDEX' file provides a categorical view of the functions in the package. This file has a very simple format

- Lines beginning with '#' are comments.
- The first non-comment line should look like this

```
toolbox >> Toolbox name
```

- Lines beginning with an alphabetical character indicates a new category of functions.
- Lines starting with a white space character indicate that the function names on the line belong to the last mentioned category.

The format can be summarized with the following example:

```
# A comment
toolbox >> Toolbox name
Category Name 1
 function1 function2 function3
 function4
Category Name 2
 function2 function5
```

If you wish to refer to a function that users might expect to find in your package but is not there, providing a work around or pointing out that the function is available elsewhere, you can use:

```
fn = workaround description
```

This workaround description will not appear when listing functions in the package with `pkg describe` but they will be published in the HTML documentation online. Workaround descriptions can use any HTML markup, but keep in mind that it will be enclosed in a bold-italic environment. For the special case of:

```
fn = use <code>alternate expression</code>
```

the bold-italic is automatically suppressed. You will need to use `<code>` even in references:

```
fn = use <a href="someothersite.html"><code>fn</code></a>
```

Sometimes functions are only partially compatible, in which case you can list the non-compatible cases separately. To refer to another function in the package, use `<f>fn</f>`. For example:

```
eig (a, b) = use <f>qz</f>
```

Since sites may have many missing functions, you can define a macro rather than typing the same link over and again.

```
$id = expansion
```

defines the macro id. You can use `$id` anywhere in the description and it will be expanded. For example:

```
$TSA = see <a href="link_to_spctools">SPC Tools</a>
arcov = $TSA <code>armcv</code>
```

id is any string of letters, numbers and _.

38.4.3 PKG_ADD and PKG_DEL Directives

If the package contains files called `PKG_ADD` or `PKG_DEL` the commands in these files will be executed when the package is added or removed from the users path. In some situations such files are a bit cumbersome to maintain, so the package manager supports automatic creation of such files. If a source file in the package contains a `PKG_ADD` or `PKG_DEL` directive they will be added to either the `PKG_ADD` or `PKG_DEL` files.

In `m`-files a `PKG_ADD` directive looks like this

```
## PKG_ADD: some_octave_command
```

Such lines should be added before the `function` keyword. In C++ files a `PKG_ADD` directive looks like this

```
// PKG_ADD: some_octave_command
```

In both cases `some_octave_command` should be replaced by the command that should be placed in the `PKG_ADD` file. `PKG_DEL` directives work in the same way, except the `PKG_ADD` keyword is replaced with `PKG_DEL` and the commands get added to the `PKG_DEL` file.

38.4.4 Missing Components

If a package relies on a component, such as another Octave package, that may not be present it may be useful to install a function which informs users what to do when a particular component is missing. The function must be written by the package maintainer and registered with Octave using `missing_component_hook`.

val = `missing_component_hook` () [Built-in Function]
old_val = `missing_component_hook` (*new_val*) [Built-in Function]
`missing_component_hook` (*new_val*, "*local*") [Built-in Function]

Query or set the internal variable that specifies the function to call when a component of Octave is missing. This can be useful for packagers that may split the Octave installation into multiple sub-packages, for example, to provide a hint to users for how to install the missing components.

When called from inside a function with the `"local"` option, the variable is changed locally for the function and any subroutines it calls. The original variable value is restored when exiting the function.

The hook function is expected to be of the form

```
fcn (component)
```

Octave will call *fcn* with the name of the function that requires the component and a string describing the missing component. The hook function should return an error message to be displayed.

See also: [missing_function_hook], page 856.

Appendix A External Code Interface

"The sum of human wisdom is not contained in any one language" —Ezra Pound

Octave is a fantastic language for solving many problems in science and engineering. However, it is not the only computer language and there are times when you may want to use code written in other languages. Good reasons for doing so include: 1) not re-inventing the wheel; existing function libraries which have been thoroughly tested and debugged or large scale simulation codebases are a good example, 2) accessing unique capabilities of a different language; for example the well-known regular expression functions of Perl (but don't do that because `regexp` already exists in Octave).

Performance should generally **not** be a reason for using compiled extensions. Although compiled extensions can run faster, particularly if they replace a loop in Octave code, this is almost never the best path to take. First, there are many techniques to speed up Octave performance while remaining within the language. Second, Octave is a high-level language that makes it easy to perform common mathematical tasks. Giving that up means shifting the focus from solving the real problem to solving a computer programming problem. It means returning to low-level constructs such as pointers, memory management, mathematical overflow/underflow, etc. Because of the low level nature, and the fact that the compiled code is executed outside of Octave, there is the very real possibility of crashing the interpreter and losing work.

Before going further, you should first determine if you really need to bother writing code outside of Octave.

- Can I get the same functionality using the Octave scripting language alone?

  Even when a function already exists outside the language, it may be better to simply reproduce the behavior in an m-file rather than attempt to interface to the outside code.

- Is the code thoroughly optimized for Octave?

  If performance is an issue you should always start with the in-language techniques for getting better performance. Chief among these is vectorization (see Chapter 19 [Vectorization and Faster Code Execution], page 457) which not only makes the code concise and more understandable but improves performance (10X-100X). If loops must be used, make sure that the allocation of space for variables takes place outside the loops using an assignment to a matrix of the right size, or zeros.

- Does the code make as much use as possible of existing built-in library routines?

  These routines are highly optimized and many do not carry the overhead of being interpreted.

- Does writing a dynamically linked function represent a useful investment of your time, relative to staying in Octave?

  It will take time to learn Octave's interface for external code and there will inevitably be issues with tools such as compilers.

With that said, Octave offers a versatile interface for including chunks of compiled code as dynamically linked extensions. These dynamically linked functions can be called from the interpreter in the same manner as any ordinary function. The interface is bi-directional and external code can call Octave functions (like `plot`) which otherwise might be very difficult to develop.

The interface is centered around supporting the languages C++, C, and Fortran. Octave itself is written in C++ and can call external C++/C code through its native oct-file interface. The C language is also supported through the mex-file interface for compatibility with MATLAB. Fortran code is easiest to reach through the oct-file interface.

Because many other languages provide C or C++ APIs it is relatively simple to build bridges between Octave and other languages. This is also a way to bridge to hardware resources which often have device drivers written in C.

A.1 Oct-Files

A.1.1 Getting Started with Oct-Files

Oct-files are pieces of C++ code that have been compiled with the Octave API into a dynamically loadable object. They take their name from the file which contains the object which has the extension '`.oct`'.

Finding a C++ compiler, using the correct switches, adding the right include paths for header files, etc. is a difficult task. Octave automates this by providing the `mkoctfile` command with which to build oct-files. The command is available from within Octave or at the shell command line.

`mkoctfile` [*-options*] *file* ... [Command]
`[`*output*`,` *status*`] = mkoctfile (...)` [Function File]

The `mkoctfile` function compiles source code written in C, C++, or Fortran. Depending on the options used with `mkoctfile`, the compiled code can be called within Octave or can be used as a stand-alone application.

`mkoctfile` can be called from the shell prompt or from the Octave prompt. Calling it from the Octave prompt simply delegates the call to the shell prompt. The output is stored in the *output* variable and the exit status in the *status* variable.

`mkoctfile` accepts the following options, all of which are optional except for the file name of the code you wish to compile:

'`-I DIR`' Add the include directory DIR to compile commands.

'`-D DEF`' Add the definition DEF to the compiler call.

'`-l LIB`' Add the library LIB to the link command.

'`-L DIR`' Add the library directory DIR to the link command.

'`-M`'
'`--depend`'
Generate dependency files (.d) for C and C++ source files.

'`-R DIR`' Add the run-time path to the link command.

'`-Wl,...`' Pass flags though the linker like "-Wl,-rpath=...". The quotes are needed since commas are interpreted as command separators.

'`-W...`' Pass flags though the compiler like "-Wa,OPTION".

'-c' Compile but do not link.

'-g' Enable debugging options for compilers.

'-o FILE'
'--output FILE'
Output file name. Default extension is .oct (or .mex if '--mex' is specified) unless linking a stand-alone executable.

'-p VAR'
'--print VAR'
Print the configuration variable VAR. Recognized variables are:

ALL_CFLAGS	INCFLAGS
ALL_CXXFLAGS	INCLUDEDIR
ALL_FFLAGS	LAPACK_LIBS
ALL_LDFLAGS	LD_CXX
AR	LDFLAGS
BLAS_LIBS	LD_STATIC_FLAG
CC	LFLAGS
CFLAGS	LIBDIR
CPICFLAG	LIBOCTAVE
CPPFLAGS	LIBOCTINTERP
CXX	LIBS
CXXFLAGS	OCTAVE_HOME
CXXPICFLAG	OCTAVE_LIBS
DEPEND_EXTRA_SED_PATTERN	OCTAVE_LINK_DEPS
DEPEND_FLAGS	OCTAVE_LINK_OPTS
DL_LD	OCTAVE_PREFIX
DL_LDFLAGS	OCTINCLUDEDIR
F77	OCTLIBDIR
F77_INTEGER8_FLAG	OCT_LINK_DEPS
FFLAGS	OCT_LINK_OPTS
FFTW3F_LDFLAGS	RANLIB
FFTW3F_LIBS	RDYNAMIC_FLAG
FFTW3_LDFLAGS	READLINE_LIBS
FFTW3_LIBS	SED
FFTW_LIBS	SPECIAL_MATH_LIB
FLIBS	XTRA_CFLAGS
FPICFLAG	XTRA_CXXFLAGS

'--link-stand-alone'
Link a stand-alone executable file.

'--mex' Assume we are creating a MEX file. Set the default output extension to ".mex".

'-s'
'--strip' Strip the output file.

'-v'
'--verbose'
Echo commands as they are executed.

'file' The file to compile or link. Recognized file types are

.c	C source
.cc	C++ source
.C	C++ source
.cpp	C++ source
.f	Fortran source (fixed form)
.F	Fortran source (fixed form)
.f90	Fortran source (free form)
.F90	Fortran source (free form)
.o	object file
.a	library file

Consider the following short example which introduces the basics of writing a C++ function that can be linked to Octave.

```
#include <octave/oct.h>

DEFUN_DLD (helloworld, args, nargout,
           "Hello World Help String")
{
  int nargin = args.length ();

  octave_stdout << "Hello World has "
                << nargin << " input arguments and "
                << nargout << " output arguments.\n";

  return octave_value_list ();
}
```

The first critical line is `#include <octave/oct.h>` which makes available most of the definitions necessary for a C++ oct-file. Note that 'octave/oct.h' is a C++ header and cannot be directly `#include`'ed in a C source file, nor any other language.

Included by 'oct.h' is a definition for the macro `DEFUN_DLD` which creates a dynamically loaded function. This macro takes four arguments:

1. The function name as it will be seen in Octave,
2. The list of arguments to the function of type `octave_value_list`,
3. The number of output arguments, which can and often is omitted if not used, and
4. The string to use for the help text of the function.

The return type of functions defined with `DEFUN_DLD` is always `octave_value_list`.

There are a couple of important considerations in the choice of function name. First, it must be a valid Octave function name and so must be a sequence of letters, digits, and underscores not starting with a digit. Second, as Octave uses the function name to define the filename it attempts to find the function in, the function name in the `DEFUN_DLD` macro must match the filename of the oct-file. Therefore, the above function should be in a file '`helloworld.cc`', and would be compiled to an oct-file using the command

```
mkoctfile helloworld.cc
```

This will create a file called '`helloworld.oct`' that is the compiled version of the function. It should be noted that it is perfectly acceptable to have more than one `DEFUN_DLD` function in a source file. However, there must either be a symbolic link to the oct-file for each of the functions defined in the source code with the `DEFUN_DLD` macro or the `autoload` (Section 11.9 [Function Files], page 178) function should be used.

The rest of the function shows how to find the number of input arguments, how to print through the Octave pager, and return from the function. After compiling this function as above, an example of its use is

```
helloworld (1, 2, 3)
-| Hello World has 3 input arguments and 0 output arguments.
```

Subsequent sections show how to use specific classes from Octave's core internals. Base classes like dMatrix (a matrix of double values) are found in the directory '`liboctave/array`'. The definitive reference for how to use a particular class is the header file itself. However, it is often enough just to study the examples in the manual in order to be able to use the class.

A.1.2 Matrices and Arrays in Oct-Files

Octave supports a number of different array and matrix classes, the majority of which are based on the Array class. The exception is the sparse matrix types discussed separately below. There are three basic matrix types

`Matrix` A double precision matrix class defined in '`dMatrix.h`',

`ComplexMatrix`
A complex matrix class defined in '`CMatrix.h`', and

`BoolMatrix`
A boolean matrix class defined in '`boolMatrix.h`'.

These are the basic two-dimensional matrix types of Octave. In addition there are a number of multi-dimensional array types including

`NDArray` A double precision array class defined in '`dNDArray.h`'

`ComplexNDarray`
A complex array class defined in '`CNDArray.h`'

`boolNDArray`
A boolean array class defined in '`boolNDArray.h`'

`int8NDArray`
`int16NDArray`
`int32NDArray`
`int64NDArray`
8, 16, 32, and 64-bit signed array classes defined in '`int8NDArray.h`', '`int16NDArray.h`', etc.

`uint8NDArray`
`uint16NDArray`
`uint32NDArray`
`uint64NDArray`
: 8, 16, 32, and 64-bit unsigned array classes defined in 'uint8NDArray.h', 'uint16NDArray.h', etc.

There are several basic ways of constructing matrices or multi-dimensional arrays. Using the class `Matrix` as an example one can

- Create an empty matrix or array with the empty constructor. For example:

```
Matrix a;
```

  This can be used for all matrix and array types.
- Define the dimensions of the matrix or array with a dim_vector which has the same characteristics as the vector returned from `size`. For example:

```
dim_vector dv (2);
dv(0) = 2; dv(1) = 3;  // 2 rows, 3 columns
Matrix a (dv);
```

  This can be used on all matrix and array types.
- Define the number of rows and columns in the matrix. For example:

```
Matrix a (2, 2)
```

  However, this constructor can only be used with matrix types.

These types all share a number of basic methods and operators. Many bear a resemblance to functions that exist in the interpreter. A selection of useful methods include

`T& operator` () (*octave_idx_type*) [Method]
`T& elem` (*octave_idx_type*) [Method]
: The `()` operator or `elem` method allow the values of the matrix or array to be read or set. These can take a single argument, which is of type `octave_idx_type`, that is the index into the matrix or array. Additionally, the matrix type allows two argument versions of the `()` operator and elem method, giving the row and column index of the value to obtain or set.

Note that these functions do significant error checking and so in some circumstances the user might prefer to access the data of the array or matrix directly through the fortran_vec method discussed below.

`octave_idx_type` *numel* (*void*) *const* [Method]
: The total number of elements in the matrix or array.

`size_t byte_size` (*void*) *const* [Method]
: The number of bytes used to store the matrix or array.

`dim_vector dims` (*void*) *const* [Method]
: The dimensions of the matrix or array in value of type dim_vector.

`int ndims` (*void*) *const* [Method]
: The number of dimensions of the matrix or array. Matrices are 2-D, but arrays can be N-dimensional.

`void resize` (*const dim_vector&*) [Method]

A method taking either an argument of type `dim_vector`, or in the case of a matrix two arguments of type `octave_idx_type` defining the number of rows and columns in the matrix.

`T* fortran_vec` (*void*) [Method]

This method returns a pointer to the underlying data of the matrix or array so that it can be manipulated directly, either within Octave or by an external library.

Operators such an +, -, or * can be used on the majority of the matrix and array types. In addition there are a number of methods that are of interest only for matrices such as `transpose`, `hermitian`, `solve`, etc.

The typical way to extract a matrix or array from the input arguments of `DEFUN_DLD` function is as follows

```
#include <octave/oct.h>

DEFUN_DLD (addtwomatrices, args, , "Add A to B")
{
  int nargin = args.length ();

  if (nargin != 2)
    print_usage ();
  else
    {
      NDArray A = args(0).array_value ();
      NDArray B = args(1).array_value ();
      if (! error_state)
        return octave_value (A + B);
    }

  return octave_value_list ();
}
```

To avoid segmentation faults causing Octave to abort this function explicitly checks that there are sufficient arguments available before accessing these arguments. It then obtains two multi-dimensional arrays of type `NDArray` and adds these together. Note that the array_value method is called without using the `is_matrix_type` type, and instead the error_state is checked before returning `A + B`. The reason to prefer this is that the arguments might be a type that is not an `NDArray`, but it would make sense to convert it to one. The `array_value` method allows this conversion to be performed transparently if possible, and sets `error_state` if it is not.

`A + B`, operating on two `NDArray`'s returns an `NDArray`, which is cast to an `octave_value` on the return from the function. An example of the use of this demonstration function is

```
addtwomatrices (ones (2, 2), eye (2, 2))
      ⇒  2  1
          1  2
```

A list of the basic `Matrix` and `Array` types, the methods to extract these from an `octave_value`, and the associated header file is listed below.

Type	**Function**	**Source Code**
RowVector	row_vector_value	'dRowVector.h'
ComplexRowVector	complex_row_vector_value	'CRowVector.h'
ColumnVector	column_vector_value	'dColVector.h'
ComplexColumnVector	complex_column_vector_value	'CColVector.h'
Matrix	matrix_value	'dMatrix.h'
ComplexMatrix	complex_matrix_value	'CMatrix.h'
boolMatrix	bool_matrix_value	'boolMatrix.h'
charMatrix	char_matrix_value	'chMatrix.h'
NDArray	array_value	'dNDArray.h'
ComplexNDArray	complex_array_value	'CNDArray.h'
boolNDArray	bool_array_value	'boolNDArray.h'
charNDArray	char_array_value	'charNDArray.h'
int8NDArray	int8_array_value	'int8NDArray.h'
int16NDArray	int16_array_value	'int16NDArray.h'
int32NDArray	int32_array_value	'int32NDArray.h'
int64NDArray	int64_array_value	'int64NDArray.h'
uint8NDArray	uint8_array_value	'uint8NDArray.h'
uint16NDArray	uint16_array_value	'uint16NDArray.h'
uint32NDArray	uint32_array_value	'uint32NDArray.h'
uint64NDArray	uint64_array_value	'uint64NDArray.h'

A.1.3 Character Strings in Oct-Files

A character string in Octave is just a special `Array` class. Consider the example:

```
#include <octave/oct.h>

DEFUN_DLD (stringdemo, args, , "String Demo")
{
  octave_value_list retval;
  int nargin = args.length ();

  if (nargin != 1)
    print_usage ();
  else
    {
      charMatrix ch = args(0).char_matrix_value ();

      if (! error_state)
        {
          retval(1) = octave_value (ch, '\'');  // Single Quote String

          octave_idx_type nr = ch.rows ();
          for (octave_idx_type i = 0; i < nr / 2; i++)
```

```
            {
              std::string tmp = ch.row_as_string (i);
              ch.insert (ch.row_as_string (nr-i-1).c_str (), i, 0);
              ch.insert (tmp.c_str (), nr-i-1, 0);
            }
          retval(0) = octave_value (ch, '"');  // Double Quote String
        }
    }
  return retval;
}
```

An example of the use of this function is

```
s0 = ["First String"; "Second String"];
[s1,s2] = stringdemo (s0)
⇒ s1 = Second String
        First String

⇒ s2 = First String
        Second String

typeinfo (s2)
⇒ sq_string
typeinfo (s1)
⇒ string
```

One additional complication of strings in Octave is the difference between single quoted and double quoted strings. To find out if an `octave_value` contains a single or double quoted string use one of the predicate tests shown below.

```
if (args(0).is_sq_string ())
  octave_stdout << "First argument is a single quoted string\n";
else if (args(0).is_dq_string ())
  octave_stdout << "First argument is a double quoted string\n";
```

Note, however, that both types of strings are represented by the `charNDArray` type, and so when assigning to an `octave_value`, the type of string should be specified. For example:

```
octave_value_list retval;
charNDArray ch;
...
// Create single quoted string
retval(1) = octave_value (ch);        // default constructor is sq_string
           OR
retval(1) = octave_value (ch, '\'');  // explicitly create sq_string

// Create a double quoted string
retval(0) = octave_value (ch, '"');
```

A.1.4 Cell Arrays in Oct-Files

Octave's cell type is also available from within oct-files. A cell array is just an array of octave_values, and thus each element of the cell array can be treated just like any other octave_value. A simple example is

```
#include <octave/oct.h>
#include <octave/Cell.h>

DEFUN_DLD (celldemo, args, , "Cell Demo")
{
  octave_value_list retval;
  int nargin = args.length ();

  if (nargin != 1)
    print_usage ();
  else
    {
      Cell c = args(0).cell_value ();
      if (! error_state)
        for (octave_idx_type i = 0; i < c.numel (); i++)
          {
            retval(i) = c(i);          // using operator syntax
            //retval(i) = c.elem (i);  // using method syntax
          }
    }

  return retval;
}
```

Note that cell arrays are used less often in standard oct-files and so the 'Cell.h' header file must be explicitly included. The rest of the example extracts the octave_values one by one from the cell array and returns them as individual return arguments. For example:

```
[b1, b2, b3] = celldemo ({1, [1, 2], "test"})
⇒
b1 =  1
b2 =

   1   2

b3 = test
```

A.1.5 Structures in Oct-Files

A structure in Octave is a map between a number of fields represented and their values. The Standard Template Library map class is used, with the pair consisting of a std::string and an Octave Cell variable.

A simple example demonstrating the use of structures within oct-files is

```
#include <octave/oct.h>
#include <octave/ov-struct.h>

DEFUN_DLD (structdemo, args, , "Struct Demo")
{
  octave_value retval;
  int nargin = args.length ();

  if (args.length () == 2)
    {
      octave_scalar_map arg0 = args(0).scalar_map_value ();
      //octave_map arg0 = args(0).map_value ();

      if (! error_state)
        {
          std::string arg1 = args(1).string_value ();

          if (! error_state)
            {
              octave_value tmp = arg0.contents (arg1);
              //octave_value tmp = arg0.contents (arg1)(0);

              if (tmp.is_defined ())
                {
                  octave_scalar_map st;

                  st.assign ("selected", tmp);

                  retval = octave_value (st);
                }
              else
                error ("structdemo: struct does not have a field named '%s'\n",
                       arg1.c_str ());
            }
          else
            error ("structdemo: ARG2 must be a character string");
        }
      else
        error ("structdemo: ARG1 must be a struct");
    }
  else
    print_usage ();

  return retval;
}
```

An example of its use is

```
x.a = 1; x.b = "test"; x.c = [1, 2];
structdemo (x, "b")
⇒ selected = test
```

The example above specifically uses the `octave_scalar_map` class which is for representing a single struct. For structure arrays the `octave_map` class is used instead. The commented code shows how the demo could be modified to handle a structure array. In that case the `contents` method returns a `Cell` which may have more than one element. Therefore, to obtain the underlying `octave_value` in this single-struct example we write

```
octave_value tmp = arg0.contents (arg1)(0);
```

where the trailing (0) is the () operator on the `Cell` object. If this were a true structure array with multiple elements we could iterate over the elements using the () operator.

Structures are a relatively complex data container and there are more functions available in '`oct-map.h`' which make coding with them easier than relying on just `contents`.

A.1.6 Sparse Matrices in Oct-Files

There are three classes of sparse objects that are of interest to the user.

`SparseMatrix`
: A double precision sparse matrix class

`SparseComplexMatrix`
: A complex sparse matrix class

`SparseBoolMatrix`
: A boolean sparse matrix class

All of these classes inherit from the `Sparse<T>` template class, and so all have similar capabilities and usage. The `Sparse<T>` class was based on Octave's `Array<T>` class, and so users familiar with Octave's `Array` classes will be comfortable with the use of the sparse classes.

The sparse classes will not be entirely described in this section, due to their similarity with the existing `Array` classes. However, there are a few differences due the different nature of sparse objects, and these will be described. First, although it is fundamentally possible to have N-dimensional sparse objects, the Octave sparse classes do not allow them at this time; All instances of the sparse classes must be 2-dimensional. This means that `SparseMatrix` is actually more similar to Octave's `Matrix` class than its `NDArray` class.

A.1.6.1 Array and Sparse Class Differences

The number of elements in a sparse matrix is considered to be the number of non-zero elements rather than the product of the dimensions. Therefore

```
SparseMatrix sm;
...
int nel = sm.nelem ();
```

returns the number of non-zero elements. If the user really requires the number of elements in the matrix, including the non-zero elements, they should use `numel` rather than `nelem`. Note that for very large matrices, where the product of the two dimensions is larger than the representation of an unsigned int, then `numel` can overflow. An example is `speye (1e6)` which will create a matrix with a million rows and columns, but only a million non-zero elements. Therefore the number of rows by the number of columns in this case is more than two hundred times the maximum value that can be represented by an unsigned int. The use of `numel` should therefore be avoided useless it is known it won't overflow.

Extreme care must be take with the elem method and the "()" operator, which perform basically the same function. The reason is that if a sparse object is non-const, then Octave will assume that a request for a zero element in a sparse matrix is in fact a request to create this element so it can be filled. Therefore a piece of code like

```
SparseMatrix sm;
...
for (int j = 0; j < nc; j++)
  for (int i = 0; i < nr; i++)
    std::cerr << " (" << i << "," << j << "): " << sm(i,j) << std::endl;
```

is a great way of turning the sparse matrix into a dense one, and a very slow way at that since it reallocates the sparse object at each zero element in the matrix.

An easy way of preventing the above from happening is to create a temporary constant version of the sparse matrix. Note that only the container for the sparse matrix will be copied, while the actual representation of the data will be shared between the two versions of the sparse matrix. So this is not a costly operation. For example, the above would become

```
SparseMatrix sm;
...
const SparseMatrix tmp (sm);
for (int j = 0; j < nc; j++)
  for (int i = 0; i < nr; i++)
    std::cerr << " (" << i << "," << j << "): " << tmp(i,j) << std::endl;
```

Finally, as the sparse types aren't represented by a contiguous block of memory, the `fortran_vec` method of the `Array<T>` is not available. It is, however, replaced by three separate methods `ridx`, `cidx` and `data`, that access the raw compressed column format that Octave sparse matrices are stored in. These methods can be used in a manner similar to `elem` to allow the matrix to be accessed or filled. However, in that case it is up to the user to respect the sparse matrix compressed column format.

A.1.6.2 Creating Sparse Matrices in Oct-Files

There are several useful alternatives for creating a sparse matrix. The first is to create three vectors representing the row index, column index, and data values, and from these create the matrix. The second alternative is to create a sparse matrix with the appropriate amount of space and then fill in the values. Both techniques have their advantages and disadvantages.

Below is an example of creating a small sparse matrix using the first technique

```
int nz, nr, nc;
nz = 4, nr = 3, nc = 4;

ColumnVector ridx (nz);
ColumnVector cidx (nz);
ColumnVector data (nz);

ridx(0) = 1; cidx(0) = 1; data(0) = 1;
ridx(1) = 2; cidx(1) = 2; data(1) = 2;
ridx(2) = 2; cidx(2) = 4; data(2) = 3;
ridx(3) = 3; cidx(3) = 4; data(3) = 4;
SparseMatrix sm (data, ridx, cidx, nr, nc);
```

which creates the matrix given in section Section 22.1.1 [Storage of Sparse Matrices], page 487. Note that the compressed matrix format is not used at the time of the creation of the matrix itself, but is used internally.

As discussed in the chapter on Sparse Matrices, the values of the sparse matrix are stored in increasing column-major ordering. Although the data passed by the user need not respect this requirement, pre-sorting the data will significantly speed up creation of the sparse matrix.

The disadvantage of this technique for creating a sparse matrix is that there is a brief time when two copies of the data exist. For extremely memory constrained problems this may not be the best technique for creating a sparse matrix.

The alternative is to first create a sparse matrix with the desired number of non-zero elements and then later fill those elements in. Sample code:

```
int nz, nr, nc;
nz = 4, nr = 3, nc = 4;
SparseMatrix sm (nr, nc, nz);
sm(0,0) = 1; sm(0,1) = 2; sm(1,3) = 3; sm(2,3) = 4;
```

This creates the same matrix as previously. Again, although not strictly necessary, it is significantly faster if the sparse matrix is created and the elements are added in column-major ordering. The reason for this is that when elements are inserted at the end of the current list of known elements then no element in the matrix needs to be moved to allow the new element to be inserted; Only the column indexes need to be updated.

There are a few further points to note about this method of creating a sparse matrix. First, it is possible to create a sparse matrix with fewer elements than are actually inserted in the matrix. Therefore,

```
int nr, nc;
nr = 3, nc = 4;
SparseMatrix sm (nr, nc, 0);
sm(0,0) = 1; sm(0,1) = 2; sm(1,3) = 3; sm(2,3) = 4;
```

is perfectly valid. However, it is a very bad idea because as each new element is added to the sparse matrix the matrix needs to request more space and reallocate memory. This is an expensive operation, that will significantly slow this means of creating a sparse matrix. Furthermore, it is possible to create a sparse matrix with too much storage, so having *nz* greater than 4 is also valid. The disadvantage is that the matrix occupies more memory than strictly needed.

It is not always possible to know the number of non-zero elements prior to filling a matrix. For this reason the additional unused storage of a sparse matrix can be removed after its creation with the `maybe_compress` function. In addition, `maybe_compress` can deallocate the unused storage, but it can also remove zero elements from the matrix. The removal of zero elements from the matrix is controlled by setting the argument of the `maybe_compress` function to be `true`. However, the cost of removing the zeros is high because it implies re-sorting the elements. If possible, it is better if the user does not add the unnecessary zeros in the first place. An example of the use of `maybe_compress` is

```
int nz, nr, nc;
nz = 6, nr = 3, nc = 4;

SparseMatrix sm1 (nr, nc, nz);
sm1(0,0) = 1; sm1(0,1) = 2; sm1(1,3) = 3; sm1(2,3) = 4;
sm1.maybe_compress ();  // No zero elements were added

SparseMatrix sm2 (nr, nc, nz);
sm2(0,0) = 1; sm2(0,1) = 2; sm(0,2) = 0; sm(1,2) = 0;
sm1(1,3) = 3; sm1(2,3) = 4;
sm2.maybe_compress (true);  // Zero elements were added
```

The use of the `maybe_compress` function should be avoided if possible as it will slow the creation of the matrix.

A third means of creating a sparse matrix is to work directly with the data in compressed row format. An example of this technique might be

```
octave_value arg;
...
int nz, nr, nc;
nz = 6, nr = 3, nc = 4;    // Assume we know the max # nz
SparseMatrix sm (nr, nc, nz);
Matrix m = arg.matrix_value ();

int ii = 0;
sm.cidx (0) = 0;
for (int j = 1; j < nc; j++)
  {
    for (int i = 0; i < nr; i++)
      {
        double tmp = foo (m(i,j));
        if (tmp != 0.)
          {
            sm.data(ii) = tmp;
            sm.ridx(ii) = i;
            ii++;
          }
      }
```

```
        sm.cidx(j+1) = ii;
     }
   sm.maybe_compress ();  // If don't know a priori the final # of nz.
```

which is probably the most efficient means of creating a sparse matrix.

Finally, it might sometimes arise that the amount of storage initially created is insufficient to completely store the sparse matrix. Therefore, the method `change_capacity` exists to reallocate the sparse memory. The above example would then be modified as

```
octave_value arg;
...
int nz, nr, nc;
nz = 6, nr = 3, nc = 4;    // Assume we know the max # nz
SparseMatrix sm (nr, nc, nz);
Matrix m = arg.matrix_value ();

int ii = 0;
sm.cidx (0) = 0;
for (int j = 1; j < nc; j++)
  {
    for (int i = 0; i < nr; i++)
      {
        double tmp = foo (m(i,j));
        if (tmp != 0.)
          {
            if (ii == nz)
              {
                nz += 2;   // Add 2 more elements
                sm.change_capacity (nz);
              }
            sm.data(ii) = tmp;
            sm.ridx(ii) = i;
            ii++;
          }
      }
    sm.cidx(j+1) = ii;
  }
sm.maybe_mutate ();  // If don't know a priori the final # of nz.
```

Note that both increasing and decreasing the number of non-zero elements in a sparse matrix is expensive as it involves memory reallocation. Also as parts of the matrix, though not its entirety, exist as old and new copies at the same time, additional memory is needed. Therefore, if possible this should be avoided.

A.1.6.3 Using Sparse Matrices in Oct-Files

Most of the same operators and functions on sparse matrices that are available from the Octave command line are also available within oct-files. The basic means of extracting a sparse matrix from an `octave_value` and returning it as an `octave_value`, can be seen in the following example.

```
octave_value_list retval;

SparseMatrix sm = args(0).sparse_matrix_value ();
SparseComplexMatrix scm =
    args(1).sparse_complex_matrix_value ();
SparseBoolMatrix sbm = args(2).sparse_bool_matrix_value ();
...
retval(2) = sbm;
retval(1) = scm;
retval(0) = sm;
```

The conversion to an `octave_value` is handled by the sparse `octave_value` constructors, and so no special care is needed.

A.1.7 Accessing Global Variables in Oct-Files

Global variables allow variables in the global scope to be accessed. Global variables can be accessed within oct-files by using the support functions `get_global_value` and `set_global_value`. `get_global_value` takes two arguments, the first is a string representing the variable name to obtain. The second argument is a boolean argument specifying what to do if no global variable of the desired name is found. An example of the use of these two functions is

```
#include <octave/oct.h>

DEFUN_DLD (globaldemo, args, , "Global Demo")
{
  octave_value retval;
  int nargin = args.length ();

  if (nargin != 1)
    print_usage ();
  else
    {
      std::string s = args(0).string_value ();
      if (! error_state)
        {
          octave_value tmp = get_global_value (s, true);
          if (tmp.is_defined ())
            retval = tmp;
          else
            retval = "Global variable not found";

          set_global_value ("a", 42.0);
        }
    }
  return retval;
}
```

An example of its use is

```
global a b
b = 10;
globaldemo ("b")
⇒ 10
globaldemo ("c")
⇒ "Global variable not found"
num2str (a)
⇒ 42
```

A.1.8 Calling Octave Functions from Oct-Files

There is often a need to be able to call another Octave function from within an oct-file, and there are many examples of such within Octave itself. For example, the `quad` function is an oct-file that calculates the definite integral by quadrature over a user supplied function.

There are also many ways in which a function might be passed. It might be passed as one of

1. Function Handle
2. Anonymous Function Handle
3. Inline Function
4. String

The example below demonstrates an example that accepts all four means of passing a function to an oct-file.

```
#include <octave/oct.h>
#include <octave/parse.h>

DEFUN_DLD (funcdemo, args, nargout, "Function Demo")
{
  octave_value_list retval;
  int nargin = args.length ();

  if (nargin < 2)
    print_usage ();
  else
    {
      octave_value_list newargs;
      for (octave_idx_type i = nargin - 1; i > 0; i--)
        newargs(i-1) = args(i);
      if (args(0).is_function_handle () || args(0).is_inline_function ())
        {
          octave_function *fcn = args(0).function_value ();
          if (! error_state)
            retval = feval (fcn, newargs, nargout);
        }
      else if (args(0).is_string ())
        {
```

```
            std::string fcn = args(0).string_value ();
            if (! error_state)
              retval = feval (fcn, newargs, nargout);
          }
        else
          error ("funcdemo: INPUT must be string, inline, or function handle");
      }
    return retval;
  }
```

The first argument to this demonstration is the user-supplied function and the remaining arguments are all passed to the user function.

```
funcdemo (@sin,1)
⇒ 0.84147
funcdemo (@(x) sin (x), 1)
⇒ 0.84147
funcdemo (inline ("sin (x)"), 1)
⇒ 0.84147
funcdemo ("sin",1)
⇒ 0.84147
funcdemo (@atan2, 1, 1)
⇒ 0.78540
```

When the user function is passed as a string the treatment of the function is different. In some cases it is necessary to have the user supplied function as an `octave_function` object. In that case the string argument can be used to create a temporary function as demonstrated below.

```
std::octave fcn_name = unique_symbol_name ("__fcn__");
std::string fcode = "function y = ";
fcode.append (fcn_name);
fcode.append ("(x) y = ");
fcn = extract_function (args(0), "funcdemo", fcn_name,
                        fcode, "; endfunction");
...
if (fcn_name.length ())
  clear_function (fcn_name);
```

There are two important things to know in this case. First, the number of input arguments to the user function is fixed, and in the above example is a single argument. Second, to avoid leaving the temporary function in the Octave symbol table it should be cleared after use. Also, by convention internal function names begin and end with the character sequence '__'.

A.1.9 Calling External Code from Oct-Files

Linking external C code to Octave is relatively simple, as the C functions can easily be called directly from C++. One possible issue is that the declarations of the external C functions may need to be explicitly defined as C functions to the compiler. If the declarations of the external C functions are in the header 'foo.h', then the tactic to ensure that the C++ compiler treats these declarations as C code is

```
#ifdef __cplusplus
extern "C"
{
#endif
#include "foo.h"
#ifdef __cplusplus
} /* end extern "C" */
#endif
```

Calling Fortran code, however, can pose more difficulties. This is due to differences in the manner in which compilers treat the linking of Fortran code with C or C++ code. Octave supplies a number of macros that allow consistent behavior across a number of compilers.

The underlying Fortran code should use the XSTOPX function to replace the Fortran STOP function. XSTOPX uses the Octave exception handler to treat failing cases in the Fortran code explicitly. Note that Octave supplies its own replacement BLAS XERBLA function, which uses XSTOPX.

If the code calls XSTOPX, then the F77_XFCN macro should be used to call the underlying Fortran function. The Fortran exception state can then be checked with the global variable f77_exception_encountered. If XSTOPX will not be called, then the F77_FCN macro should be used instead to call the Fortran code.

There is no great harm in using F77_XFCN in all cases, except that for Fortran code that is short running and executes a large number of times, there is potentially an overhead in doing so. However, if F77_FCN is used with code that calls XSTOP, Octave can generate a segmentation fault.

An example of the inclusion of a Fortran function in an oct-file is given in the following example, where the C++ wrapper is

```
#include <octave/oct.h>
#include <octave/f77-fcn.h>

extern "C"
{
  F77_RET_T
  F77_FUNC (fortransub, FORTSUB)
    (const int&, double*, F77_CHAR_ARG_DECL F77_CHAR_ARG_LEN_DECL);
}

DEFUN_DLD (fortrandemo, args, , "Fortran Demo")
{
  octave_value_list retval;
  int nargin = args.length ();

  if (nargin != 1)
    print_usage ();
  else
```

```
      {
        NDArray a = args(0).array_value ();
        if (! error_state)
          {
            double *av = a.fortran_vec ();
            octave_idx_type na = a.numel ();
            OCTAVE_LOCAL_BUFFER (char, ctmp, 128);

            F77_XFCN (fortransub, FORTSUB,
                      (na, av, ctmp F77_CHAR_ARG_LEN (128)));

            retval(1) = std::string (ctmp);
            retval(0) = a;
          }
      }
    return retval;
  }
```

and the Fortran function is

```
      subroutine fortransub (n, a, s)
      implicit none
      character*(*) s
      real*8 a(*)
      integer*4 i, n, ioerr
      do i = 1, n
        if (a(i) .eq. 0d0) then
          call xstopx ('fortransub: divide by zero')
        else
          a(i) = 1d0 / a(i)
        endif
      enddo
      write (unit = s, fmt = '(a,i3,a,a)', iostat = ioerr)
     $       'There are ', n,
     $       ' values in the input vector', char(0)
      if (ioerr .ne. 0) then
        call xstopx ('fortransub: error writing string')
      endif
      return
      end
```

This example demonstrates most of the features needed to link to an external Fortran function, including passing arrays and strings, as well as exception handling. Both the Fortran and C++ files need to be compiled in order for the example to work.

```
mkoctfile fortrandemo.cc fortransub.f
[b, s] = fortrandemo (1:3)
⇒
  b = 1.00000   0.50000   0.33333
  s = There are   3 values in the input vector
[b, s] = fortrandemo (0:3)
error: fortrandemo: fortransub: divide by zero
```

A.1.10 Allocating Local Memory in Oct-Files

Allocating memory within an oct-file might seem easy as the C++ new/delete operators can be used. However, in that case great care must be taken to avoid memory leaks. The preferred manner in which to allocate memory for use locally is to use the `OCTAVE_LOCAL_BUFFER` macro. An example of its use is

```
OCTAVE_LOCAL_BUFFER (double, tmp, len)
```

that returns a pointer `tmp` of type `double *` of length `len`.

In this case Octave itself will worry about reference counting and variable scope and will properly free memory without programmer intervention.

A.1.11 Input Parameter Checking in Oct-Files

As oct-files are compiled functions they open up the possibility of crashing Octave through careless function calls or memory faults. It is quite important that each and every function have a sufficient level of parameter checking to ensure that Octave behaves well.

The minimum requirement, as previously discussed, is to check the number of input arguments before using them to avoid referencing a non-existent argument. However, in some cases this might not be sufficient as the underlying code imposes further constraints. For example, an external function call might be undefined if the input arguments are not integers, or if one of the arguments is zero, or if the input is complex and a real value was expected. Therefore, oct-files often need additional input parameter checking.

There are several functions within Octave that can be useful for the purposes of parameter checking. These include the methods of the octave_value class like `is_real_matrix`, `is_numeric_type`, etc. Often, with a knowledge of the Octave m-file language, you can guess at what the corresponding C++ routine will. In addition there are some more specialized input validation functions of which a few are demonstrated below.

```
#include <octave/oct.h>

DEFUN_DLD (paramdemo, args, nargout, "Parameter Check Demo")
{
  octave_value retval;
  int nargin = args.length ();

  if (nargin != 1)
    print_usage ();
  else if (nargout != 0)
    error ("paramdemo: OUTPUT argument required");
  else
```

```
    {
      NDArray m = args(0).array_value ();
      double min_val = -10.0;
      double max_val = 10.0;
      octave_stdout << "Properties of input array:\n";
      if (m.any_element_is_negative ())
        octave_stdout << "  includes negative values\n";
      if (m.any_element_is_inf_or_nan ())
        octave_stdout << "  includes Inf or NaN values\n";
      if (m.any_element_not_one_or_zero ())
        octave_stdout << "  includes other values than 1 and 0\n";
      if (m.all_elements_are_int_or_inf_or_nan ())
        octave_stdout << "  includes only int, Inf or NaN values\n";
      if (m.all_integers (min_val, max_val))
        octave_stdout << "  includes only integers in [-10,10]\n";
    }
  return retval;
}
```

An example of its use is:

```
paramdemo ([1, 2, NaN, Inf])
⇒ Properties of input array:
     includes Inf or NaN values
     includes other values than 1 and 0
     includes only int, Inf or NaN values
```

A.1.12 Exception and Error Handling in Oct-Files

Another important feature of Octave is its ability to react to the user typing CONTROL-C even during calculations. This ability is based on the C++ exception handler, where memory allocated by the C++ new/delete methods are automatically released when the exception is treated. When writing an oct-file, to allow Octave to treat the user typing CONTROL-C, the `OCTAVE_QUIT` macro is supplied. For example:

```
for (octave_idx_type i = 0; i < a.nelem (); i++)
  {
    OCTAVE_QUIT;
    b.elem (i) = 2. * a.elem (i);
  }
```

The presence of the `OCTAVE_QUIT` macro in the inner loop allows Octave to treat the user request with the CONTROL-C. Without this macro, the user must either wait for the function to return before the interrupt is processed, or press CONTROL-C three times to force Octave to exit.

The `OCTAVE_QUIT` macro does impose a very small speed penalty, and so for loops that are known to be small it might not make sense to include `OCTAVE_QUIT`.

When creating an oct-file that uses an external libraries, the function might spend a significant portion of its time in the external library. It is not generally possible to use the `OCTAVE_QUIT` macro in this case. The alternative in this case is

```
BEGIN_INTERRUPT_IMMEDIATELY_IN_FOREIGN_CODE;
...  some code that calls a "foreign" function ...
END_INTERRUPT_IMMEDIATELY_IN_FOREIGN_CODE;
```

The disadvantage of this is that if the foreign code allocates any memory internally, then this memory might be lost during an interrupt, without being deallocated. Therefore, ideally Octave itself should allocate any memory that is needed by the foreign code, with either the fortran_vec method or the `OCTAVE_LOCAL_BUFFER` macro.

The Octave unwind_protect mechanism (Section 10.8 [The unwind_protect Statement], page 164) can also be used in oct-files. In conjunction with the exception handling of Octave, it is important to enforce that certain code is run to allow variables, etc. to be restored even if an exception occurs. An example of the use of this mechanism is

```
#include <octave/oct.h>
#include <octave/unwind-prot.h>

void
my_err_handler (const char *fmt, ...)
{
  // Do nothing!!
}

DEFUN_DLD (unwinddemo, args, nargout, "Unwind Demo")
{
  octave_value retval;
  int nargin = args.length ();

  if (nargin < 2)
    print_usage ();
  else
    {
      NDArray a = args(0).array_value ();
      NDArray b = args(1).array_value ();

      if (! error_state)
        {
          // Declare unwind_protect frame which lasts as long as
          // the variable frame has scope.
          unwind_protect frame;
          frame.protect_var (current_liboctave_warning_handler);

          set_liboctave_warning_handler (my_err_handler);
          retval = octave_value (quotient (a, b));
        }
    }
  return retval;
}
```

As can be seen in the example:

```
unwinddemo (1, 0)
⇒ Inf
1 / 0
⇒ warning: division by zero
   Inf
```

The warning for division by zero (and in fact all warnings) are disabled in the `unwinddemo` function.

A.1.13 Documentation and Test of Oct-Files

The documentation of an oct-file is the fourth string parameter of the `DEFUN_DLD` macro. This string can be formatted in the same manner as the help strings for user functions (see Section C.4 [Documentation Tips], page 811), however there are some issue that are particular to the formatting of help strings within oct-files.

The major issue is that the help string will typically be longer than a single line of text, and so the formatting of long help strings needs to be taken into account. There are several possible solutions, but the most common is illustrated in the following example,

```
DEFUN_DLD (do_what_i_want, args, nargout,
  "-*- texinfo -*-\n\
@deftypefn {Function File} {} do_what_i_say (@var{n})\n\
A function that does what the user actually wants rather\n\
than what they requested.\n\
@end deftypefn")
{
...
}
```

where, as can be seen, each line of text is terminated by `\n\` which is an embedded new-line in the string together with a C++ string continuation character. Note that the final `\` must be the last character on the line.

Octave also includes the ability to embed test and demonstration code for a function within the code itself (see Appendix B [Test and Demo Functions], page 799). This can be used from within oct-files (or in fact any file) with certain provisos. First, the test and demo functions of Octave look for `%!` as the first two characters of a line to identify test and demonstration code. This is a requirement for oct-files as well. In addition, the test and demonstration code must be wrapped in a comment block to avoid it being interpreted by the compiler. Finally, the Octave test and demonstration code must have access to the original source code of the oct-file and not just the compiled code as the tests are stripped from the compiled code. An example in an oct-file might be

```
/*
%!assert (sin ([1,2]), [sin(1),sin(2)])
%!error (sin ())
%!error (sin (1,1))
*/
```

A.2 Mex-Files

Octave includes an interface to allow legacy mex-files to be compiled and used with Octave. This interface can also be used to share code between Octave and MATLAB users. However, as mex-files expose MATLAB's internal API, and the internal structure of Octave is different, a mex-file can never have the same performance in Octave as the equivalent oct-file. In particular, to support the manner in which variables are passed to mex functions there are a significant number of additional copies of memory blocks when calling or returning from a mex-file function. For this reason, it is recommended that any new code be written with the oct-file interface previously discussed.

A.2.1 Getting Started with Mex-Files

The basic command to build a mex-file is either `mkoctfile --mex` or `mex`. The first command can be used either from within Octave or from the command line. However, to avoid issues with MATLAB's own `mex` command, the use of the command `mex` is limited to within Octave. Compiled mex-files have the extension '`.mex`'.

`mex` [*options*] *file* ... [Command]

Compile source code written in C, C++, or Fortran, to a MEX file. This is equivalent to `mkoctfile --mex [options] file`.

See also: [mkoctfile], page 760.

`mexext` () [Function File]

Return the filename extension used for MEX files.

See also: [mex], page 784.

Consider the following short example:

```
#include "mex.h"

void
mexFunction (int nlhs, mxArray *plhs[],
             int nrhs, const mxArray *prhs[])
{
  mexPrintf ("Hello, World!\n");

  mexPrintf ("I have %d inputs and %d outputs\n", nrhs, nlhs);
}
```

The first line `#include "mex.h"` makes available all of the definitions necessary for a mex-file. One important difference between Octave and MATLAB is that the header file `"matrix.h"` is implicitly included through the inclusion of `"mex.h"`. This is necessary to avoid a conflict with the Octave file `"Matrix.h"` for operating systems and compilers that don't distinguish between filenames in upper and lower case.

The entry point into the mex-file is defined by `mexFunction`. The function takes four arguments:

1. The number of return arguments (# of left-hand side args).

2. An array of pointers to return arguments.
3. The number of input arguments (# of right-hand side args).
4. An array of pointers to input arguments.

Note that the function name definition is not explicitly included in `mexFunction` and so there can only be a single `mexFunction` entry point per file. Instead, the name of the function as seen in Octave is determined by the name of the mex-file itself minus the extension. Therefore, if the above function is in the file 'myhello.c', it can be compiled with

```
mkoctfile --mex myhello.c
```

which creates a file 'myhello.mex'. The function can then be run from Octave as

```
myhello (1,2,3)
⇒ Hello, World!
⇒ I have 3 inputs and 0 outputs
```

It should be noted that the mex-file contains no help string for the functions it contains. To document mex-files, there should exist an m-file in the same directory as the mex-file itself. Taking the above as an example, we would therefore have a file 'myhello.m' that might contain the text

```
%MYHELLO Simple test of the functionality of a mex-file.
```

In this case, the function that will be executed within Octave will be given by the mex-file, while the help string will come from the m-file. This can also be useful to allow a sample implementation of the mex-file within the Octave language itself for testing purposes.

Although there cannot be multiple entry points in a single mex-file, one can use the `mexFunctionName` function to determine what name the mex-file was called with. This can be used to alter the behavior of the mex-file based on the function name. For example, if

```
#include "mex.h"

void
mexFunction (int nlhs, mxArray *plhs[],
             int nrhs, const mxArray *prhs[])
{
  const char *nm;

  nm = mexFunctionName ();
  mexPrintf ("You called function: %s\n", nm);
  if (strcmp (nm, "myfunc") == 0)
    mexPrintf ("This is the principal function\n", nm);

  return;
}
```

is in file 'myfunc.c', and it is compiled with

```
mkoctfile --mex myfunc.c
ln -s myfunc.mex myfunc2.mex
```

then as can be seen by

```
myfunc ()
⇒ You called function: myfunc
    This is the principal function
myfunc2 ()
⇒ You called function: myfunc2
```

the behavior of the mex-file can be altered depending on the functions name.

Although the user should only include 'mex.h' in their code, Octave declares additional functions, typedefs, etc., available to the user to write mex-files in the headers 'mexproto.h' and 'mxarray.h'.

A.2.2 Working with Matrices and Arrays in Mex-Files

The basic mex type of all variables is mxArray. Any object, such as a matrix, cell array, or structure is stored in this basic type. As such, mxArray serves basically the same purpose as the octave_value class in oct-files in that it acts as a container for the more specialized types.

The mxArray structure contains at a minimum, the name of the variable it represents, its dimensions, its type, and whether the variable is real or complex. It can also contain a number of additional fields depending on the type of the mxArray. There are a number of functions to create mxArray structures, including mxCreateDoubleMatrix, mxCreateCellArray, mxCreateSparse, and the generic mxCreateNumericArray.

The basic function to access the data contained in an array is mxGetPr. As the mex interface assumes that real and imaginary parts of a complex array are stored separately, there is an equivalent function mxGetPi that gets the imaginary part. Both of these functions are only for use with double precision matrices. The generic functions mxGetData and mxGetImagData perform the same operation on all matrix types. For example:

```
mxArray *m;
mwSize *dims;
UINT32_T *pr;

dims = (mwSize *) mxMalloc (2 * sizeof (mwSize));
dims[0] = 2; dims[1] = 2;
m = mxCreateNumericArray (2, dims, mxUINT32_CLASS, mxREAL);
pr = (UINT32_T *) mxGetData (m);
```

There are also the functions mxSetPr, etc., that perform the inverse, and set the data of an array to use the block of memory pointed to by the argument of mxSetPr.

Note the type mwSize used above, and also mwIndex, are defined as the native precision of the indexing in Octave on the platform on which the mex-file is built. This allows both 32- and 64-bit platforms to support mex-files. mwSize is used to define array dimensions and the maximum number or elements, while mwIndex is used to define indexing into arrays.

An example that demonstrates how to work with arbitrary real or complex double precision arrays is given by the file 'mypow2.c' shown below.

```
#include "mex.h"

void
mexFunction (int nlhs, mxArray* plhs[],
```

```
              int nrhs, const mxArray* prhs[])
{
  mwSize n;
  mwIndex i;
  double *vri, *vro;

  if (nrhs != 1 || ! mxIsNumeric (prhs[0]))
    mexErrMsgTxt ("ARG1 must be a matrix");

  n = mxGetNumberOfElements (prhs[0]);
  plhs[0] = mxCreateNumericArray (mxGetNumberOfDimensions (prhs[0]),
                                  mxGetDimensions (prhs[0]),
                                  mxGetClassID (prhs[0]),
                                  mxIsComplex (prhs[0]));
  vri = mxGetPr (prhs[0]);
  vro = mxGetPr (plhs[0]);

  if (mxIsComplex (prhs[0]))
    {
      double *vii, *vio;
      vii = mxGetPi (prhs[0]);
      vio = mxGetPi (plhs[0]);

      for (i = 0; i < n; i++)
        {
          vro[i] = vri[i] * vri[i] - vii[i] * vii[i];
          vio[i] = 2 * vri[i] * vii[i];
        }
    }
  else
    {
      for (i = 0; i < n; i++)
        vro[i] = vri[i] * vri[i];
    }
}
```

with an example of its use

```
b = randn (4,1) + 1i * randn (4,1);
all (b.^2 == mypow2 (b))
⇒ 1
```

The example above uses the functions `mxGetDimensions`, `mxGetNumberOfElements`, and `mxGetNumberOfDimensions` to work with the dimensions of multi-dimensional arrays. The functions `mxGetM`, and `mxGetN` are also available to find the number of rows and columns in a 2-D matrix.

A.2.3 Character Strings in Mex-Files

As mex-files do not make the distinction between single and double quoted strings within Octave, there is perhaps less complexity in the use of strings and character matrices in mex-files. An example of their use that parallels the demo in 'stringdemo.cc' is given in the file 'mystring.c', as shown below.

```
#include <string.h>
#include "mex.h"

void
mexFunction (int nlhs, mxArray *plhs[],
             int nrhs, const mxArray *prhs[])
{
  mwSize m, n;
  mwIndex i, j;
  mxChar *pi, *po;

  if (nrhs != 1 || ! mxIsChar (prhs[0])
      || mxGetNumberOfDimensions (prhs[0]) > 2)
    mexErrMsgTxt ("ARG1 must be a char matrix");

  m = mxGetM (prhs[0]);
  n = mxGetN (prhs[0]);
  pi = mxGetChars (prhs[0]);
  plhs[0] = mxCreateNumericMatrix (m, n, mxCHAR_CLASS, mxREAL);
  po = mxGetChars (plhs[0]);

  for (j = 0; j < n; j++)
    for (i = 0; i < m; i++)
      po[j*m + m - 1 - i] = pi[j*m + i];
}
```

An example of its expected output is

```
mystring (["First String"; "Second String"])
⇒ Second String
   First String
```

Other functions in the mex interface for handling character strings are `mxCreateString`, `mxArrayToString`, and `mxCreateCharMatrixFromStrings`. In a mex-file, a character string is considered to be a vector rather than a matrix. This is perhaps an arbitrary distinction as the data in the mxArray for the matrix is consecutive in any case.

A.2.4 Cell Arrays with Mex-Files

One can perform exactly the same operations on Cell arrays in mex-files as in oct-files. An example that reduplicates the function of the 'celldemo.cc' oct-file in a mex-file is given by 'mycell.c' as shown below.

```
#include "mex.h"
```

```
void
mexFunction (int nlhs, mxArray* plhs[],
             int nrhs, const mxArray* prhs[])
{
  mwSize n;
  mwIndex i;

  if (nrhs != 1 || ! mxIsCell (prhs[0]))
    mexErrMsgTxt ("ARG1 must be a cell");

  n = mxGetNumberOfElements (prhs[0]);
  n = (n > nlhs ? nlhs : n);

  for (i = 0; i < n; i++)
    plhs[i] = mxDuplicateArray (mxGetCell (prhs[0], i));
}
```

The output is identical to the oct-file version as well.

```
[b1, b2, b3] = mycell ({1, [1, 2], "test"})
⇒
b1 =  1
b2 =

   1   2

b3 = test
```

Note in the example the use of the `mxDuplicateArray` function. This is needed as the `mxArray` pointer returned by `mxGetCell` might be deallocated. The inverse function to `mxGetCell`, used for setting Cell values, is `mxSetCell` and is defined as

```
void mxSetCell (mxArray *ptr, int idx, mxArray *val);
```

Finally, to create a cell array or matrix, the appropriate functions are

```
mxArray *mxCreateCellArray (int ndims, const int *dims);
mxArray *mxCreateCellMatrix (int m, int n);
```

A.2.5 Structures with Mex-Files

The basic function to create a structure in a mex-file is `mxCreateStructMatrix` which creates a structure array with a two dimensional matrix, or `mxCreateStructArray`.

```
mxArray *mxCreateStructArray (int ndims, int *dims,
                              int num_keys,
                              const char **keys);
mxArray *mxCreateStructMatrix (int rows, int cols,
                               int num_keys,
                               const char **keys);
```

Accessing the fields of the structure can then be performed with **mxGetField** and **mxSetField** or alternatively with the **mxGetFieldByNumber** and **mxSetFieldByNumber** functions.

```
mxArray *mxGetField (const mxArray *ptr, mwIndex index,
                     const char *key);
mxArray *mxGetFieldByNumber (const mxArray *ptr,
                             mwIndex index, int key_num);
void mxSetField (mxArray *ptr, mwIndex index,
                 const char *key, mxArray *val);
void mxSetFieldByNumber (mxArray *ptr, mwIndex index,
                         int key_num, mxArray *val);
```

A difference between the oct-file interface to structures and the mex-file version is that the functions to operate on structures in mex-files directly include an `index` over the elements of the arrays of elements per `field`; Whereas, the oct-file structure includes a Cell Array per field of the structure.

An example that demonstrates the use of structures in a mex-file can be found in the file 'mystruct.c' shown below.

```
#include "mex.h"

void
mexFunction (int nlhs, mxArray* plhs[],
             int nrhs, const mxArray* prhs[])
{
  int i;
  mwIndex j;
  mxArray *v;
  const char *keys[] = { "this", "that" };

  if (nrhs != 1 || ! mxIsStruct (prhs[0]))
    mexErrMsgTxt ("expects struct");

  for (i = 0; i < mxGetNumberOfFields (prhs[0]); i++)
    for (j = 0; j < mxGetNumberOfElements (prhs[0]); j++)
      {
        mexPrintf ("field %s(%d) = ", mxGetFieldNameByNumber (prhs[0], i), j);
        v = mxGetFieldByNumber (prhs[0], j, i);
        mexCallMATLAB (0, NULL, 1, &v, "disp");
      }

  v = mxCreateStructMatrix (2, 2, 2, keys);

  mxSetFieldByNumber (v, 0, 0, mxCreateString ("this1"));
  mxSetFieldByNumber (v, 0, 1, mxCreateString ("that1"));
  mxSetFieldByNumber (v, 1, 0, mxCreateString ("this2"));
  mxSetFieldByNumber (v, 1, 1, mxCreateString ("that2"));
  mxSetFieldByNumber (v, 2, 0, mxCreateString ("this3"));
```

```
  mxSetFieldByNumber (v, 2, 1, mxCreateString ("that3"));
  mxSetFieldByNumber (v, 3, 0, mxCreateString ("this4"));
  mxSetFieldByNumber (v, 3, 1, mxCreateString ("that4"));

  if (nlhs)
    plhs[0] = v;
}
```

An example of the behavior of this function within Octave is then

```
a(1).f1 = "f11"; a(1).f2 = "f12";
a(2).f1 = "f21"; a(2).f2 = "f22";
b = mystruct (a);
⇒   field f1(0) = f11
    field f1(1) = f21
    field f2(0) = f12
    field f2(1) = f22
b
⇒ 2x2 struct array containing the fields:

     this
     that

b(3)
⇒ scalar structure containing the fields:

     this = this3
     that = that3
```

A.2.6 Sparse Matrices with Mex-Files

The Octave format for sparse matrices is identical to the mex format in that it is a compressed column sparse format. Also in both, sparse matrices are required to be two-dimensional. The only difference is that the real and imaginary parts of the matrix are stored separately.

The mex-file interface, in addition to using `mxGetM`, `mxGetN`, `mxSetM`, `mxSetN`, `mxGetPr`, `mxGetPi`, `mxSetPr`, and `mxSetPi`, also supplies the following functions.

```
mwIndex *mxGetIr (const mxArray *ptr);
mwIndex *mxGetJc (const mxArray *ptr);
mwSize mxGetNzmax (const mxArray *ptr);

void mxSetIr (mxArray *ptr, mwIndex *ir);
void mxSetJc (mxArray *ptr, mwIndex *jc);
void mxSetNzmax (mxArray *ptr, mwSize nzmax);
```

`mxGetNzmax` gets the maximum number of elements that can be stored in the sparse matrix. This is not necessarily the number of non-zero elements in the sparse matrix. `mxGetJc` returns an array with one additional value than the number of columns in the sparse matrix.

The difference between consecutive values of the array returned by **mxGetJc** define the number of non-zero elements in each column of the sparse matrix. Therefore,

```
mwSize nz, n;
mwIndex *Jc;
mxArray *m;
...
n = mxGetN (m);
Jc = mxGetJc (m);
nz = Jc[n];
```

returns the actual number of non-zero elements stored in the matrix in **nz**. As the arrays returned by **mxGetPr** and **mxGetPi** only contain the non-zero values of the matrix, we also need a pointer to the rows of the non-zero elements, and this is given by **mxGetIr**. A complete example of the use of sparse matrices in mex-files is given by the file '**mysparse.c**' shown below.

```
#include "mex.h"

void
mexFunction (int nlhs, mxArray *plhs[],
             int nrhs, const mxArray *prhs[])
{
  mwSize m, n, nz;
  mxArray *v;
  mwIndex i;
  double *pr, *pi;
  double *pr2, *pi2;
  mwIndex *ir, *jc;
  mwIndex *ir2, *jc2;

  if (nrhs != 1 || ! mxIsSparse (prhs[0]))
    mexErrMsgTxt ("ARG1 must be a sparse matrix");

  m = mxGetM (prhs[0]);
  n = mxGetN (prhs[0]);
  nz = mxGetNzmax (prhs[0]);

  if (mxIsComplex (prhs[0]))
    {
      mexPrintf ("Matrix is %d-by-%d complex sparse matrix", m, n);
      mexPrintf (" with %d elements\n", nz);

      pr = mxGetPr (prhs[0]);
      pi = mxGetPi (prhs[0]);
      ir = mxGetIr (prhs[0]);
      jc = mxGetJc (prhs[0]);

      i = n;
```

```
    while (jc[i] == jc[i-1] && i != 0) i--;

    mexPrintf ("last non-zero element (%d, %d) = (%g, %g)\n",
               ir[nz-1]+ 1, i, pr[nz-1], pi[nz-1]);

    v = mxCreateSparse (m, n, nz, mxCOMPLEX);
    pr2 = mxGetPr (v);
    pi2 = mxGetPi (v);
    ir2 = mxGetIr (v);
    jc2 = mxGetJc (v);

    for (i = 0; i < nz; i++)
      {
        pr2[i] = 2 * pr[i];
        pi2[i] = 2 * pi[i];
        ir2[i] = ir[i];
      }
    for (i = 0; i < n + 1; i++)
      jc2[i] = jc[i];

    if (nlhs > 0)
      plhs[0] = v;
  }
else if (mxIsLogical (prhs[0]))
  {
    mxLogical *pbr, *pbr2;
    mexPrintf ("Matrix is %d-by-%d logical sparse matrix", m, n);
    mexPrintf (" with %d elements\n", nz);

    pbr = mxGetLogicals (prhs[0]);
    ir = mxGetIr (prhs[0]);
    jc = mxGetJc (prhs[0]);

    i = n;
    while (jc[i] == jc[i-1] && i != 0) i--;
    mexPrintf ("last non-zero element (%d, %d) = %d\n",
               ir[nz-1]+ 1, i, pbr[nz-1]);

    v = mxCreateSparseLogicalMatrix (m, n, nz);
    pbr2 = mxGetLogicals (v);
    ir2 = mxGetIr (v);
    jc2 = mxGetJc (v);

    for (i = 0; i < nz; i++)
      {
        pbr2[i] = pbr[i];
        ir2[i] = ir[i];
```

```
          }
        for (i = 0; i < n + 1; i++)
          jc2[i] = jc[i];

        if (nlhs > 0)
          plhs[0] = v;
      }
    else
      {
        mexPrintf ("Matrix is %d-by-%d real sparse matrix", m, n);
        mexPrintf (" with %d elements\n", nz);

        pr = mxGetPr (prhs[0]);
        ir = mxGetIr (prhs[0]);
        jc = mxGetJc (prhs[0]);

        i = n;
        while (jc[i] == jc[i-1] && i != 0) i--;
        mexPrintf ("last non-zero element (%d, %d) = %g\n",
                   ir[nz-1]+ 1, i, pr[nz-1]);

        v = mxCreateSparse (m, n, nz, mxREAL);
        pr2 = mxGetPr (v);
        ir2 = mxGetIr (v);
        jc2 = mxGetJc (v);

        for (i = 0; i < nz; i++)
          {
            pr2[i] = 2 * pr[i];
            ir2[i] = ir[i];
          }
        for (i = 0; i < n + 1; i++)
          jc2[i] = jc[i];

        if (nlhs > 0)
          plhs[0] = v;
      }
}
```

A sample usage of mysparse is

```
sm = sparse ([1, 0; 0, pi]);
mysparse (sm)
⇒
Matrix is 2-by-2 real sparse matrix with 2 elements
last non-zero element (2, 2) = 3.14159
```

A.2.7 Calling Other Functions in Mex-Files

It is possible to call other Octave functions from within a mex-file using `mexCallMATLAB`. An example of the use of `mexCallMATLAB` can be see in the example below.

```
#include "mex.h"

void
mexFunction (int nlhs, mxArray* plhs[],
             int nrhs, const mxArray* prhs[])
{
  char *str;

  mexPrintf ("Starting file myfeval.mex\n");

  mexPrintf ("I have %d inputs and %d outputs\n", nrhs, nlhs);

  if (nrhs < 1 || ! mxIsString (prhs[0]))
    mexErrMsgTxt ("ARG1 must be a function name");

  str = mxArrayToString (prhs[0]);

  mexPrintf ("I'm going to call the function %s\n", str);

  if (nlhs == 0)
    nlhs = 1;  // Octave's automatic 'ans' variable

  /* Cast prhs just to get rid of 'const' qualifier and stop compile warning */
  mexCallMATLAB (nlhs, plhs, nrhs-1, (mxArray**)prhs+1, str);

  mxFree (str);
}
```

If this code is in the file '`myfeval.c`', and is compiled to '`myfeval.mex`', then an example of its use is

```
a = myfeval ("sin", 1)
⇒ Starting file myfeval.mex
   I have 2 inputs and 1 outputs
   I'm going to call the interpreter function sin
   a =  0.84147
```

Note that it is not possible to use function handles or inline functions within a mex-file.

A.3 Standalone Programs

The libraries Octave itself uses can be utilized in standalone applications. These applications then have access, for example, to the array and matrix classes, as well as to all of the Octave algorithms. The following C++ program, uses class Matrix from '`liboctave.a`' or '`liboctave.so`'.

```
#include <iostream>
#include <octave/oct.h>

int
main (void)
{
  std::cout << "Hello Octave world!\n";

  int n = 2;
  Matrix a_matrix = Matrix (n, n);

  for (octave_idx_type i = 0; i < n; i++)
    for (octave_idx_type j = 0; j < n; j++)
      a_matrix(i,j) = (i + 1) * 10 + (j + 1);

  std::cout << a_matrix;

  return 0;
}
```

mkoctfile can be used to build a standalone application with a command like

```
$ mkoctfile --link-stand-alone standalone.cc -o standalone
$ ./standalone
Hello Octave world!
  11 12
  21 22
$
```

Note that the application standalone will be dynamically linked against the Octave libraries and any Octave support libraries. The above allows the Octave math libraries to be used by an application. It does not, however, allow the script files, oct-files, or built-in functions of Octave to be used by the application. To do that the Octave interpreter needs to be initialized first. An example of how to do this can then be seen in the code

```
#include <iostream>
#include <octave/oct.h>
#include <octave/octave.h>
#include <octave/parse.h>
#include <octave/toplev.h>

int
main (void)
{
  string_vector argv (2);
  argv(0) = "embedded";
  argv(1) = "-q";

  octave_main (2, argv.c_str_vec (), 1);
```

```
    octave_idx_type n = 2;
    octave_value_list in;

    for (octave_idx_type i = 0; i < n; i++)
      in(i) = octave_value (5 * (i + 2));

    octave_value_list out = feval ("gcd", in, 1);

    if (! error_state && out.length () > 0)
      std::cout << "GCD of ["
                << in(0).int_value ()
                << ", "
                << in(1).int_value ()
                << "] is " << out(0).int_value ()
                << std::endl;
    else
      std::cout << "invalid\n";

    clean_up_and_exit (0);
  }
```

which, as before, is compiled and run as a standalone application with

```
$ mkoctfile --link-stand-alone embedded.cc -o embedded
$ ./embedded
GCD of [10, 15] is 5
$
```

It is worth noting that, if only built-in functions are to be called from a C++ standalone program, then it does not need to initialize the interpreter to do so. The general rule is that, for a built-in function named `function_name` in the interpreter, there will be a C++ function named `Ffunction_name` (note the prepended capital `F`) accessible in the C++ API. The declarations for all built-in functions are collected in the header file `builtin-defun-decls.h`. This feature should be used with care as the list of built-in functions can change. No guarantees can be made that a function that is currently built in won't be implemented as a .m file or as a dynamically linked function in the future. An example of how to call built-in functions from C++ can be seen in the code

```
#include <iostream>
#include <octave/oct.h>
#include <octave/builtin-defun-decls.h>

int
main (void)
{
  int n = 2;
  Matrix a_matrix = Matrix (n, n);
```

```
  for (octave_idx_type i = 0; i < n; i++)
    for (octave_idx_type j = 0; j < n; j++)
      a_matrix(i,j) = (i + 1) * 10 + (j + 1);

  std::cout << "This is a matrix:" << std::endl
            << a_matrix            << std::endl;

  octave_value_list in;
  in(0) = a_matrix;

  octave_value_list out = Fnorm (in, 1);
  double norm_of_the_matrix = out(0).double_value ();

  std::cout << "This is the norm of the matrix:" << std::endl
            << norm_of_the_matrix                << std::endl;

  return 0;
}
```

which, again, is compiled and run as a standalone application with

```
$ mkoctfile --link-stand-alone standalonebuiltin.cc -o standalonebuiltin
$ ./standalonebuiltin
This is a matrix:
 11 12
 21 22

This is the norm of the matrix:
34.4952
$
```

Appendix B Test and Demo Functions

Octave includes a number of functions to allow the integration of testing and demonstration code in the source code of the functions themselves.

B.1 Test Functions

`test` *name* [Command]
`test` *name* *quiet* | *normal* | *verbose* [Command]
`test ("`*name*`", "`*quiet* | *normal* | *verbose*`",` *fid*`)` [Function File]
`test ([], "`*explain*`",` *fid*`)` [Function File]
success `= test (...)` [Function File]
`[`*n*`,` *max*`] = test (...)` [Function File]
`[`*code*`,` *idx*`] = test ("`*name*`", "`*grabdemo*`")` [Function File]

Perform tests from the first file in the loadpath matching *name*. `test` can be called as a command or as a function. Called with a single argument *name*, the tests are run interactively and stop after the first error is encountered.

With a second argument the tests which are performed and the amount of output is selected.

`"quiet"` Don't report all the tests as they happen, just the errors.

`"normal"` Report all tests as they happen, but don't do tests which require user interaction.

`"verbose"`
Do tests which require user interaction.

The argument *fid* can be used to allow batch processing. Errors can be written to the already open file defined by *fid*, and hopefully when Octave crashes this file will tell you what was happening when it did. You can use `stdout` if you want to see the results as they happen. You can also give a file name rather than an *fid*, in which case the contents of the file will be replaced with the log from the current test.

Called with a single output argument *success*, `test` returns true if all of the tests were successful. Called with two output arguments *n* and *max*, the number of successful tests and the total number of tests in the file *name* are returned.

If the second argument is the string `"grabdemo"`, the contents of the demo blocks are extracted but not executed. Code for all code blocks is concatenated and returned as *code* with *idx* being a vector of positions of the ends of the demo blocks.

If the second argument is `"explain"`, then *name* is ignored and an explanation of the line markers used is written to the file *fid*.

See also: [assert], page 804, [fail], page 804, [error], page 197, [demo], page 805, [example], page 806.

`test` scans the named script file looking for lines which start with the identifier '`%!`'. The prefix is stripped off and the rest of the line is processed through the Octave interpreter. If the code generates an error, then the test is said to fail.

Since `eval()` will stop at the first error it encounters, you must divide your tests up into blocks, with anything in a separate block evaluated separately. Blocks are introduced by valid keywords like `test`, `function`, or `assert` immediately following '`%!`'. A block is defined by indentation as in Python. Lines beginning with '`%!<whitespace>`' are part of the preceeding block.

For example:

```
%!test error ("this test fails!");
%!test "test doesn't fail. it doesn't generate an error";
```

When a test fails, you will see something like:

```
  ***** test error ("this test fails!")
!!!!! test failed
this test fails!
```

Generally, to test if something works, you want to assert that it produces a correct value. A real test might look something like

```
%!test
%! a = [1, 2, 3; 4, 5, 6]; B = [1; 2];
%! expect = [ a ; 2*a ];
%! get = kron (b, a);
%! if (any (size (expect) != size (get)))
%!   error ("wrong size: expected %d,%d but got %d,%d",
%!          size (expect), size (get));
%! elseif (any (any (expect != get)))
%!   error ("didn't get what was expected.");
%! endif
```

To make the process easier, use the `assert` function. For example, with `assert` the previous test is reduced to:

```
%!test
%! a = [1, 2, 3; 4, 5, 6]; b = [1; 2];
%! assert (kron (b, a), [ a; 2*a ]);
```

`assert` can accept a tolerance so that you can compare results absolutely or relatively. For example, the following all succeed:

```
%!test assert (1+eps, 1, 2*eps)          # absolute error
%!test assert (100+100*eps, 100, -2*eps) # relative error
```

You can also do the comparison yourself, but still have assert generate the error:

```
%!test assert (isempty ([]))
%!test assert ([1, 2; 3, 4] > 0)
```

Because `assert` is so frequently used alone in a test block, there is a shorthand form:

```
%!assert (...)
```

which is equivalent to:

```
%!test assert (...)
```

Occasionally a block of tests will depend on having optional functionality in Octave. Before testing such blocks the availability of the required functionality must be checked. A `%!testif HAVE_XXX` block will only be run if Octave was compiled with functionality

'HAVE_XXX'. For example, the sparse single value decomposition, svds(), depends on having the ARPACK library. All of the tests for svds begin with

```
%!testif HAVE_ARPACK
```

Review 'config.h' or octave_config_info ("features") to see some of the possible values to check.

Sometimes during development there is a test that should work but is known to fail. You still want to leave the test in because when the final code is ready the test should pass, but you may not be able to fix it immediately. To avoid unnecessary bug reports for these known failures, mark the block with xtest rather than test:

```
%!xtest assert (1==0)
%!xtest fail ("success=1", "error")
```

In this case, the test will run and any failure will be reported. However, testing is not aborted and subsequent test blocks will be processed normally. Another use of xtest is for statistical tests which should pass most of the time but are known to fail occasionally.

Each block is evaluated in its own function environment, which means that variables defined in one block are not automatically shared with other blocks. If you do want to share variables, then you must declare them as shared before you use them. For example, the following declares the variable *a*, gives it an initial value (default is empty), and then uses it in several subsequent tests.

```
%!shared a
%! a = [1, 2, 3; 4, 5, 6];
%!assert (kron ([1; 2], a), [ a; 2*a ]);
%!assert (kron ([1, 2], a), [ a, 2*a ]);
%!assert (kron ([1,2; 3,4], a), [ a,2*a; 3*a,4*a ]);
```

You can share several variables at the same time:

```
%!shared a, b
```

You can also share test functions:

```
%!function a = fn (b)
%!  a = 2*b;
%!endfunction
%!assert (fn(2), 4);
```

Note that all previous variables and values are lost when a new shared block is declared.

Remember that %!function begins a new block and that %!endfunction ends this block. Be aware that until a new block is started, lines starting with '%!<space>' will be discarded as comments. The following is nearly identical to the example above, but does nothing.

```
%!function a = fn (b)
%!  a = 2*b;
%!endfunction
%! assert (fn(2), 4);
```

Because there is a space after '%!' the assert statement does not begin a new block and this line is treated as a comment.

Error and warning blocks are like test blocks, but they only succeed if the code generates an error. You can check the text of the error is correct using an optional regular expression `<pattern>`. For example:

```
%!error <passes!> error ("this test passes!");
```

If the code doesn't generate an error, the test fails. For example:

```
%!error "this is an error because it succeeds.";
```

produces

```
  ***** error "this is an error because it succeeds.";
!!!!! test failed: no error
```

It is important to automate the tests as much as possible, however some tests require user interaction. These can be isolated into demo blocks, which if you are in batch mode, are only run when called with `demo` or the `verbose` option to `test`. The code is displayed before it is executed. For example,

```
%!demo
%! t = [0:0.01:2*pi]; x = sin (t);
%! plot (t, x);
%! # you should now see a sine wave in your figure window
```

produces

```
funcname example 1:
 t = [0:0.01:2*pi]; x = sin (t);
 plot (t, x);
 # you should now see a sine wave in your figure window

Press <enter> to continue:
```

Note that demo blocks cannot use any shared variables. This is so that they can be executed by themselves, ignoring all other tests.

If you want to temporarily disable a test block, put `#` in place of the block type. This creates a comment block which is echoed in the log file but not executed. For example:

```
%!#demo
%! t = [0:0.01:2*pi]; x = sin (t);
%! plot (t, x);
%! # you should now see a sine wave in your figure window
```

The following trivial code snippet provides examples for the use of fail, assert, error and xtest:

```
function output = must_be_zero (input)
  if (input != 0)
    error ("Non-zero input!")
  endif
  output = input;
endfunction

%!fail ("must_be_zero (1)");
%!assert (must_be_zero (0), 0);
%!error <Non-zero> must_be_zero (1);
%!xtest error ("This code generates an error");
```

When putting this a file 'must_be_zero.m', and running the test, we see

```
test must_be_zero verbose

⇒
>>>>> /path/to/must_be_zero.m
  ***** fail ("must_be_zero (1)");
  ***** assert (must_be_zero (0), 0);
  ***** error <Non-zero> must_be_zero (1);
  ***** xtest error ("This code generates an error");
!!!!! known failure
This code generates an error
PASSES 4 out of 4 tests (1 expected failures)
```

Block type summary:

`%!test` check that entire block is correct

`%!testif HAVE_XXX`
check block only if Octave was compiled with feature HAVE_XXX.

`%!xtest` check block, report a test failure but do not abort testing.

`%!error` check for correct error message

`%!warning`
check for correct warning message

`%!demo` demo only executes in interactive mode

`%!#` comment: ignore everything within the block

`%!shared x,y,z`
declare variables for use in multiple tests

`%!function`
define a function for use in multiple tests

`%!endfunction`
close a function definition

`%!assert (x, y, tol)`
shorthand for `%!test assert (x, y, tol)`

You can also create test scripts for built-in functions and your own C++ functions. To do so, put a file with the bare function name (no .m extension) in a directory in the load path and it will be discovered by the `test` function. Alternatively, you can embed tests directly in your C++ code:

```
/*
%!test disp ("this is a test")
*/
```

or

```
#if 0
%!test disp ("this is a test")
#endif
```

However, in this case the raw source code will need to be on the load path and the user will have to remember to type `test ("funcname.cc")`.

`assert` (*cond*) [Function File]
`assert` (*cond*, *errmsg*, ...) [Function File]
`assert` (*cond*, *msg_id*, *errmsg*, ...) [Function File]
`assert` (*observed*, *expected*) [Function File]
`assert` (*observed*, *expected*, *tol*) [Function File]

Produce an error if the specified condition is not met. `assert` can be called in three different ways.

`assert (`*cond*`)`
`assert (`*cond*`,` *errmsg*`, ...)`
`assert (`*cond*`,` *msg_id*`,` *errmsg*`, ...)`

Called with a single argument *cond*, `assert` produces an error if *cond* is zero. When called with more than one argument the additional arguments are passed to the `error` function.

`assert (`*observed*`,` *expected*`)`

Produce an error if observed is not the same as expected. Note that *observed* and *expected* can be scalars, vectors, matrices, strings, cell arrays, or structures.

`assert (`*observed*`,` *expected*`,` *tol*`)`

Produce an error if observed is not the same as expected but equality comparison for numeric data uses a tolerance *tol*. If *tol* is positive then it is an absolute tolerance which will produce an error if `abs (`*observed* `-` *expected*`) > abs (`*tol*`)`. If *tol* is negative then it is a relative tolerance which will produce an error if `abs (`*observed* `-` *expected*`) > abs (`*tol* `*` *expected*`)`. If *expected* is zero *tol* will always be interpreted as an absolute tolerance. If *tol* is not scalar its dimensions must agree with those of *observed* and *expected* and tests are performed on an element-wise basis.

See also: [test], page 799, [fail], page 804, [error], page 197.

`fail` (*code*) [Function File]
`fail` (*code*, *pattern*) [Function File]

fail (*code*, "*warning*", *pattern*) [Function File]

Return true if *code* fails with an error message matching *pattern*, otherwise produce an error. Note that *code* is a string and if *code* runs successfully, the error produced is:

```
expected error <.> but got none
```

Code must be in the form of a string that may be passed by `fail` to the Octave interpreter via the `evalin` function, that is, a (quoted) string constant or a string variable.

If called with two arguments, the behavior is similar to `fail (code)`, except the return value will only be true if code fails with an error message containing pattern (case sensitive). If the code fails with a different error to that given in pattern, the message produced is:

```
expected <pattern>
but got <text of actual error>
```

The angle brackets are not part of the output.

Called with three arguments, the behavior is similar to `fail (code, pattern)`, but produces an error if no warning is given during code execution or if the code fails.

See also: [assert], page 804.

B.2 Demonstration Functions

demo *name* [Command]
demo *name n* [Command]
demo ("*name*") [Function File]
demo ("*name*", *n*) [Function File]

Run example code block *n* associated with the function *name*. If *n* is not specified, all examples are run.

Examples are stored in the script file, or in a file with the same name but no extension located on Octave's load path. To keep examples separate from regular script code, all lines are prefixed by `%!`. Each example must also be introduced by the keyword `"demo"` flush left to the prefix with no intervening spaces. The remainder of the example can contain arbitrary Octave code. For example:

```
%!demo
%! t = 0:0.01:2*pi;
%! x = sin (t);
%! plot (t, x);
%! %-------------------------------------------------
%! % the figure window shows one cycle of a sine wave
```

Note that the code is displayed before it is executed, so a simple comment at the end suffices for labeling what is being shown. It is generally not necessary to use `disp` or `printf` within the demo.

Demos are run in a function environment with no access to external variables. This means that every demo must have separate initialization code. Alternatively, all demos can be combined into a single large demo with the code

```
%! input("Press <enter> to continue: ","s");
```

between the sections, but this is discouraged. Other techniques to avoid multiple initialization blocks include using multiple plots with a new `figure` command between each plot, or using `subplot` to put multiple plots in the same window.

Also, because demo evaluates within a function context, you cannot define new functions inside a demo. If you must have function blocks, rather than just anonymous functions or inline functions, you will have to use `eval (example ("function",n))` to see them. Because eval only evaluates one line, or one statement if the statement crosses multiple lines, you must wrap your demo in "`if 1 <demo stuff> endif`" with the "`if`" on the same line as "`demo`". For example:

```
%!demo if 1
%!  function y=f(x)
%!    y=x;
%!  endfunction
%!  f(3)
%! endif
```

See also: [test], page 799, [example], page 806.

`example` *name* [Command]
`example` *name n* [Command]
`example ("`*name*`")` [Function File]
`example ("`*name*`",` *n*`)` [Function File]
`[`*s*`,` *idx*`] = example (...)` [Function File]

Display the code for example *n* associated with the function *name*, but do not run it. If *n* is not specified, all examples are displayed.

When called with output arguments, the examples are returned in the form of a string *s*, with *idx* indicating the ending position of the various examples.

See `demo` for a complete explanation.

See also: [demo], page 805, [test], page 799.

`rundemos ()` [Function File]
`rundemos (`*directory*`)` [Function File]

Execute built-in demos for all function files in the specified directory. Also executes demos in any C++ source files found in the directory, for use with dynamically linked functions.

If no directory is specified, operate on all directories in Octave's search path for functions.

See also: [runtests], page 806, [path], page 181.

`runtests ()` [Function File]
`runtests (`*directory*`)` [Function File]

Execute built-in tests for all function files in the specified directory. Also executes tests in any C++ source files found in the directory, for use with dynamically linked functions.

If no directory is specified, operate on all directories in Octave's search path for functions.

See also: [rundemos], page 806, [path], page 181.

`speed (f, init, max_n, f2, tol)` [Function File]
`[order, n, T_f, T_f2] = speed (...)` [Function File]

Determine the execution time of an expression (*f*) for various input values (*n*). The *n* are log-spaced from 1 to *max_n*. For each *n*, an initialization expression (*init*) is computed to create any data needed for the test. If a second expression (*f2*) is given then the execution times of the two expressions are compared. When called without output arguments the results are printed to stdout and displayed graphically.

f The code expression to evaluate.

max_n The maximum test length to run. The default value is 100. Alternatively, use `[min_n, max_n]` or specify the *n* exactly with `[n1, n2, ..., nk]`.

init Initialization expression for function argument values. Use *k* for the test number and *n* for the size of the test. This should compute values for all variables used by *f*. Note that *init* will be evaluated first for $k = 0$, so things which are constant throughout the test series can be computed once. The default value is `x = randn (n, 1)`.

f2 An alternative expression to evaluate, so that the speed of two expressions can be directly compared. The default is `[]`.

tol Tolerance used to compare the results of expression *f* and expression *f2*. If *tol* is positive, the tolerance is an absolute one. If *tol* is negative, the tolerance is a relative one. The default is `eps`. If *tol* is `Inf`, then no comparison will be made.

order The time complexity of the expression $O(a * n^p)$. This is a structure with fields `a` and `p`.

n The values *n* for which the expression was calculated **AND** the execution time was greater than zero.

T_f The nonzero execution times recorded for the expression *f* in seconds.

T_f2 The nonzero execution times recorded for the expression *f2* in seconds. If required, the mean time ratio is simply `mean (T_f ./ T_f2)`.

The slope of the execution time graph shows the approximate power of the asymptotic running time $O(n^p)$. This power is plotted for the region over which it is approximated (the latter half of the graph). The estimated power is not very accurate, but should be sufficient to determine the general order of an algorithm. It should indicate if, for example, the implementation is unexpectedly $O(n^2)$ rather than $O(n)$ because it extends a vector each time through the loop rather than pre-allocating storage. In the current version of Octave, the following is not the expected $O(n)$.

```
speed ("for i = 1:n, y{i} = x(i); endfor", "", [1000, 10000])
```

But it is if you preallocate the cell array y:

```
speed ("for i = 1:n, y{i} = x(i); endfor", ...
       "x = rand (n, 1); y = cell (size (x));", [1000, 10000])
```

An attempt is made to approximate the cost of individual operations, but it is wildly inaccurate. You can improve the stability somewhat by doing more work for each `n`. For example:

```
speed ("airy(x)", "x = rand (n, 10)", [10000, 100000])
```

When comparing two different expressions (*f*, *f2*), the slope of the line on the speedup ratio graph should be larger than 1 if the new expression is faster. Better algorithms have a shallow slope. Generally, vectorizing an algorithm will not change the slope of the execution time graph, but will shift it relative to the original. For example:

```
speed ("sum (x)", "", [10000, 100000], ...
       "v = 0; for i = 1:length (x), v += x(i); endfor")
```

The following is a more complex example. If there was an original version of `xcorr` using for loops and a second version using an FFT, then one could compare the run speed for various lags as follows, or for a fixed lag with varying vector lengths as follows:

```
speed ("xcorr (x, n)", "x = rand (128, 1);", 100,
       "xcorr_orig (x, n)", -100*eps)
speed ("xcorr (x, 15)", "x = rand (20+n, 1);", 100,
       "xcorr_orig (x, n)", -100*eps)
```

Assuming one of the two versions is in xcorr_orig, this would compare their speed and their output values. Note that the FFT version is not exact, so one must specify an acceptable tolerance on the comparison `100*eps`. In this case, the comparison should be computed relatively, as `abs ((x - y) ./ y)` rather than absolutely as `abs (x - y)`.

Type `example ("speed")` to see some real examples or `demo ("speed")` to run them.

Appendix C Tips and Standards

This chapter describes no additional features of Octave. Instead it gives advice on making effective use of the features described in the previous chapters.

C.1 Writing Clean Octave Programs

Here are some tips for avoiding common errors in writing Octave code intended for widespread use:

- Since all global variables share the same name space, and all functions share another name space, you should choose a short word to distinguish your program from other Octave programs. Then take care to begin the names of all global variables, constants, and functions with the chosen prefix. This helps avoid name conflicts.

  If you write a function that you think ought to be added to Octave under a certain name, such as `fiddle_matrix`, don't call it by that name in your program. Call it `mylib_fiddle_matrix` in your program, and send mail to `maintainers@octave.org` suggesting that it be added to Octave. If and when it is, the name can be changed easily enough.

  If one prefix is insufficient, your package may use two or three alternative common prefixes, so long as they make sense.

  Separate the prefix from the rest of the symbol name with an underscore '`_`'. This will be consistent with Octave itself and with most Octave programs.

- When you encounter an error condition, call the function `error` (or `usage`). The `error` and `usage` functions do not return. See Section 2.5 [Errors], page 34.

- Please put a copyright notice on the file if you give copies to anyone. Use the same lines that appear at the top of the function files distributed with Octave. If you have not signed papers to assign the copyright to anyone else, then place your name in the copyright notice.

C.2 Tips on Writing Comments

Here are the conventions to follow when writing comments.

'`#`' Comments that start with a single sharp-sign, '`#`', should all be aligned to the same column on the right of the source code. Such comments usually explain how the code on the same line does its job. In the Emacs mode for Octave, the *M-;* (`indent-for-comment`) command automatically inserts such a '`#`' in the right place, or aligns such a comment if it is already present.

'`##`' Comments that start with a double sharp-sign, '`##`', should be aligned to the same level of indentation as the code. Such comments usually describe the purpose of the following lines or the state of the program at that point.

The indentation commands of the Octave mode in Emacs, such as *M-;* (`indent-for-comment`) and *TAB* (`octave-indent-line`) automatically indent comments according to these conventions, depending on the number of semicolons. See Section "Manipulating Comments" in *The GNU Emacs Manual*.

C.3 Conventional Headers for Octave Functions

Octave has conventions for using special comments in function files to give information such as who wrote them. This section explains these conventions.

The top of the file should contain a copyright notice, followed by a block of comments that can be used as the help text for the function. Here is an example:

```
## Copyright (C) 1996, 1997, 2007 John W. Eaton
##
## This file is part of Octave.
##
## Octave is free software; you can redistribute it and/or
## modify it under the terms of the GNU General Public
## License as published by the Free Software Foundation;
## either version 3 of the License, or (at your option) any
## later version.
##
## Octave is distributed in the hope that it will be useful,
## but WITHOUT ANY WARRANTY; without even the implied
## warranty of MERCHANTABILITY or FITNESS FOR A PARTICULAR
## PURPOSE.  See the GNU General Public License for more
## details.
##
## You should have received a copy of the GNU General Public
## License along with Octave; see the file COPYING.  If not,
## see <http://www.gnu.org/licenses/>.

## usage: [IN, OUT, PID] = popen2 (COMMAND, ARGS)
##
## Start a subprocess with two-way communication.  COMMAND
## specifies the name of the command to start.  ARGS is an
## array of strings containing options for COMMAND.  IN and
## OUT are the file ids of the input and streams for the
## subprocess, and PID is the process id of the subprocess,
## or -1 if COMMAND could not be executed.
##
## Example:
##
##  [in, out, pid] = popen2 ("sort", "-nr");
##  fputs (in, "these\nare\nsome\nstrings\n");
##  fclose (in);
##  while (ischar (s = fgets (out)))
##    fputs (stdout, s);
##  endwhile
##  fclose (out);
```

Octave uses the first block of comments in a function file that do not appear to be a copyright notice as the help text for the file. For Octave to recognize the first comment block as a copyright notice, it must start with the word 'Copyright' after stripping the leading comment characters.

After the copyright notice and help text come several *header comment* lines, each beginning with '`## header-name:`'. For example,

```
## Author: jwe
## Keywords: subprocesses input-output
## Maintainer: jwe
```

Here is a table of the conventional possibilities for *header-name*:

'`Author`' This line states the name and net address of at least the principal author of the library.

```
## Author: John W. Eaton <jwe@octave.org>
```

'`Maintainer`'
This line should contain a single name/address as in the Author line, or an address only, or the string '`jwe`'. If there is no maintainer line, the person(s) in the Author field are presumed to be the maintainers. The example above is mildly bogus because the maintainer line is redundant.

The idea behind the '`Author`' and '`Maintainer`' lines is to make possible a function to "send mail to the maintainer" without having to mine the name out by hand.

Be sure to surround the network address with '`<...>`' if you include the person's full name as well as the network address.

'`Created`' This optional line gives the original creation date of the file. For historical interest only.

'`Version`' If you wish to record version numbers for the individual Octave program, put them in this line.

'`Adapted-By`'
In this header line, place the name of the person who adapted the library for installation (to make it fit the style conventions, for example).

'`Keywords`'
This line lists keywords. Eventually, it will be used by an apropos command to allow people will find your package when they're looking for things by topic area. To separate the keywords, you can use spaces, commas, or both.

Just about every Octave function ought to have the '`Author`' and '`Keywords`' header comment lines. Use the others if they are appropriate. You can also put in header lines with other header names—they have no standard meanings, so they can't do any harm.

C.4 Tips for Documentation Strings

As noted above, documentation is typically in a commented header block on an Octave function following the copyright statement. The help string shown above is an unformatted string and will be displayed as is by Octave. Here are some tips for the writing of documentation strings.

- Every command, function, or variable intended for users to know about should have a documentation string.
- An internal variable or subroutine of an Octave program might as well have a documentation string.
- The first line of the documentation string should consist of one or two complete sentences that stand on their own as a summary.

  The documentation string can have additional lines that expand on the details of how to use the function or variable. The additional lines should also be made up of complete sentences.
- For consistency, phrase the verb in the first sentence of a documentation string as an infinitive with "to" omitted. For instance, use "Return the frob of A and B." in preference to "Returns the frob of A and B." Usually it looks good to do likewise for the rest of the first paragraph. Subsequent paragraphs usually look better if they have proper subjects.
- Write documentation strings in the active voice, not the passive, and in the present tense, not the future. For instance, use "Return a list containing A and B." instead of "A list containing A and B will be returned."
- Avoid using the word "cause" (or its equivalents) unnecessarily. Instead of, "Cause Octave to display text in boldface," just write "Display text in boldface."
- Use two spaces between the period marking the end of a sentence and the word which opens the next sentence. This convention has no effect for typeset formats like TEX, but improves the readability of the documentation in fixed-width environments such as the Info reader.
- Do not start or end a documentation string with whitespace.
- Format the documentation string so that it fits within an 80-column screen. It is a good idea for most lines to be no wider than 60 characters.

  However, rather than simply filling the entire documentation string, you can make it much more readable by choosing line breaks with care. Use blank lines between topics if the documentation string is long.
- **Do not** indent subsequent lines of a documentation string so that the text is lined up in the source code with the text of the first line. This looks nice in the source code, but looks bizarre when users view the documentation. Remember that the indentation before the starting double-quote is not part of the string!
- When choosing variable names try to adhere to the following guidelines.

  vectors : x,y,z,t,w

  matrices : A,B,M

  strings : str,s

  filenames : fname

  cells,cellstrs :
  c,cstr
- The documentation string for a variable that is a yes-or-no flag should start with words such as "Nonzero means...", to make it clear that all nonzero values are equivalent and indicate explicitly what zero and nonzero mean.

- When a function's documentation string mentions the value of an argument of the function, use the argument name in capital letters as if it were a name for that value. Thus, the documentation string of the operator / refers to its second argument as 'DIVISOR', because the actual argument name is divisor.

  Also use all caps for meta-syntactic variables, such as when you show the decomposition of a list or vector into subunits, some of which may vary.

Octave also allows extensive formatting of the help string of functions using Texinfo. The effect on the online documentation is relatively small, but makes the help string of functions conform to the help of Octave's own functions. However, the effect on the appearance of printed or online documentation will be greatly improved.

The fundamental building block of Texinfo documentation strings is the Texinfo-macro `@deftypefn`, which takes three arguments: The class the function is in, its output arguments, and the function's signature. Typical classes for functions include `Function File` for standard Octave functions, and `Loadable Function` for dynamically linked functions. A skeletal Texinfo documentation string therefore looks like this

```
-*- texinfo -*-
@deftypefn {Function File} {@var{ret} =} fn (...)
@cindex index term
Help text in Texinfo format.  Code samples should be marked
like @code{sample of code} and variables should be marked
as @var{variable}.
@seealso{fn2, fn3}
@end deftypefn
```

This help string must be commented in user functions, or in the help string of the `DEFUN_DLD` macro for dynamically loadable functions. The important aspects of the documentation string are

-*- texinfo -*-
: This string signals Octave that the following text is in Texinfo format, and should be the first part of any help string in Texinfo format.

@deftypefn {class} ... @end deftypefn
: The entire help string should be enclosed within the block defined by deftypefn.

@cindex index term
: This generates an index entry, and can be useful when the function is included as part of a larger piece of documentation. It is ignored within Octave's help viewer. Only one index term may appear per line but multiple @cindex lines are valid if the function should be filed under different terms.

@var{variable}
: All variables should be marked with this macro. The markup of variables is then changed appropriately for display.

@code{sample of code}
: All samples of code should be marked with this macro for the same reasons as the @var macro.

@qcode{"sample_code"}
@qcode{'sample_code'}

All samples of code which are quoted should use this more specialized macro. This happens frequently when discussing graphics properties such as "position" or options such as "on"/"off".

@seealso{function2, function3}

This is a comma separated list of function names that allows cross referencing from one function documentation string to another.

Texinfo format has been designed to generate output for online viewing with text terminals as well as generating high-quality printed output. To these ends, Texinfo has commands which control the diversion of parts of the document into a particular output processor. Three formats are of importance: info, HTML, and TeX. These are selected with

```
@ifinfo
Text area for info only
@end ifinfo
```

```
@ifhtml
Text area for HTML only
@end ifhtml
```

```
@tex
Text area for TeX only
@end tex
```

Note that often TeX output can be used in HTML documents and so often the `@ifhtml` blocks are unnecessary. If no specific output processor is chosen, by default, the text goes into all output processors. It is usual to have the above blocks in pairs to allow the same information to be conveyed in all output formats, but with a different markup. Currently, most Octave documentation only makes a distinction between TeX and all other formats. Therefore, the following construct is seen repeatedly.

```
@tex
text for TeX only
@end tex
@ifnottex
text for info, HTML, plaintext
@end ifnottex
```

Another important feature of Texinfo that is often used in Octave help strings is the `@example` environment. An example of its use is

```
@example
@group
@code{2 * 2}
@result{} 4
@end group
@end example
```

which produces

```
2 * 2
⇒ 4
```

The `@group` block prevents the example from being split across a page boundary, while the `@result{}` macro produces a right arrow signifying the result of a command. If your example is larger than 20 lines it is better *NOT* to use grouping so that a reasonable page boundary can be calculated.

In many cases a function has multiple ways in which it can be called, and the `@deftypefnx` macro can be used to give alternatives. For example

```
-*- texinfo -*-
@deftypefn  {Function File} {@var{a} =} fn (@var{x}, ...)
@deftypefnx {Function File} {@var{a} =} fn (@var{y}, ...)
Help text in Texinfo format.
@end deftypefn
```

Many complete examples of Texinfo documentation can be taken from the help strings for the Octave functions themselves. A relatively complete example of which is the `nchoosek` function. The Texinfo documentation string for `nchoosek` is

```
-*- texinfo -*-
@deftypefn  {Function File} {@var{c} =} nchoosek (@var{n}, @var{k})
@deftypefnx {Function File} {@var{c} =} nchoosek (@var{set}, @var{k})

Compute the binomial coefficient or all combinations of a set of items.

If @var{n} is a scalar then calculate the binomial coefficient
of @var{n} and @var{k} which is defined as
@tex
$$
 {n \choose k} = {n (n-1) (n-2) \cdots (n-k+1) \over k!}
               = {n! \over k! (n-k)!}
$$
@end tex
@ifnottex

@example
@group
 /   \
 | n |    n (n-1) (n-2) @dots{} (n-k+1)       n!
 |   |  = ------------------------- =  ---------
 | k |               k!                k! (n-k)!
 \   /
@end group
@end example

@end ifnottex
@noindent
This is the number of combinations of @var{n} items taken in groups of
size @var{k}.

If the first argument is a vector, @var{set}, then generate all
```

```
combinations of the elements of @var{set}, taken @var{k} at a time, with
one row per combination.  The result @var{c} has @var{k} columns and
@w{@code{nchoosek (length (@var{set}), @var{k})}} rows.

For example:

How many ways can three items be grouped into pairs?

@example
@group
nchoosek (3, 2)
   @result{} 3
@end group
@end example

What are the possible pairs?

@example
@group
nchoosek (1:3, 2)
   @result{}  1   2
       1   3
       2   3
@end group
@end example

@code{nchoosek} works only for non-negative, integer arguments.  Use
@code{bincoeff} for non-integer and negative scalar arguments, or for
computing many binomial coefficients at once with vector inputs
for @var{n} or @var{k}.

@seealso{bincoeff, perms}
@end deftypefn
```

which demonstrates most of the concepts discussed above. This documentation string renders as

```
 -- Function File: C = nchoosek (N, K)
 -- Function File: C = nchoosek (SET, K)
     Compute the binomial coefficient or all combinations of a set of
     items.

     If N is a scalar then calculate the binomial coefficient of N and
     K which is defined as

           /   \
           | n |    n (n-1) (n-2) ... (n-k+1)       n!
           |   |  = ------------------------- =  ---------
```

```
         | k |                k!                 k! (n-k)!
         \   /

    This is the number of combinations of N items taken in groups of
    size K.

    If the first argument is a vector, SET, then generate all
    combinations of the elements of SET, taken K at a time, with one
    row per combination.  The result C has K columns and
    'nchoosek (length (SET), K)' rows.

    For example:

    How many ways can three items be grouped into pairs?

         nchoosek (3, 2)
            => 3

    What are the possible pairs?

         nchoosek (1:3, 2)
            =>  1   2
                1   3
                2   3

    'nchoosek' works only for non-negative, integer arguments.  Use
    'bincoeff' for non-integer and negative scalar arguments, or for
    computing many binomial coefficients at once with vector inputs
    for N or K.

    See also: bincoeff, perms
```

using info, whereas in a printed documentation using TeX it will appear as

c = **nchoosek** (*n*, *k*) [Function File]
c = **nchoosek** (*set*, *k*) [Function File]

Compute the binomial coefficient or all combinations of a set of items.

If n is a scalar then calculate the binomial coefficient of n and k which is defined as

$$\binom{n}{k} = \frac{n(n-1)(n-2)\cdots(n-k+1)}{k!} = \frac{n!}{k!(n-k)!}$$

This is the number of combinations of n items taken in groups of size k.

If the first argument is a vector, *set*, then generate all combinations of the elements of *set*, taken k at a time, with one row per combination. The result c has k columns and `nchoosek (length (set), k)` rows.

For example:

How many ways can three items be grouped into pairs?

```
nchoosek (3, 2)
   ⇒ 3
```

What are the possible pairs?

```
nchoosek (1:3, 2)
   ⇒ 1  2
      1  3
      2  3
```

`nchoosek` works only for non-negative, integer arguments. Use `bincoeff` for non-integer and negative scalar arguments, or for computing many binomial coefficients at once with vector inputs for *n* or *k*.

See also: bincoeff, perms.

Appendix D Contributing Guidelines

This chapter is dedicated to those who wish to contribute code to Octave.

D.1 How to Contribute

The mailing list for Octave development discussions is maintainers@octave.org. Patches should be submitted to Octave's patch tracker. This concerns the development of Octave core, i.e., code that goes in to Octave directly. You may consider developing and publishing a package instead; a great place for this is the allied Octave-Forge project (http://octave.sourceforge.net). Note that the Octave core project is inherently more conservative and follows narrower rules.

D.2 Building the Development Sources

The directions for building from the development sources change from time to time, so you should read the resources for developers on the web or in the development sources archive. Start here: http://www.octave.org/get-involved.html.

D.3 Basics of Generating a Changeset

The best way to contribute is to create a Mercurial changeset and submit it to the bug or patch trackers[1]. Mercurial is the source code management system currently used to develop Octave. Other forms of contributions (e.g., simple diff patches) are also acceptable, but they slow down the review process. If you want to make more contributions, you should really get familiar with Mercurial. A good place to start is http://www.selenic.com/mercurial/wiki/index.cgi/Tutorial. There you will also find help about how to install Mercurial.

A simple contribution sequence could look like this:

```
hg clone http://www.octave.org/hg/octave
                             # make a local copy of the octave
                             # source repository
cd octave
# change some sources...
hg commit -m "make Octave the coolest software ever"
                             # commit the changeset into your
                             # local repository
hg export -o ../cool.diff tip
                             # export the changeset to a diff
                             # file
# attach ../cool.diff to your bug report
```

You may want to get familiar with Mercurial queues to manage your changesets. To work with queues you must activate the extension mq with the following entry in Mercurial's configuration file '.hgrc' (or 'Mercurial.ini' on Windows):

[1] Please use the patch tracker only for patches which add new features. If you have a patch to submit that fixes a bug, you should use the bug tracker instead.

```
[extensions]
mq=
```

Here is a slightly more complex example using Mercurial queues, where work on two unrelated changesets is done in parallel and one of the changesets is updated after discussion on the bug tracker:

```
hg qnew nasty_bug             # create a new patch
# change sources...
hg qref                       # save the changes into the patch
# change even more...
hg qref -m "solution to nasty bug!"
                              # save again with commit message
hg export -o ../nasty.diff tip
                              # export the patch
# attach ../nasty.diff to your bug report
hg qpop                       # undo the application of the patch
                              # and remove the changes from the
                              # source tree
hg qnew doc_improvements      # create an unrelated patch
# change doc sources...
hg qref -m "could not find myfav.m in the doc"
                              # save the changes into the patch
hg export -o ../doc.diff tip
                              # export the second patch
# attach ../doc.diff to your bug report
hg qpop
# discussion in the bug tracker ...
hg qpush nasty_bug            # apply the patch again
# change sources yet again ...
hg qref
hg export -o ../nasty2.diff tip
# attach ../nasty2.diff to your bug report
```

Mercurial has a few more useful extensions that really should be enabled. They are not enabled by default due to a number of factors (mostly because they don't work in all terminal types).

The following entries in the '`.hgrc`' are recommended

```
[extensions]
graphlog=
color=
progress=
pager=
```

For the color extension, default color and formatting of `hg status` can be modified by

```
[color]
status.modified = magenta bold
status.added = green bold
status.removed = red bold
status.deleted = cyan bold
status.unknown = black  bold
status.ignored = black bold
```

Sometimes a few further improvements for the pager extension are necessary. The following options should not be enabled unless paging is not working correctly.

```
[pager]
# Some options for the less pager, see less(1) for their meaning.
pager = LESS='FSRX' less

# Some commands that aren't paged by default; also enable paging
# for them
attend = tags, help, annotate, cat, diff, export, status, \
         outgoing, incoming
```

Enabling the described extensions should immediately lead to a difference when using the command line version of hg. Of these options, the only one that enables a new command is graphlog. It is recommanded that to use the command `hg glog`, instead of `hg log`, for a better feel about what commits are being based on.

D.4 General Guidelines

All Octave's sources are distributed under the GNU General Public License (GPL). Currently, Octave uses GPL version 3. For details about this license, see `http://www.gnu.org/licenses/gpl.html`. Therefore, whenever you create a new source file, it should have the following comment header (use appropriate year, name and comment marks):

```
## Copyright (C) 1996-2013 John W. Eaton <jwe@octave.org>
##
## This file is part of Octave.
##
## Octave is free software; you can redistribute it and/or modify it
## under the terms of the GNU General Public License as published by
## the Free Software Foundation; either version 3 of the License, or
## (at your option) any later version.
##
## Octave is distributed in the hope that it will be useful, but
## WITHOUT ANY WARRANTY; without even the implied warranty of
## MERCHANTABILITY or FITNESS FOR A PARTICULAR PURPOSE.  See the
## GNU General Public License for more details.
##
## You should have received a copy of the GNU General Public License
## along with Octave; see the file COPYING.  If not,
## see <http://www.gnu.org/licenses/>.
```

Always include commit messages in changesets. After making your source changes, record and briefly describe the changes in your commit message. You should have previously configured your '`.hgrc`' (or '`Mercurial.ini`' on Windows) with your name and email, which will be automatically added to your commit message. Your commit message should have a brief one-line explanation of what the commit does. If you are patching a bug, this one-line explanation should mention the bug number at the end. If your change is small and only touches one file then this is typically sufficient. If you are modifying several files, or several parts of one file, you should enumerate your changes roughly following the GNU coding standards for changelogs, as in the following example:

```
look for methods before constructors

* symtab.cc (symbol_table::fcn_info::fcn_info_rep::find):
Look for class methods before constructors, contrary to MATLAB
documentation.

* test/ctor-vs-method: New directory of test classes.
* test/test_ctor_vs_method.m: New file.
* test/Makefile.am: Include ctor-vs-method/module.mk.
(FCN_FILES): Include test_ctor_vs_method.m in the list.
```

In this example, the names of the file changed is listed first, and in parentheses the name of the function in that file that was modified. There is no need to mention the function for m-files that only contain one function. The commit message should describe what was changed, not why it was changed. Any explanation for why a change is needed should appear as comments in the code, particularly if there is something that might not be obvious to someone reading it later.

When submitting code which addresses a known bug on the Octave bug tracker (`http://bugs.octave.org`), please add '(bug #XXXXX)' to the first line of the commit messages. For example:

```
Fix bug for complex input for gradient (bug #34292).
```

The preferred comment mark for places that may need further attention is `FIXME:`.

D.5 Octave Sources (m-files)

Don't use tabs. Tabs cause trouble. If you are used to them, set up your editor so that it converts tabs to spaces. Indent the bodies of statement blocks. The recommended indent is 2 spaces. When calling functions, put spaces after commas and before the calling parentheses, like this:

```
x = max (sin (y+3), 2);
```

An exception are matrix or cell constructors:

```
[sin(x), cos(x)]
{sin(x), cos(x)}
```

Here, putting spaces after `sin`, `cos` would result in a parse error. For an indexing expression, do not put a space after the identifier (this differentiates indexing and function calls nicely). The space after a comma is not necessary if index expressions are simple, i.e., you may write

```
A(:,i,j)
```

but

```
A([1:i-1;i+1:n], XI(:,2:n-1))
```

Use lowercase names if possible. Uppercase is acceptable for variable names consisting of 1-2 letters. Do not use mixed case names. Function names must be lowercase. Function names are global, so choose them wisely.

Always use a specific end-of-block statement (like `endif`, `endswitch`) rather than the generic `end`.

Enclose the `if`, `while`, `until`, and `switch` conditions in parentheses, as in C:

```
if (isvector (a))
  s = sum (a);
endif
```

Do not do this, however, with the iteration counter portion of a `for` statement. Write:

```
for i = 1:n
  b(i) = sum (a(:,i));
endfor
```

D.6 C++ Sources

Don't use tabs. Tabs cause trouble. If you are used to them, set up your editor so that it converts tabs to spaces. Format function headers like this:

```
static bool
matches_patterns (const string_vector& patterns, int pat_idx,
                  int num_pat, const std::string& name)
```

The function name should start in column 1, and multi-line argument lists should be aligned on the first char after the open parenthesis. You should put a space before the left open parenthesis and after commas, for both function definitions and function calls.

The recommended indent is 2 spaces. When indenting, indent the statement after control structures (like `if`, `while`, etc.). If there is a compound statement, indent *both* the curly braces and the body of the statement (so that the body gets indented by *two* indents). Example:

```
if (have_args)
  {
    idx.push_back (first_args);
    have_args = false;
  }
else
  idx.push_back (make_value_list (*p_args, *p_arg_nm, &tmp));
```

If you have nested `if` statements, use extra braces for extra clarification.

Split long expressions in such a way that a continuation line starts with an operator rather than identifier. If the split occurs inside braces, continuation should be aligned with the first char after the innermost braces enclosing the split. Example:

```
SVD::type type = ((nargout == 0 || nargout == 1)
                  ? SVD::sigma_only
                  : (nargin == 2) ? SVD::economy : SVD::std);
```

Consider putting extra braces around a multi-line expression to make it more readable, even if they are not necessary. Also, do not hesitate to put extra braces anywhere if it improves clarity.

Declare variables just before they are needed. Use local variables of blocks—it helps optimization. Don't write a multi-line variable declaration with a single type specification and multiple variables. If the variables don't fit on single line, repeat the type specification. Example:

```
octave_value retval;

octave_idx_type nr = b.rows ();
octave_idx_type nc = b.cols ();

double d1, d2;
```

Use lowercase names if possible. Uppercase is acceptable for variable names consisting of 1-2 letters. Do not use mixed case names.

Use Octave's types and classes if possible. Otherwise, use the C++ standard library. Use of STL containers and algorithms is encouraged. Use templates wisely to reduce code duplication. Avoid comma expressions, labels and gotos, and explicit typecasts. If you need to typecast, use the modern C++ casting operators. In functions, minimize the number of `return` statements—use nested `if` statements if possible.

D.7 Other Sources

Apart from C++ and Octave language (m-files), Octave's sources include files written in C, Fortran, M4, Perl, Unix shell, AWK, Texinfo, and TEX. There are not many rules to follow when using these other languages; some of them are summarized below. In any case, the golden rule is: if you modify a source file, try to follow any conventions you can detect in the file or other similar files.

For C you should obviously follow all C++ rules that can apply.

If you modify a Fortran file, you should stay within Fortran 77 with common extensions like `END DO`. Currently, we want all sources to be compilable with the f2c and g77 compilers, without special flags if possible. This usually means that non-legacy compilers also accept the sources.

The M4 macro language is mainly used for Autoconf configuration files. You should follow normal M4 rules when contributing to these files. Some M4 files come from external source, namely the Autoconf archive `http://autoconf-archive.cryp.to`.

If you give a code example in the documentation written in Texinfo with the `@example` environment, you should be aware that the text within such an environment will not be wrapped. It is recommended that you keep the lines short enough to fit on pages in the generated pdf or ps documents. Here is a ruler (in an `@example` environment) for finding the appropriate line width:

```
         1         2         3         4         5         6
123456789012345678901234567890123456789012345678901234567890
```

Appendix E Obsolete Functions

After being marked as deprecated for two major releases, the following functions have been removed from Octave. The third column of the table shows the version of Octave in which the function was removed. Prior to removal, each function in the list was marked as deprecated for at least two major releases. All deprecated functions issue warnings explaining that they will be removed in a future version of Octave, and which function should be used instead.

Replacement functions do not always accept precisely the same arguments as the obsolete function, but should provide equivalent functionality.

Obsolete Function	Replacement	Version
beta_cdf	betacdf	3.4.0
beta_inv	betainv	3.4.0
beta_pdf	betapdf	3.4.0
beta_rnd	betarnd	3.4.0
binomial_cdf	binocdf	3.4.0
binomial_inv	binoinv	3.4.0
binomial_pdf	binopdf	3.4.0
binomial_rnd	binornd	3.4.0
chisquare_cdf	chi2cdf	3.4.0
chisquare_inv	chi2inv	3.4.0
chisquare_pdf	chi2pdf	3.4.0
chisquare_rnd	chi2rnd	3.4.0
clearplot	clf	3.4.0
com2str	num2str	3.4.0
exponential_cdf	expcdf	3.4.0
exponential_inv	expinv	3.4.0
exponential_pdf	exppdf	3.4.0
exponential_rnd	exprnd	3.4.0
f_cdf	fcdf	3.4.0
f_inv	finv	3.4.0
f_pdf	fpdf	3.4.0
f_rnd	frnd	3.4.0
gamma_cdf	gamcdf	3.4.0
gamma_inv	gaminv	3.4.0
gamma_pdf	gampdf	3.4.0
gamma_rnd	gamrnd	3.4.0
geometric_cdf	geocdf	3.4.0
geometric_inv	geoinv	3.4.0
geometric_pdf	geopdf	3.4.0
geometric_rnd	geornd	3.4.0
hypergeometric_cdf	hygecdf	3.4.0
hypergeometric_inv	hygeinv	3.4.0
hypergeometric_pdf	hygepdf	3.4.0
hypergeometric_rnd	hygernd	3.4.0

intersection	intersect	3.4.0
is_bool	isbool	3.4.0
is_complex	iscomplex	3.4.0
is_list	islist	3.4.0
is_matrix	ismatrix	3.4.0
is_scalar	isscalar	3.4.0
is_square	issquare	3.4.0
is_stream	isstream	3.4.0
is_struct	isstruct	3.4.0
is_symmetric	issymmetric	3.4.0
is_vector	isvector	3.4.0
lognormal_cdf	logncdf	3.4.0
lognormal_inv	logninv	3.4.0
lognormal_pdf	lognpdf	3.4.0
lognormal_rnd	lognrnd	3.4.0
meshdom	meshgrid	3.4.0
normal_cdf	normcdf	3.4.0
normal_inv	norminv	3.4.0
normal_pdf	normpdf	3.4.0
normal_rnd	normrnd	3.4.0
pascal_cdf	nbincdf	3.4.0
pascal_inv	nbininv	3.4.0
pascal_pdf	nbinpdf	3.4.0
pascal_rnd	nbinrnd	3.4.0
poisson_cdf	poisscdf	3.4.0
poisson_inv	poissinv	3.4.0
poisson_pdf	poisspdf	3.4.0
poisson_rnd	poissrnd	3.4.0
polyinteg	polyint	3.4.0
struct_contains	isfield	3.4.0
struct_elements	fieldnames	3.4.0
t_cdf	tcdf	3.4.0
t_inv	tinv	3.4.0
t_pdf	tpdf	3.4.0
t_rnd	trnd	3.4.0
uniform_cdf	unifcdf	3.4.0
uniform_inv	unifinv	3.4.0
uniform_pdf	unifpdf	3.4.0
uniform_rnd	unifrnd	3.4.0
weibull_cdf	wblcdf	3.4.0
weibull_inv	wblinv	3.4.0
weibull_pdf	wblpdf	3.4.0
weibull_rnd	wblrnd	3.4.0
wiener_rnd	wienrnd	3.4.0
create_set	unique	3.6.0
dmult	diag (A) * B	3.6.0
iscommand	None	3.6.0

`israwcommand`	None	3.6.0
`lchol`	`chol (..., "lower")`	3.6.0
`loadimage`	`load` or `imread`	3.6.0
`mark_as_command`	None	3.6.0
`mark_as_rawcommand`	None	3.6.0
`spatan2`	`atan2`	3.6.0
`spchol`	`chol`	3.6.0
`spchol2inv`	`chol2inv`	3.6.0
`spcholinv`	`cholinv`	3.6.0
`spcumprod`	`cumprod`	3.6.0
`spcumsum`	`cumsum`	3.6.0
`spdet`	`det`	3.6.0
`spdiag`	`sparse (diag (...))`	3.6.0
`spfind`	`find`	3.6.0
`sphcat`	`horzcat`	3.6.0
`spinv`	`inv`	3.6.0
`spkron`	`kron`	3.6.0
`splchol`	`chol (..., "lower")`	3.6.0
`split`	`char (strsplit (s, t))`	3.6.0
`splu`	`lu`	3.6.0
`spmax`	`max`	3.6.0
`spmin`	`min`	3.6.0
`spprod`	`prod`	3.6.0
`spqr`	`qr`	3.6.0
`spsum`	`sum`	3.6.0
`spsumsq`	`sumsq`	3.6.0
`spvcat`	`vertcat`	3.6.0
`str2mat`	`char`	3.6.0
`unmark_command`	None	3.6.0
`unmark_rawcommand`	None	3.6.0

Appendix F Known Causes of Trouble

This section describes known problems that affect users of Octave. Most of these are not Octave bugs per se—if they were, we would fix them. But the result for a user may be like the result of a bug.

Some of these problems are due to bugs in other software, some are missing features that are too much work to add, and some are places where people's opinions differ as to what is best.

F.1 Actual Bugs We Haven't Fixed Yet

- Output that comes directly from Fortran functions is not sent through the pager and may appear out of sequence with other output that is sent through the pager. One way to avoid this is to force pending output to be flushed before calling a function that will produce output from within Fortran functions. To do this, use the command

  ```
  fflush (stdout)
  ```

  Another possible workaround is to use the command

  ```
  page_screen_output (false);
  ```

  to turn the pager off.

A list of ideas for future enhancements is distributed with Octave. See the file 'PROJECTS' in the top level directory in the source distribution.

F.2 Reporting Bugs

Your bug reports play an essential role in making Octave reliable.

When you encounter a problem, the first thing to do is to see if it is already known. See Appendix F [Trouble], page 829. If it isn't known, then you should report the problem.

Reporting a bug may help you by bringing a solution to your problem, or it may not. In any case, the principal function of a bug report is to help the entire community by making the next version of Octave work better. Bug reports are your contribution to the maintenance of Octave.

In order for a bug report to serve its purpose, you must include the information that makes it possible to fix the bug.

F.2.1 Have You Found a Bug?

If you are not sure whether you have found a bug, here are some guidelines:

- If Octave gets a fatal signal, for any input whatever, that is a bug. Reliable interpreters never crash.
- If Octave produces incorrect results, for any input whatever, that is a bug.
- Some output may appear to be incorrect when it is in fact due to a program whose behavior is undefined, which happened by chance to give the desired results on another system. For example, the range operator may produce different results because of differences in the way floating point arithmetic is handled on various systems.
- If Octave produces an error message for valid input, that is a bug.

- If Octave does not produce an error message for invalid input, that is a bug. However, you should note that your idea of "invalid input" might be my idea of "an extension" or "support for traditional practice".
- If you are an experienced user of programs like Octave, your suggestions for improvement are welcome in any case.

F.2.2 Where to Report Bugs

To report a bug in Octave, submit a bug report to the Octave bug tracker `http://bugs.octave.org`.

Do not send bug reports to '`help-octave`'. Most users of Octave do not want to receive bug reports.

F.2.3 How to Report Bugs

Submit bug reports for Octave to the Octave bug tracker `http://bugs.octave.org`.

The fundamental principle of reporting bugs usefully is this: **report all the facts**. If you are not sure whether to state a fact or leave it out, state it!

Often people omit facts because they think they know what causes the problem and they conclude that some details don't matter. Thus, you might assume that the name of the variable you use in an example does not matter. Well, probably it doesn't, but one cannot be sure. Perhaps the bug is a stray memory reference which happens to fetch from the location where that name is stored in memory; perhaps, if the name were different, the contents of that location would fool the interpreter into doing the right thing despite the bug. Play it safe and give a specific, complete example.

Keep in mind that the purpose of a bug report is to enable someone to fix the bug if it is not known. Always write your bug reports on the assumption that the bug is not known.

Sometimes people give a few sketchy facts and ask, "Does this ring a bell?" This cannot help us fix a bug. It is better to send a complete bug report to begin with.

Try to make your bug report self-contained. If we have to ask you for more information, it is best if you include all the previous information in your response, as well as the information that was missing.

To enable someone to investigate the bug, you should include all these things:

- The version of Octave. You can get this by noting the version number that is printed when Octave starts, or running it with the '`-v`' option.
- A complete input file that will reproduce the bug.

  A single statement may not be enough of an example—the bug might depend on other details that are missing from the single statement where the error finally occurs.
- The command arguments you gave Octave to execute that example and observe the bug. To guarantee you won't omit something important, list all the options.

  If we were to try to guess the arguments, we would probably guess wrong and then we would not encounter the bug.
- The type of machine you are using, and the operating system name and version number.
- The command-line arguments you gave to the `configure` command when you installed the interpreter.

- A complete list of any modifications you have made to the interpreter source.

  Be precise about these changes—show a context diff for them.
- Details of any other deviations from the standard procedure for installing Octave.
- A description of what behavior you observe that you believe is incorrect. For example, "The interpreter gets a fatal signal," or, "The output produced at line 208 is incorrect."

  Of course, if the bug is that the interpreter gets a fatal signal, then one can't miss it. But if the bug is incorrect output, we might not notice unless it is glaringly wrong.

  Even if the problem you experience is a fatal signal, you should still say so explicitly. Suppose something strange is going on, such as, your copy of the interpreter is out of sync, or you have encountered a bug in the C library on your system. Your copy might crash and the copy here would not. If you said to expect a crash, then when the interpreter here fails to crash, we would know that the bug was not happening. If you don't say to expect a crash, then we would not know whether the bug was happening. We would not be able to draw any conclusion from our observations.

  Often the observed symptom is incorrect output when your program is run. Unfortunately, this is not enough information unless the program is short and simple. It is very helpful if you can include an explanation of the expected output, and why the actual output is incorrect.
- If you wish to suggest changes to the Octave source, send them as context diffs. If you even discuss something in the Octave source, refer to it by context, not by line number, because the line numbers in the development sources probably won't match those in your sources.

Here are some things that are not necessary:

- A description of the envelope of the bug.

  Often people who encounter a bug spend a lot of time investigating which changes to the input file will make the bug go away and which changes will not affect it. Such information is usually not necessary to enable us to fix bugs in Octave, but if you can find a simpler example to report *instead* of the original one, that is a convenience. Errors in the output will be easier to spot, running under the debugger will take less time, etc. Most Octave bugs involve just one function, so the most straightforward way to simplify an example is to delete all the function definitions except the one in which the bug occurs.

  However, simplification is not vital; if you don't want to do this, report the bug anyway and send the entire test case you used.
- A patch for the bug. Patches can be helpful, but if you find a bug, you should report it, even if you cannot send a fix for the problem.

F.2.4 Sending Patches for Octave

If you would like to write bug fixes or improvements for Octave, that is very helpful. When you send your changes, please follow these guidelines to avoid causing extra work for us in studying the patches.

If you don't follow these guidelines, your information might still be useful, but using it will take extra work. Maintaining Octave is a lot of work in the best of circumstances, and we can't keep up unless you do your best to help.

- Send an explanation with your changes of what problem they fix or what improvement they bring about. For a bug fix, just include a copy of the bug report, and explain why the change fixes the bug.
- Always include a proper bug report for the problem you think you have fixed. We need to convince ourselves that the change is right before installing it. Even if it is right, we might have trouble judging it if we don't have a way to reproduce the problem.
- Include all the comments that are appropriate to help people reading the source in the future understand why this change was needed.
- Don't mix together changes made for different reasons. Send them *individually*.

  If you make two changes for separate reasons, then we might not want to install them both. We might want to install just one.
- Use '`diff -c`' to make your diffs. Diffs without context are hard for us to install reliably. More than that, they make it hard for us to study the diffs to decide whether we want to install them. Unified diff format is better than contextless diffs, but not as easy to read as '`-c`' format.

  If you have GNU diff, use '`diff -cp`', which shows the name of the function that each change occurs in.
- Write the change log entries for your changes.

  Read the '`ChangeLog`' file to see what sorts of information to put in, and to learn the style that we use. The purpose of the change log is to show people where to find what was changed. So you need to be specific about what functions you changed; in large functions, it's often helpful to indicate where within the function the change was made.

  On the other hand, once you have shown people where to find the change, you need not explain its purpose. Thus, if you add a new function, all you need to say about it is that it is new. If you feel that the purpose needs explaining, it probably does—but the explanation will be much more useful if you put it in comments in the code.

  If you would like your name to appear in the header line for who made the change, send us the header line.

F.3 How To Get Help with Octave

The mailing list `help@octave.org` exists for the discussion of matters related to using and installing Octave. If would like to join the discussion, please send a short note to `help-request@octave.org`.

Please do not send requests to be added or removed from the mailing list, or other administrative trivia to the list itself.

If you think you have found a bug in Octave or in the installation procedure, however, you should submit a complete bug report to the Octave bug tracker at `http://bugs.octave.org`. But before you submit a bug report, please read `http://www.octave.org/bugs.html` to learn how to submit a useful bug report.

Appendix G Installing Octave

The procedure for installing Octave from source on a Unix-like system is described next. Building on other platforms will follow similar steps. Note that this description applies to Octave releases. Building the development sources from the Mercurial archive requires additional steps as described in Section D.2 [Building the Development Sources], page 819.

G.1 Build Dependencies

Octave is a fairly large program with many build dependencies. You may be able to find pre-packaged versions of the dependencies distributed as part of your system, or you may have to build some or all of them yourself.

G.1.1 Obtaining the Dependencies Automatically

On some systems you can obtain many of Octave's build dependencies automatically. The commands for doing this vary by system. Similarly, the names of pre-compiled packages vary by system and do not always match exactly the names listed in Section G.1.2 [Build Tools], page 833 and Section G.1.3 [External Packages], page 834.

You will usually need the development version of an external dependency so that you get the libraries and header files for building software, not just for running already compiled programs. These packages typically have names that end with the suffix `-dev` or `-devel`.

On systems with `apt-get` (Debian, Ubuntu, etc.), you may be able to install most of the tools and external packages using a command similar to

```
apt-get build-dep octave
```

The specific package name may be `octave3.2` or `octave3.4`. The set of required tools and external dependencies does not change frequently, so it is not important that the version match exactly, but you should use the most recent one available.

On systems with `yum` (Fedora, Red Hat, etc.), you may be able to install most of the tools and external packages using a command similar to

```
yum-builddep octave
```

The `yum-builddep` utility is part of the `yum-utils` package.

For either type of system, the package name may include a version number. The set of required tools and external dependencies does not change frequently, so it is not important that the version exactly match the version you are installing, but you should use the most recent one available.

G.1.2 Build Tools

The following tools are required:

C++, C, and Fortran compilers

: The Octave sources are primarily written in C++, but some portions are also written in C and Fortran. The Octave sources are intended to be portable. Recent versions of the GNU compiler collection (GCC) should work (`http://gcc.gnu.org`). If you use GCC, you should avoid mixing versions. For example, be sure that you are not using the obsolete `g77` Fortran compiler with modern versions of `gcc` and `g++`.

GNU Make
: Tool for building software (`http://www.gnu.org/software/make`). Octave's build system requires GNU Make. Other versions of Make will not work. Fortunately, GNU Make is highly portable and easy to install.

AWK, sed, and other Unix utilities
: Basic Unix system utilities are required for building Octave. All will be available with any modern Unix system and also on Windows with either Cygwin or MinGW and MSYS.

Additionally, the following tools may be needed:

Bison
: Parser generator (`http://www.gnu.org/software/bison`). You will need Bison if you modify the **`oct-parse.yy`** source file or if you delete the files that are generated from it.

Flex
: Lexer analyzer (`http://www.gnu.org/software/flex`). You will need Flex if you modify the **`lex.ll`** source file or if you delete the files that are generated from it.

Autoconf
: Package for software configuration (`http://www.gnu.org/software/autoconf`). Autoconf is required if you modify Octave's **`configure.ac`** file or other files that it requires.

Automake
: Package for Makefile generation (`http://www.gnu.org/software/automake`). Automake is required if you modify Octave's **`Makefile.am`** files or other files that they depend on.

Libtool
: Package for building software libraries (`http://www.gnu.org/software/libtool`). Libtool is required by Automake.

gperf
: Perfect hash function generator (`http://www.gnu.org/software/gperf`). You will need gperf if you modify the **`octave.gperf`** file or if you delete the file that is generated from it.

Texinfo
: Package for generating online and print documentation (`http://www.gnu.org/software/te`
You will need Texinfo to build Octave's documentation or if you modify the documentation source files or the docstring of any Octave function.

G.1.3 External Packages

The following external packages are required:

BLAS
: Basic Linear Algebra Subroutine library (`http://www.netlib.org/blas`). Accelerated BLAS libraries such as ATLAS (`http://math-atlas.sourceforge.net`) are recommeded for better performance.

LAPACK
: Linear Algebra Package (`http://www.netlib.org/lapack`).

PCRE
: The Perl Compatible Regular Expression library (`http://www.pcre.org`).

The following external package is optional but strongly recommended:

GNU Readline
: Command-line editing library (`www.gnu.org/s/readline`).

If you wish to build Octave without GNU readline installed, you must use the '--disable-readline' option when running the configure script.

The following external software packages are optional but recommended:

ARPACK
: Library for the solution of large-scale eigenvalue problems (http://forge.scilab.org/index.php/p/arpack-ng). ARPACK is required to provide the functions **eigs** and **svds**.

cURL
: Library for transferring data with URL syntax (http://curl.haxx.se). cURL is required to provide the **urlread** and **urlwrite** functions and the **ftp** class.

FFTW3
: Library for computing discrete Fourier transforms (http://www.fftw.org). FFTW3 is used to provide better performance for functions that compute discrete Fourier transforms (**fft**, **ifft**, **fft2**, etc.)

FLTK
: Portable GUI toolkit (http://www.fltk.org). FLTK is currently used to provide windows for Octave's OpenGL-based graphics functions.

fontconfig
: Library for configuring and customizing font access (http://www.freedesktop.org/wiki/Softw Fontconfig is used to manage fonts for Octave's OpenGL-based graphics functions.

FreeType
: Portable font engine (http://www.freetype.org). FreeType is used to perform font rendering for Octave's OpenGL-based graphics functions.

GLPK
: GNU Linear Programming Kit (http://www.gnu.org/software/glpk). GPLK is required for the function **glpk**.

gl2ps
: OpenGL to PostScript printing library (http://www.geuz.org/gl2ps/). gl2ps is required for printing when using the FLTK toolkit.

gnuplot
: Interactive graphics program (http://www.gnuplot.info). gnuplot is currently the default graphics renderer for Octave.

GraphicsMagick++
: Image processing library (http://www.graphicsmagick.org). GraphicsMagick++ is used to provide the **imread** and **imwrite** functions.

HDF5
: Library for manipulating portable data files (http://www.hdfgroup.org/HDF5). HDF5 is required for Octave's **load** and **save** commands to read and write HDF data files.

Java Development Kit
: Java programming language compiler and libraries. The OpenJDK free software implementation is recommended (http://openjdk.java.net/), although other JDK implementations may work. Java is required to be able to call Java functions from within Octave.

LLVM
: Compiler framework, (http://www.llvm.org). LLVM is required for Octave's experimental just-in-time (JIT) compilation for speeding up the interpreter.

OpenGL
: API for portable 2-D and 3-D graphics (http://www.opengl.org). An OpenGL implementation is required to provide Octave's OpenGL-based graphics functions. Octave's OpenGL-based graphics functions usually

outperform the gnuplot-based graphics functions because plot data can be rendered directly instead of sending data and commands to gnuplot for interpretation and rendering.

Qhull
: Computational geometry library (`http://www.qhull.org`). Qhull is required to provide the functions **`convhull`**, **`convhulln`**, **`delaunay`**, **`delaunay3`**, **`delaunayn`**, **`voronoi`**, and **`voronoin`**.

QRUPDATE
: QR factorization updating library (`http://sourceforge.net/projects/qrupdate`). QRUPDATE is used to provide improved performance for the functions **`qrdelete`**, **`qrinsert`**, **`qrshift`**, and **`qrupdate`**.

QScintilla
: Source code highlighter and manipulator; a Qt port of Scintilla (`http://www.riverbankcomputing.co.uk/software/qscintilla`). QScintilla is used for syntax highlighting and code completion in the GUI.

Qt
: GUI and utility libraries (). Qt is required for building the GUI. It is a large framework, but the only components required are the GUI, core, and network modules.

SuiteSparse
: Sparse matrix factorization library (`http://www.cise.ufl.edu/research/sparse/SuiteSparse`) SuiteSparse is required to provide sparse matrix factorizations and solution of linear equations for sparse systems.

zlib
: Data compression library (`http://zlib.net`). The zlib library is required for Octave's **`load`** and **`save`** commands to handle compressed data, including MATLAB v5 MAT files.

G.2 Running Configure and Make

- Run the shell script '**`configure`**'. This will determine the features your system has (or doesn't have) and create a file named '**`Makefile`**' from each of the files named '**`Makefile.in`**'.

  Here is a summary of the configure options that are most frequently used when building Octave:

  `--help` Print a summary of the options recognized by the configure script.

  `--prefix=`*`prefix`*
  : Install Octave in subdirectories below *prefix*. The default value of *prefix* is '**`/usr/local`**'.

  `--srcdir=`*`dir`*
  : Look for Octave sources in the directory *dir*.

  `--enable-64`
  : This is an **experimental** option to enable Octave to use 64-bit integers for array dimensions and indexing on 64-bit platforms. You probably don't want to use this option unless you know what you are doing. See Section G.3 [Compiling Octave with 64-bit Indexing], page 840, for more details about building Octave with this option.

--enable-bounds-check
: Enable bounds checking for indexing operators in the internal array classes. This option is primarily used for debugging Octave. Building Octave with this option has a negative impact on performance and is not recommended for general use.

--disable-docs
: Disable building all forms of the documentation (Info, PDF, HTML). The default is to build documentation, but your system will need functioning Texinfo and TeX installs for this to succeed.

--enable-float-truncate
: This option allows for truncation of intermediate floating point results in calculations. It is only necessary for certain platforms.

--enable-readline
: Use the readline library to provide for editing of the command line in terminal environments. This option is on by default.

--enable-shared
: Create shared libraries (this is the default). If you are planning to use the dynamic loading features, you will probably want to use this option. It will make your '.oct' files much smaller and on some systems it may be necessary to build shared libraries in order to use dynamically linked functions.

  You may also want to build a shared version of libstdc++, if your system doesn't already have one.

--enable-dl
: Use dlopen and friends to make Octave capable of dynamically linking externally compiled functions (this is the default if '--enable-shared' is specified). This option only works on systems that actually have these functions. If you plan on using this feature, you should probably also use '--enable-shared' to reduce the size of your '.oct' files.

--with-blas=<lib>
: By default, configure looks for the best BLAS matrix libraries on your system, including optimized implementations such as the free ATLAS 3.0, as well as vendor-tuned libraries. (The use of an optimized BLAS will generally result in several-times faster matrix operations.) Use this option to specify a particular BLAS library that Octave should use.

--with-lapack=<lib>
: By default, configure looks for the best LAPACK matrix libraries on your system, including optimized implementations such as the free ATLAS 3.0, as well as vendor-tuned libraries. (The use of an optimized LAPACK will generally result in several-times faster matrix operations.) Use this option to specify a particular LAPACK library that Octave should use.

--with-magick=<lib>
: Select the library to use for image I/O. The two possible values are "GraphicsMagick" (default) or "ImageMagick".

--with-sepchar=<char>
: Use <char> as the path separation character. This option can help when running Octave on non-Unix systems.

--without-amd
: Don't use AMD, disable some sparse matrix functionality.

--without-camd
: Don't use CAMD, disable some sparse matrix functionality.

--without-colamd
: Don't use COLAMD, disable some sparse matrix functionality.

--without-ccolamd
: Don't use CCOLAMD, disable some sparse matrix functionality.

--without-cholmod
: Don't use CHOLMOD, disable some sparse matrix functionality.

--without-curl
: Don't use the cURL library, disable the ftp objects, `urlread` and `urlwrite` functions.

--without-cxsparse
: Don't use CXSPARSE, disable some sparse matrix functionality.

--without-fftw3
: Use the included FFTPACK library for computing Fast Fourier Transforms instead of the FFTW3 library.

--without-fftw3f
: Use the included FFTPACK library for computing Fast Fourier Transforms instead of the FFTW3 library when operating on single precision (float) values.

--without-glpk
: Don't use the GLPK library for linear programming.

--without-hdf5
: Don't use the HDF5 library, disable reading and writing of HDF5 files.

--without-opengl
: Don't use OpenGL, disable native graphics toolkit for plotting. You will need `gnuplot` installed in order to make plots.

--without-qhull
: Don't use Qhull, disable `delaunay`, `convhull`, and related functions.

--without-qrupdate
: Don't use QRUPDATE, disable QR and Cholesky update functions.

--without-umfpack
: Don't use UMFPACK, disable some sparse matrix functionality.

--without-zlib
: Don't use the zlib library, disable data file compression and support for recent MAT file formats.

`--without-framework-carbon`
: Don't use framework Carbon headers, libraries, or specific source code even if the configure test succeeds (the default is to use Carbon framework if available). This is a platform specific configure option for Mac systems.

`--without-framework-opengl`
: Don't use framework OpenGL headers, libraries, or specific source code even if the configure test succeeds. If this option is given then OpenGL headers and libraries in standard system locations are tested (the default value is '`--with-framework-opengl`'). This is a platform specific configure option for Mac systems.

See the file '`INSTALL`' for more general information about the command line options used by configure. That file also contains instructions for compiling in a directory other than the one where the source is located.

- Run make.

  You will need a recent version of GNU Make as Octave relies on certain features not generally available in all versions of make. Modifying Octave's makefiles to work with other make programs is probably not worth your time; instead, we simply recommend installing GNU Make.

  There are currently two options for plotting in Octave: (1) the external program gnuplot, or (2) the internal graphics engine using OpenGL and FLTK. Gnuplot is a command-driven interactive function plotting program. Gnuplot is copyrighted, but freely distributable. As of Octave release 3.4, gnuplot is the default option for plotting. But, the internal graphics engine is nearly 100% compatible, certainly for most ordinary plots, and users are encouraged to test it. It is anticipated that the internal engine will become the default option at the next major release of Octave.

  To compile Octave, you will need a recent version of `g++` or other ANSI C++ compiler. In addition, you will need a Fortran 77 compiler or `f2c`. If you use `f2c`, you will need a script like `fort77` that works like a normal Fortran compiler by combining `f2c` with your C compiler in a single script.

  If you plan to modify the parser you will also need GNU `bison` and `flex`. If you modify the documentation, you will need GNU Texinfo.

  GNU Make, `gcc` (and `libstdc++`), `gnuplot`, `bison`, `flex`, and Texinfo are all available from many anonymous ftp archives. The primary site is `ftp.gnu.org`, but it is often very busy. A list of sites that mirror the software on `ftp.gnu.org` is available by anonymous ftp from `ftp://ftp.gnu.org/pub/gnu/GNUinfo/FTP`.

  Octave requires approximately 1.4 GB of disk storage to unpack and compile from source (significantly less, 400 MB, if you don't compile with debugging symbols). To compile without debugging symbols try the command

  ```
  make CFLAGS=-O CXXFLAGS=-O LDFLAGS=
  ```

  instead of just `make`.

- If you encounter errors while compiling Octave, first see Section G.4 [Installation Problems], page 843 for a list of known problems and if there is a workaround or solution for your problem. If not, see Appendix F [Trouble], page 829 for information about how to report bugs.

- Once you have successfully compiled Octave, run `make install`.

  This will install a copy of Octave, its libraries, and its documentation in the destination directory. As distributed, Octave is installed in the following directories. In the table below, *prefix* defaults to '`/usr/local`', *version* stands for the current version number of the interpreter, and *arch* is the type of computer on which Octave is installed (for example, '`i586-unknown-gnu`').

  '`prefix/bin`'
  : Octave and other binaries that people will want to run directly.

  '`prefix/lib/octave-version`'
  : Libraries like liboctave.a and liboctinterp.a.

  '`prefix/octave-version/include/octave`'
  : Include files distributed with Octave.

  '`prefix/share`'
  : Architecture-independent data files.

  '`prefix/share/man/man1`'
  : Unix-style man pages describing Octave.

  '`prefix/share/info`'
  : Info files describing Octave.

  '`prefix/share/octave/version/m`'
  : Function files distributed with Octave. This includes the Octave version, so that multiple versions of Octave may be installed at the same time.

  '`prefix/libexec/octave/version/exec/arch`'
  : Executables to be run by Octave rather than the user.

  '`prefix/lib/octave/version/oct/arch`'
  : Object files that will be dynamically loaded.

  '`prefix/share/octave/version/imagelib`'
  : Image files that are distributed with Octave.

G.3 Compiling Octave with 64-bit Indexing

Note: the following only applies to systems that have 64-bit pointers. Configuring Octave with '`--enable-64`' cannot magically make a 32-bit system have a 64-bit address space.

On 64-bit systems, Octave is limited to (approximately) the following array sizes when using the default 32-bit indexing mode:

```
double:          16GB
single:           8GB
uint64, int64:   16GB
uint32, int32:    8GB
uint16, int16:    4GB
uint8, int8:      2GB
```

In each case, the limit is really (approximately) 2^{31} elements because of the default type of the value used for indexing arrays (signed 32-bit integer, corresponding to the size of a Fortran INTEGER value).

Trying to create larger arrays will produce the following error:

```
octave:1> a = zeros (1024*1024*1024*3, 1, 'int8');
error: memory exhausted or requested size too large
       for range of Octave's index type --
       trying to return to prompt
```

You will obtain this error even if your system has enough memory to create this array (4 GB in the above case).

To use arrays larger than 2 GB, Octave has to be configured with the option '`--enable-64`'. This option is experimental and you are encouraged to submit bug reports if you find a problem. With this option, Octave will use 64-bit integers internally for array dimensions and indexing. However, all numerical libraries used by Octave will **also** need to use 64-bit integers for array dimensions and indexing. In most cases, this means they will need to be compiled from source since most (all?) distributions which package these libraries compile them with the default Fortran integer size, which is normally 32-bits wide.

The following instructions were tested with the development version of Octave and GCC 4.3.4 on an x86_64 Debian system.

The versions listed below are the versions used for testing. If newer versions of these packages are available, you should try to use them, although there may be some differences.

All libraries and header files will be installed in subdirectories of `$prefix64` (you must choose the location of this directory).

- BLAS and LAPACK (`http://www.netlib.org/lapack`)

  Reference versions for both libraries are included in the reference LAPACK 3.2.1 distribution from `netlib.org`.

  - Copy the file '`make.inc.example`' and name it '`make.inc`'. The options '`-fdefault-integer-8`' and '`-fPIC`' (on 64-bit CPU) have to be added to the variable `OPTS` and `NOOPT`.
  - Once you have compiled this library make sure that you use it for compiling Suite Sparse and Octave. In the following we assume that you installed the LAPACK library as $prefix64/lib/liblapack.a.

- QRUPDATE (`http://sourceforge.net/projects/qrupdate`)

  In the '`Makeconf`' file:

  - Add '`-fdefault-integer-8`' to `FFLAGS`.
  - Adjust the BLAS and LAPACK variables as needed if your 64-bit aware BLAS and LAPACK libraries are in a non-standard location.
  - Set `PREFIX` to the top-level directory of your install tree.
  - Run `make solib` to make a shared library.
  - Run `make install` to install the library.

- SuiteSparse (`http://www.cise.ufl.edu/research/sparse/SuiteSparse`)

  Pass the following options to `make` to enable 64-bit integers for BLAS library calls. On 64-bit Windows systems, use `-DLONGBLAS="long long"` instead.

  ```
  CFLAGS='-DLONGBLAS=long'
  CXXFLAGS='-DLONGBLAS=long'
  ```

The SuiteSparse makefiles don't generate shared libraries. On some systems, you can generate them by doing something as simple as

```
top=$(pwd)
for f in *.a; do
  mkdir tmp
  cd tmp
  ar vx ../$f
  gcc -shared -o ../${f%%.a}.so *.o
  cd $top
  rm -rf tmp
done
```

Other systems may require a different solution.

- ATLAS instead of reference BLAS and LAPACK

  Suggestions on how to compile ATLAS would be most welcome.
- GLPK
- Qhull (http://www.qhull.org)

  Both GLPK and Qhull use `int` internally so maximum problem sizes may be limited.
- Octave

  Octave's 64-bit index support is activated with the configure option '`--enable-64`'.

  ```
  ./configure \
    LD_LIBRARY_PATH="$prefix64/lib" \
    CPPFLAGS="-I$prefix64/include" LDFLAGS="-L$prefix64/lib" \
    --enable-64
  ```

  You must ensure that all Fortran sources except those in the '`liboctave/cruft/ranlib`' directory are compiled such that INTEGERS are 8-bytes wide. If you are using gfortran, the configure script should automatically set the Makefile variable `F77_INTEGER_8_FLAG` to '`-fdefault-integer-8`'. If you are using another compiler, you must set this variable yourself. You should NOT set this flag in `FFLAGS`, otherwise the files in '`liboctave/cruft/ranlib`' will be miscompiled.
- Other dependencies

  Probably nothing special needs to be done for the following dependencies. If you discover that something does need to be done, please submit a bug report.

  - pcre
  - zlib
  - hdf5
  - fftw3
  - cURL
  - GraphicsMagick++
  - OpenGL
  - freetype
  - fontconfig
  - fltk

G.4 Installation Problems

This section contains a list of problems (and some apparent problems that don't really mean anything is wrong) that may show up during installation of Octave.

- On some SCO systems, `info` fails to compile if `HAVE_TERMIOS_H` is defined in '`config.h`'. Simply removing the definition from '`info/config.h`' should allow it to compile.
- If `configure` finds `dlopen`, `dlsym`, `dlclose`, and `dlerror`, but not the header file '`dlfcn.h`', you need to find the source for the header file and install it in the directory '`usr/include`'. This is reportedly a problem with Slackware 3.1. For Linux/GNU systems, the source for '`dlfcn.h`' is in the `ldso` package.
- Building '`.oct`' files doesn't work.

  You should probably have a shared version of `libstdc++`. A patch is needed to build shared versions of version 2.7.2 of `libstdc++` on the HP-PA architecture. You can find the patch at ftp://ftp.cygnus.com/pub/g++/libg++-2.7.2-hppa-gcc-fix.
- On some DEC alpha systems there may be a problem with the `libdxml` library, resulting in floating point errors and/or segmentation faults in the linear algebra routines called by Octave. If you encounter such problems, then you should modify the configure script so that `SPECIAL_MATH_LIB` is not set to `-ldxml`.
- On FreeBSD systems Octave may hang while initializing some internal constants. The fix appears to be to use

  ```
  options       GPL_MATH_EMULATE
  ```

  rather than

  ```
  options       MATH_EMULATE
  ```

  in the kernel configuration files (typically found in the directory '`/sys/i386/conf`'. After making this change, you'll need to rebuild the kernel, install it, and reboot.
- If you encounter errors like

  ```
  passing 'void (*)()' as argument 2 of
    'octave_set_signal_handler(int, void (*)(int))'
  ```

  or

  ```
  warning: ANSI C++ prohibits conversion from '(int)'
           to '(...)'
  ```

  while compiling '`sighandlers.cc`', you may need to edit some files in the `gcc` include subdirectory to add proper prototypes for functions there. For example, Ultrix 4.2 needs proper declarations for the `signal` function and the `SIG_IGN` macro in the file '`signal.h`'.

  On some systems the `SIG_IGN` macro is defined to be something like this:

  ```
  #define  SIG_IGN  (void (*)())1
  ```

  when it should really be something like:

  ```
  #define  SIG_IGN  (void (*)(int))1
  ```

  to match the prototype declaration for the `signal` function. This change should also be made for the `SIG_DFL` and `SIG_ERR` symbols. It may be necessary to change the definitions in '`sys/signal.h`' as well.

The gcc fixincludes and fixproto scripts should probably fix these problems when gcc installs its modified set of header files, but I don't think that's been done yet.

You should not change the files in '/usr/include'. You can find the gcc include directory tree by running the command

```
gcc -print-libgcc-file-name
```

The directory of gcc include files normally begins in the same directory that contains the file 'libgcc.a'.

- Some of the Fortran subroutines may fail to compile with older versions of the Sun Fortran compiler. If you get errors like

  ```
  zgemm.f:
          zgemm:
  warning: unexpected parent of complex expression subtree
  zgemm.f, line 245: warning: unexpected parent of complex
    expression subtree
  warning: unexpected parent of complex expression subtree
  zgemm.f, line 304: warning: unexpected parent of complex
    expression subtree
  warning: unexpected parent of complex expression subtree
  zgemm.f, line 327: warning: unexpected parent of complex
    expression subtree
  pcc_binval: missing IR_CONV in complex op
  make[2]: *** [zgemm.o] Error 1
  ```

  when compiling the Fortran subroutines in the 'liboctave/cruft' subdirectory, you should either upgrade your compiler or try compiling with optimization turned off.

- On NeXT systems, if you get errors like this:

  ```
  /usr/tmp/cc007458.s:unknown:Undefined local
        symbol LBB7656
  /usr/tmp/cc007458.s:unknown:Undefined local
        symbol LBE7656
  ```

  when compiling 'Array.cc' and 'Matrix.cc', try recompiling these files without '-g'.

- Some people have reported that calls to system() and the pager do not work on SunOS systems. This is apparently due to having G_HAVE_SYS_WAIT defined to be 0 instead of 1 when compiling libg++.

- On NeXT systems, linking to 'libsys_s.a' may fail to resolve the following functions

  ```
  _tcgetattr
  _tcsetattr
  _tcflow
  ```

  which are part of 'libposix.a'. Unfortunately, linking Octave with '-posix' results in the following undefined symbols.

```
.destructors_used
.constructors_used
_objc_msgSend
_NXGetDefaultValue
_NXRegisterDefaults
.objc_class_name_NXStringTable
.objc_class_name_NXBundle
```

One kluge around this problem is to extract 'termios.o' from 'libposix.a', put it in Octave's 'src' directory, and add it to the list of files to link together in the makefile. Suggestions for better ways to solve this problem are welcome!

- If Octave crashes immediately with a floating point exception, it is likely that it is failing to initialize the IEEE floating point values for infinity and NaN.

  If your system actually does support IEEE arithmetic, you should be able to fix this problem by modifying the function `octave_ieee_init` in the file 'lo-ieee.cc' to correctly initialize Octave's internal infinity and NaN variables.

  If your system does not support IEEE arithmetic but Octave's configure script incorrectly determined that it does, you can work around the problem by editing the file 'config.h' to not define `HAVE_ISINF`, `HAVE_FINITE`, and `HAVE_ISNAN`.

  In any case, please report this as a bug since it might be possible to modify Octave's configuration script to automatically determine the proper thing to do.

- If Octave is unable to find a header file because it is installed in a location that is not normally searched by the compiler, you can add the directory to the include search path by specifying (for example) `CPPFLAGS=-I/some/nonstandard/directory` as an argument to `configure`. Other variables that can be specified this way are `CFLAGS`, `CXXFLAGS`, `FFLAGS`, and `LDFLAGS`. Passing them as options to the configure script also records them in the 'config.status' file. By default, `CPPFLAGS` and `LDFLAGS` are empty, `CFLAGS` and `CXXFLAGS` are set to `"-g -O"` and `FFLAGS` is set to `"-O"`.

Appendix H Emacs Octave Support

The development of Octave code can greatly be facilitated using Emacs with Octave mode, a major mode for editing Octave files which can e.g. automatically indent the code, do some of the typing (with Abbrev mode) and show keywords, comments, strings, etc. in different faces (with Font-lock mode on devices that support it).

It is also possible to run Octave from within Emacs, either by directly entering commands at the prompt in a buffer in Inferior Octave mode, or by interacting with Octave from within a file with Octave code. This is useful in particular for debugging Octave code.

Finally, you can convince Octave to use the Emacs info reader for `help -i`.

All functionality is provided by the Emacs Lisp package EOS (for "Emacs Octave Support"). This chapter describes how to set up and use this package.

Please contact `Kurt.Hornik@wu-wien.ac.at` if you have any questions or suggestions on using EOS.

H.1 Installing EOS

The Emacs package EOS consists of the three files '`octave-mod.el`', '`octave-inf.el`', and '`octave-hlp.el`'. These files, or better yet their byte-compiled versions, should be somewhere in your Emacs load-path.

If you have GNU Emacs with a version number at least as high as 19.35, you are all set up, because EOS is respectively will be part of GNU Emacs as of version 19.35.

Otherwise, copy the three files from the '`emacs`' subdirectory of the Octave distribution to a place where Emacs can find them (this depends on how your Emacs was installed). Byte-compile them for speed if you want.

H.2 Using Octave Mode

If you are lucky, your sysadmins have already arranged everything so that Emacs automatically goes into Octave mode whenever you visit an Octave code file as characterized by its extension '`.m`'. If not, proceed as follows.

1. To begin using Octave mode for all '`.m`' files you visit, add the following lines to a file loaded by Emacs at startup time, typically your '`~/.emacs`' file:

   ```
   (autoload 'octave-mode "octave-mod" nil t)
   (setq auto-mode-alist
         (cons '("\\.m$" . octave-mode) auto-mode-alist))
   ```

2. Finally, to turn on the abbrevs, auto-fill and font-lock features automatically, also add the following lines to one of the Emacs startup files:

   ```
   (add-hook 'octave-mode-hook
             (lambda ()
               (abbrev-mode 1)
               (auto-fill-mode 1)
               (if (eq window-system 'x)
                   (font-lock-mode 1))))
   ```

   See the Emacs manual for more information about how to customize Font-lock mode.

In Octave mode, the following special Emacs commands can be used in addition to the standard Emacs commands.

C-h m
: Describe the features of Octave mode.

LFD
: Reindent the current Octave line, insert a newline and indent the new line (`octave-reindent-then-newline-and-indent`). An abbrev before point is expanded if `abbrev-mode` is non-`nil`.

TAB
: Indents current Octave line based on its contents and on previous lines (`indent-according-to-mode`).

;
: Insert an "electric" semicolon (`octave-electric-semi`). If `octave-auto-indent` is non-`nil`, reindent the current line. If `octave-auto-newline` is non-`nil`, automagically insert a newline and indent the new line.

'
: Start entering an abbreviation (`octave-abbrev-start`). If Abbrev mode is turned on, typing *'C-h* or *'?* lists all abbrevs. Any other key combination is executed normally. Note that all Octave abbrevs start with a grave accent.

M-LFD
: Break line at point and insert continuation marker and alignment (`octave-split-line`).

M-TAB
: Perform completion on Octave symbol preceding point, comparing that symbol against Octave's reserved words and built-in variables (`octave-complete-symbol`).

M-C-a
: Move backward to the beginning of a function (`octave-beginning-of-defun`). With prefix argument *N*, do it that many times if *N* is positive; otherwise, move forward to the *N*-th following beginning of a function.

M-C-e
: Move forward to the end of a function (`octave-end-of-defun`). With prefix argument *N*, do it that many times if *N* is positive; otherwise, move back to the *N*-th preceding end of a function.

M-C-h
: Puts point at beginning and mark at the end of the current Octave function, i.e., the one containing point or following point (`octave-mark-defun`).

M-C-q
: Properly indents the Octave function which contains point (`octave-indent-defun`).

M-;
: If there is no comment already on this line, create a code-level comment (started by two comment characters) if the line is empty, or an in-line comment (started by one comment character) otherwise (`octave-indent-for-comment`). Point is left after the start of the comment which is properly aligned.

C-c ;
: Puts the comment character '`#`' (more precisely, the string value of `octave-comment-start`) at the beginning of every line in the region (`octave-comment-region`). With just *C-u* prefix argument, uncomment each line in the region. A numeric prefix argument *N* means use *N* comment characters.

C-c :
: Uncomments every line in the region (`octave-uncomment-region`).

C-c C-p
: Move one line of Octave code backward, skipping empty and comment lines (`octave-previous-code-line`). With numeric prefix argument *N*, move that many code lines backward (forward if *N* is negative).

C-c C-n Move one line of Octave code forward, skipping empty and comment lines (`octave-next-code-line`). With numeric prefix argument *N*, move that many code lines forward (backward if *N* is negative).

C-c C-a Move to the 'real' beginning of the current line (`octave-beginning-of-line`). If point is in an empty or comment line, simply go to its beginning; otherwise, move backwards to the beginning of the first code line which is not inside a continuation statement, i.e., which does not follow a code line ending in '`...`' or '\', or is inside an open parenthesis list.

C-c C-e Move to the 'real' end of the current line (`octave-end-of-line`). If point is in a code line, move forward to the end of the first Octave code line which does not end in '`...`' or '\' or is inside an open parenthesis list. Otherwise, simply go to the end of the current line.

C-c M-C-n Move forward across one balanced begin-end block of Octave code (`octave-forward-block`). With numeric prefix argument *N*, move forward across *n* such blocks (backward if *N* is negative).

C-c M-C-p Move back across one balanced begin-end block of Octave code (`octave-backward-block`). With numeric prefix argument *N*, move backward across *N* such blocks (forward if *N* is negative).

C-c M-C-d Move forward down one begin-end block level of Octave code (`octave-down-block`). With numeric prefix argument, do it that many times; a negative argument means move backward, but still go down one level.

C-c M-C-u Move backward out of one begin-end block level of Octave code (`octave-backward-up-block`). With numeric prefix argument, do it that many times; a negative argument means move forward, but still to a less deep spot.

C-c M-C-h Put point at the beginning of this block, mark at the end (`octave-mark-block`). The block marked is the one that contains point or follows point.

C-c] Close the current block on a separate line (`octave-close-block`). An error is signaled if no block to close is found.

C-c f Insert a function skeleton, prompting for the function's name, arguments and return values which have to be entered without parentheses (`octave-insert-defun`).

C-c C-h Search the function, operator and variable indices of all info files with documentation for Octave for entries (`octave-help`). If used interactively, the entry is prompted for with completion. If multiple matches are found, one can cycle through them using the standard '`,`' (`Info-index-next`) command of the Info reader.

The variable `octave-help-files` is a list of files to search through and defaults to `'("octave")`. If there is also an Octave Local Guide with corresponding info file, say, '`octave-LG`', you can have `octave-help` search both files by

```
(setq octave-help-files '("octave" "octave-LG"))
```

in one of your Emacs startup files.

A common problem is that the RET key does *not* indent the line to where the new text should go after inserting the newline. This is because the standard Emacs convention is that RET (aka *C-m*) just adds a newline, whereas LFD (aka *C-j*) adds a newline and indents it. This is particularly inconvenient for users with keyboards which do not have a special LFD key at all; in such cases, it is typically more convenient to use RET as the LFD key (rather than typing *C-j*).

You can make RET do this by adding

```
(define-key octave-mode-map "\C-m"
  'octave-reindent-then-newline-and-indent)
```

to one of your Emacs startup files. Another, more generally applicable solution is

```
(defun RET-behaves-as-LFD ()
  (let ((x (key-binding "\C-j")))
    (local-set-key "\C-m" x)))
(add-hook 'octave-mode-hook 'RET-behaves-as-LFD)
```

(this works for all modes by adding to the startup hooks, without having to know the particular binding of RET in that mode!). Similar considerations apply for using M-RET as M-LFD. As Barry A. Warsaw `bwarsaw@cnri.reston.va.us` says in the documentation for his `cc-mode`, "This is a very common question. `:-)` If you want this to be the default behavior, don't lobby me, lobby RMS!"

The following variables can be used to customize Octave mode.

`octave-auto-indent`
: Non-`nil` means auto-indent the current line after a semicolon or space. Default is `nil`.

`octave-auto-newline`
: Non-`nil` means auto-insert a newline and indent after semicolons are typed. The default value is `nil`.

`octave-blink-matching-block`
: Non-`nil` means show matching begin of block when inserting a space, newline or '`;`' after an else or end keyword. Default is `t`. This is an extremely useful feature for automatically verifying that the keywords match—if they don't, an error message is displayed.

`octave-block-offset`
: Extra indentation applied to statements in block structures. Default is 2.

`octave-continuation-offset`
: Extra indentation applied to Octave continuation lines. Default is 4.

`octave-continuation-string`
: String used for Octave continuation lines. Normally '`\`'.

`octave-mode-startup-message`
: If `t` (default), a startup message is displayed when Octave mode is called.

If Font Lock mode is enabled, Octave mode will display

- strings in `font-lock-string-face`
- comments in `font-lock-comment-face`

- the Octave reserved words (such as all block keywords) and the text functions (such as 'cd' or 'who') which are also reserved using `font-lock-keyword-face`
- the built-in operators ('&&', '==', ...) using `font-lock-reference-face`
- and the function names in function declarations in `font-lock-function-name-face`.

There is also rudimentary support for Imenu (currently, function names can be indexed).

You can generate TAGS files for Emacs from Octave '.m' files using the shell script `octave-tags` that is installed alongside your copy of Octave.

Customization of Octave mode can be performed by modification of the variable `octave-mode-hook`. If the value of this variable is non-`nil`, turning on Octave mode calls its value.

If you discover a problem with Octave mode, you can conveniently send a bug report using *C-c C-b* (`octave-submit-bug-report`). This automatically sets up a mail buffer with version information already added. You just need to add a description of the problem, including a reproducible test case and send the message.

H.3 Running Octave from Within Emacs

The package '`octave`' provides commands for running an inferior Octave process in a special Emacs buffer. Use

```
M-x run-octave
```

to directly start an inferior Octave process. If Emacs does not know about this command, add the line

```
(autoload 'run-octave "octave-inf" nil t)
```

to your '`.emacs`' file.

This will start Octave in a special buffer the name of which is specified by the variable `inferior-octave-buffer` and defaults to `"*Inferior Octave*"`. From within this buffer, you can interact with the inferior Octave process 'as usual', i.e., by entering Octave commands at the prompt. The buffer is in Inferior Octave mode, which is derived from the standard Comint mode, a major mode for interacting with an inferior interpreter. See the documentation for `comint-mode` for more details, and use *C-h b* to find out about available special keybindings.

You can also communicate with an inferior Octave process from within files with Octave code (i.e., buffers in Octave mode), using the following commands.

C-c i l Send the current line to the inferior Octave process (`octave-send-line`). With positive prefix argument *N*, send that many lines. If `octave-send-line-auto-forward` is non-`nil`, go to the next unsent code line.

C-c i b Send the current block to the inferior Octave process (`octave-send-block`).

C-c i f Send the current function to the inferior Octave process (`octave-send-defun`).

C-c i r Send the region to the inferior Octave process (`octave-send-region`).

C-c i s Make sure that 'inferior-octave-buffer' is displayed (`octave-show-process-buffer`).

C-c i h Delete all windows that display the inferior Octave buffer (`octave-hide-process-buffer`).

C-c i k Kill the inferior Octave process and its buffer (`octave-kill-process`).

The effect of the commands which send code to the Octave process can be customized by the following variables.

`octave-send-echo-input`
: Non-`nil` means echo input sent to the inferior Octave process. Default is `t`.

`octave-send-show-buffer`
: Non-`nil` means display the buffer running the Octave process after sending a command (but without selecting it). Default is `t`.

If you send code and there is no inferior Octave process yet, it will be started automatically.

The startup of the inferior Octave process is highly customizable. The variable `inferior-octave-startup-args` can be used for specifying command lines arguments to be passed to Octave on startup as a list of strings. For example, to suppress the startup message and use 'traditional' mode, set this to `'("-q" "--traditional")`. You can also specify a startup file of Octave commands to be loaded on startup; note that these commands will not produce any visible output in the process buffer. Which file to use is controlled by the variable `inferior-octave-startup-file`. If this is `nil`, the file '`~/.emacs-octave`' is used if it exists.

And finally, `inferior-octave-mode-hook` is run after starting the process and putting its buffer into Inferior Octave mode. Hence, if you like the up and down arrow keys to behave in the interaction buffer as in the shell, and you want this buffer to use nice colors, add

```
(add-hook 'inferior-octave-mode-hook
          (lambda ()
            (turn-on-font-lock)
            (define-key inferior-octave-mode-map [up]
              'comint-previous-input)
            (define-key inferior-octave-mode-map [down]
              'comint-next-input)))
```

to your '`.emacs`' file. You could also swap the roles of *C-a* (`beginning-of-line`) and `C-c C-a` (`comint-bol`) using this hook.

> **Note** that if you set your Octave prompts to something different from the defaults, make sure that `inferior-octave-prompt` matches them. Otherwise, *nothing* will work, because Emacs will not know when Octave is waiting for input, or done sending output.

H.4 Using the Emacs Info Reader for Octave

You may also use the Emacs Info reader with Octave's `doc` function. For this, the package '`gnuserv`' needs to be installed.

If '`gnuserv`' is installed, add the lines

```
(autoload 'octave-help "octave-hlp" nil t)
(require 'gnuserv)
(gnuserv-start)
```

to your '`.emacs`' file.

You can use either 'plain' Emacs Info or the function `octave-help` as your Octave info reader (for '`help -i`'). In the former case, use `info_program ("info-emacs-info")`. The latter is perhaps more attractive because it allows to look up keys in the indices of *several* info files related to Octave (provided that the Emacs variable `octave-help-files` is set correctly). In this case, use `info_program ("info-emacs-octave-help")`.

If you use Octave from within Emacs, it is best to add these settings to your '`~/.emacs-octave`' startup file (or the file pointed to by the Emacs variable `inferior-octave-startup-file`).

Appendix I Grammar and Parser

This appendix will eventually contain a semi-formal description of Octave's language.

I.1 Keywords

The following identifiers are keywords, and may not be used as variable or function names:

`__FILE__`	`__LINE__`	`break`
`case`	`catch`	`classdef`
`continue`	`do`	`else`
`elseif`	`end`	`end_try_catch`
`end_unwind_protect`	`endclassdef`	`endenumeration`
`endevents`	`endfor`	`endfunction`
`endif`	`endmethods`	`endparfor`
`endproperties`	`endswitch`	`endwhile`
`enumeration`	`events`	`for`
`function`	`global`	`if`
`methods`	`otherwise`	`parfor`
`persistent`	`properties`	`return`
`static`	`switch`	`try`
`until`	`unwind_protect`	`unwind_protect_cleanup`
`while`		

The function `iskeyword` can be used to quickly check whether an identifier is reserved by Octave.

`iskeyword ()` [Built-in Function]
`iskeyword (`*name*`)` [Built-in Function]

Return true if *name* is an Octave keyword. If *name* is omitted, return a list of keywords.

See also: [isvarname], page 121, [exist], page 128.

I.2 Parser

The parser has a number of variables that affect its internal operation. These variables are generally documented in the manual alongside the code that they affect. For example, `allow_noninteger_range_as_index` is discussed in the section on index expressions.

In addition, there are three non-specific parser customization functions. `add_input_event_hook` can be used to schedule a user function for periodic evaluation. `remove_input_event_hook` will stop a user function from being evaluated periodically.

id `= add_input_event_hook (`*fcn*`)` [Built-in Function]
id `= add_input_event_hook (`*fcn*`, `*data*`)` [Built-in Function]

Add the named function or function handle *fcn* to the list of functions to call periodically when Octave is waiting for input. The function should have the form

```
fcn (data)
```

If *data* is omitted, Octave calls the function without any arguments.

The returned identifier may be used to remove the function handle from the list of input hook functions.

See also: [remove_input_event_hook], page 856.

`remove_input_event_hook (`*name*`)` [Built-in Function]
`remove_input_event_hook (`*fcn_id*`)` [Built-in Function]

Remove the named function or function handle with the given identifier from the list of functions to call periodically when Octave is waiting for input.

See also: [add_input_event_hook], page 855.

Finally, when the parser cannot identify an input token it calls a particular function to handle this. By default, this is the internal function `"__unimplemented__"` which makes suggestions about possible Octave substitutes for MATLAB functions.

val `= missing_function_hook ()` [Built-in Function]
old_val `= missing_function_hook (`*new_val*`)` [Built-in Function]
`missing_function_hook (`*new_val*, "*local*"`)` [Built-in Function]

Query or set the internal variable that specifies the function to call when an unknown identifier is requested.

When called from inside a function with the `"local"` option, the variable is changed locally for the function and any subroutines it calls. The original variable value is restored when exiting the function.

See also: [missing_component_hook], page 758.

Appendix J GNU GENERAL PUBLIC LICENSE

Preamble

The GNU General Public License is a free, copyleft license for software and other kinds of works.

The licenses for most software and other practical works are designed to take away your freedom to share and change the works. By contrast, the GNU General Public License is intended to guarantee your freedom to share and change all versions of a program—to make sure it remains free software for all its users. We, the Free Software Foundation, use the GNU General Public License for most of our software; it applies also to any other work released this way by its authors. You can apply it to your programs, too.

When we speak of free software, we are referring to freedom, not price. Our General Public Licenses are designed to make sure that you have the freedom to distribute copies of free software (and charge for them if you wish), that you receive source code or can get it if you want it, that you can change the software or use pieces of it in new free programs, and that you know you can do these things.

To protect your rights, we need to prevent others from denying you these rights or asking you to surrender the rights. Therefore, you have certain responsibilities if you distribute copies of the software, or if you modify it: responsibilities to respect the freedom of others.

For example, if you distribute copies of such a program, whether gratis or for a fee, you must pass on to the recipients the same freedoms that you received. You must make sure that they, too, receive or can get the source code. And you must show them these terms so they know their rights.

Developers that use the GNU GPL protect your rights with two steps: (1) assert copyright on the software, and (2) offer you this License giving you legal permission to copy, distribute and/or modify it.

For the developers' and authors' protection, the GPL clearly explains that there is no warranty for this free software. For both users' and authors' sake, the GPL requires that modified versions be marked as changed, so that their problems will not be attributed erroneously to authors of previous versions.

Some devices are designed to deny users access to install or run modified versions of the software inside them, although the manufacturer can do so. This is fundamentally incompatible with the aim of protecting users' freedom to change the software. The systematic pattern of such abuse occurs in the area of products for individuals to use, which is precisely where it is most unacceptable. Therefore, we have designed this version of the GPL to prohibit the practice for those products. If such problems arise substantially in other domains, we stand ready to extend this provision to those domains in future versions of the GPL, as needed to protect the freedom of users.

Finally, every program is threatened constantly by software patents. States should not allow patents to restrict development and use of software on general-purpose computers, but in those that do, we wish to avoid the special danger that patents applied to a free program could make it effectively proprietary. To prevent this, the GPL assures that patents cannot be used to render the program non-free.

The precise terms and conditions for copying, distribution and modification follow.

TERMS AND CONDITIONS

0. Definitions.

   "This License" refers to version 3 of the GNU General Public License.

   "Copyright" also means copyright-like laws that apply to other kinds of works, such as semiconductor masks.

   "The Program" refers to any copyrightable work licensed under this License. Each licensee is addressed as "you". "Licensees" and "recipients" may be individuals or organizations.

   To "modify" a work means to copy from or adapt all or part of the work in a fashion requiring copyright permission, other than the making of an exact copy. The resulting work is called a "modified version" of the earlier work or a work "based on" the earlier work.

   A "covered work" means either the unmodified Program or a work based on the Program.

   To "propagate" a work means to do anything with it that, without permission, would make you directly or secondarily liable for infringement under applicable copyright law, except executing it on a computer or modifying a private copy. Propagation includes copying, distribution (with or without modification), making available to the public, and in some countries other activities as well.

   To "convey" a work means any kind of propagation that enables other parties to make or receive copies. Mere interaction with a user through a computer network, with no transfer of a copy, is not conveying.

   An interactive user interface displays "Appropriate Legal Notices" to the extent that it includes a convenient and prominently visible feature that (1) displays an appropriate copyright notice, and (2) tells the user that there is no warranty for the work (except to the extent that warranties are provided), that licensees may convey the work under this License, and how to view a copy of this License. If the interface presents a list of user commands or options, such as a menu, a prominent item in the list meets this criterion.

1. Source Code.

   The "source code" for a work means the preferred form of the work for making modifications to it. "Object code" means any non-source form of a work.

   A "Standard Interface" means an interface that either is an official standard defined by a recognized standards body, or, in the case of interfaces specified for a particular programming language, one that is widely used among developers working in that language.

The “System Libraries” of an executable work include anything, other than the work as a whole, that (a) is included in the normal form of packaging a Major Component, but which is not part of that Major Component, and (b) serves only to enable use of the work with that Major Component, or to implement a Standard Interface for which an implementation is available to the public in source code form. A “Major Component”, in this context, means a major essential component (kernel, window system, and so on) of the specific operating system (if any) on which the executable work runs, or a compiler used to produce the work, or an object code interpreter used to run it.

The “Corresponding Source” for a work in object code form means all the source code needed to generate, install, and (for an executable work) run the object code and to modify the work, including scripts to control those activities. However, it does not include the work's System Libraries, or general-purpose tools or generally available free programs which are used unmodified in performing those activities but which are not part of the work. For example, Corresponding Source includes interface definition files associated with source files for the work, and the source code for shared libraries and dynamically linked subprograms that the work is specifically designed to require, such as by intimate data communication or control flow between those subprograms and other parts of the work.

The Corresponding Source need not include anything that users can regenerate automatically from other parts of the Corresponding Source.

The Corresponding Source for a work in source code form is that same work.

2. Basic Permissions.

   All rights granted under this License are granted for the term of copyright on the Program, and are irrevocable provided the stated conditions are met. This License explicitly affirms your unlimited permission to run the unmodified Program. The output from running a covered work is covered by this License only if the output, given its content, constitutes a covered work. This License acknowledges your rights of fair use or other equivalent, as provided by copyright law.

   You may make, run and propagate covered works that you do not convey, without conditions so long as your license otherwise remains in force. You may convey covered works to others for the sole purpose of having them make modifications exclusively for you, or provide you with facilities for running those works, provided that you comply with the terms of this License in conveying all material for which you do not control copyright. Those thus making or running the covered works for you must do so exclusively on your behalf, under your direction and control, on terms that prohibit them from making any copies of your copyrighted material outside their relationship with you.

   Conveying under any other circumstances is permitted solely under the conditions stated below. Sublicensing is not allowed; section 10 makes it unnecessary.

3. Protecting Users' Legal Rights From Anti-Circumvention Law.

   No covered work shall be deemed part of an effective technological measure under any applicable law fulfilling obligations under article 11 of the WIPO copyright treaty adopted on 20 December 1996, or similar laws prohibiting or restricting circumvention of such measures.

When you convey a covered work, you waive any legal power to forbid circumvention of technological measures to the extent such circumvention is effected by exercising rights under this License with respect to the covered work, and you disclaim any intention to limit operation or modification of the work as a means of enforcing, against the work's users, your or third parties' legal rights to forbid circumvention of technological measures.

4. Conveying Verbatim Copies.

   You may convey verbatim copies of the Program's source code as you receive it, in any medium, provided that you conspicuously and appropriately publish on each copy an appropriate copyright notice; keep intact all notices stating that this License and any non-permissive terms added in accord with section 7 apply to the code; keep intact all notices of the absence of any warranty; and give all recipients a copy of this License along with the Program.

   You may charge any price or no price for each copy that you convey, and you may offer support or warranty protection for a fee.

5. Conveying Modified Source Versions.

   You may convey a work based on the Program, or the modifications to produce it from the Program, in the form of source code under the terms of section 4, provided that you also meet all of these conditions:

   a. The work must carry prominent notices stating that you modified it, and giving a relevant date.
   b. The work must carry prominent notices stating that it is released under this License and any conditions added under section 7. This requirement modifies the requirement in section 4 to "keep intact all notices".
   c. You must license the entire work, as a whole, under this License to anyone who comes into possession of a copy. This License will therefore apply, along with any applicable section 7 additional terms, to the whole of the work, and all its parts, regardless of how they are packaged. This License gives no permission to license the work in any other way, but it does not invalidate such permission if you have separately received it.
   d. If the work has interactive user interfaces, each must display Appropriate Legal Notices; however, if the Program has interactive interfaces that do not display Appropriate Legal Notices, your work need not make them do so.

   A compilation of a covered work with other separate and independent works, which are not by their nature extensions of the covered work, and which are not combined with it such as to form a larger program, in or on a volume of a storage or distribution medium, is called an "aggregate" if the compilation and its resulting copyright are not used to limit the access or legal rights of the compilation's users beyond what the individual works permit. Inclusion of a covered work in an aggregate does not cause this License to apply to the other parts of the aggregate.

6. Conveying Non-Source Forms.

   You may convey a covered work in object code form under the terms of sections 4 and 5, provided that you also convey the machine-readable Corresponding Source under the terms of this License, in one of these ways:

a. Convey the object code in, or embodied in, a physical product (including a physical distribution medium), accompanied by the Corresponding Source fixed on a durable physical medium customarily used for software interchange.

b. Convey the object code in, or embodied in, a physical product (including a physical distribution medium), accompanied by a written offer, valid for at least three years and valid for as long as you offer spare parts or customer support for that product model, to give anyone who possesses the object code either (1) a copy of the Corresponding Source for all the software in the product that is covered by this License, on a durable physical medium customarily used for software interchange, for a price no more than your reasonable cost of physically performing this conveying of source, or (2) access to copy the Corresponding Source from a network server at no charge.

c. Convey individual copies of the object code with a copy of the written offer to provide the Corresponding Source. This alternative is allowed only occasionally and noncommercially, and only if you received the object code with such an offer, in accord with subsection 6b.

d. Convey the object code by offering access from a designated place (gratis or for a charge), and offer equivalent access to the Corresponding Source in the same way through the same place at no further charge. You need not require recipients to copy the Corresponding Source along with the object code. If the place to copy the object code is a network server, the Corresponding Source may be on a different server (operated by you or a third party) that supports equivalent copying facilities, provided you maintain clear directions next to the object code saying where to find the Corresponding Source. Regardless of what server hosts the Corresponding Source, you remain obligated to ensure that it is available for as long as needed to satisfy these requirements.

e. Convey the object code using peer-to-peer transmission, provided you inform other peers where the object code and Corresponding Source of the work are being offered to the general public at no charge under subsection 6d.

A separable portion of the object code, whose source code is excluded from the Corresponding Source as a System Library, need not be included in conveying the object code work.

A "User Product" is either (1) a "consumer product", which means any tangible personal property which is normally used for personal, family, or household purposes, or (2) anything designed or sold for incorporation into a dwelling. In determining whether a product is a consumer product, doubtful cases shall be resolved in favor of coverage. For a particular product received by a particular user, "normally used" refers to a typical or common use of that class of product, regardless of the status of the particular user or of the way in which the particular user actually uses, or expects or is expected to use, the product. A product is a consumer product regardless of whether the product has substantial commercial, industrial or non-consumer uses, unless such uses represent the only significant mode of use of the product.

"Installation Information" for a User Product means any methods, procedures, authorization keys, or other information required to install and execute modified versions of a covered work in that User Product from a modified version of its Corresponding Source.

The information must suffice to ensure that the continued functioning of the modified object code is in no case prevented or interfered with solely because modification has been made.

If you convey an object code work under this section in, or with, or specifically for use in, a User Product, and the conveying occurs as part of a transaction in which the right of possession and use of the User Product is transferred to the recipient in perpetuity or for a fixed term (regardless of how the transaction is characterized), the Corresponding Source conveyed under this section must be accompanied by the Installation Information. But this requirement does not apply if neither you nor any third party retains the ability to install modified object code on the User Product (for example, the work has been installed in ROM).

The requirement to provide Installation Information does not include a requirement to continue to provide support service, warranty, or updates for a work that has been modified or installed by the recipient, or for the User Product in which it has been modified or installed. Access to a network may be denied when the modification itself materially and adversely affects the operation of the network or violates the rules and protocols for communication across the network.

Corresponding Source conveyed, and Installation Information provided, in accord with this section must be in a format that is publicly documented (and with an implementation available to the public in source code form), and must require no special password or key for unpacking, reading or copying.

7. Additional Terms.

"Additional permissions" are terms that supplement the terms of this License by making exceptions from one or more of its conditions. Additional permissions that are applicable to the entire Program shall be treated as though they were included in this License, to the extent that they are valid under applicable law. If additional permissions apply only to part of the Program, that part may be used separately under those permissions, but the entire Program remains governed by this License without regard to the additional permissions.

When you convey a copy of a covered work, you may at your option remove any additional permissions from that copy, or from any part of it. (Additional permissions may be written to require their own removal in certain cases when you modify the work.) You may place additional permissions on material, added by you to a covered work, for which you have or can give appropriate copyright permission.

Notwithstanding any other provision of this License, for material you add to a covered work, you may (if authorized by the copyright holders of that material) supplement the terms of this License with terms:

a. Disclaiming warranty or limiting liability differently from the terms of sections 15 and 16 of this License; or
b. Requiring preservation of specified reasonable legal notices or author attributions in that material or in the Appropriate Legal Notices displayed by works containing it; or
c. Prohibiting misrepresentation of the origin of that material, or requiring that modified versions of such material be marked in reasonable ways as different from the original version; or

d. Limiting the use for publicity purposes of names of licensors or authors of the material; or
e. Declining to grant rights under trademark law for use of some trade names, trademarks, or service marks; or
f. Requiring indemnification of licensors and authors of that material by anyone who conveys the material (or modified versions of it) with contractual assumptions of liability to the recipient, for any liability that these contractual assumptions directly impose on those licensors and authors.

All other non-permissive additional terms are considered "further restrictions" within the meaning of section 10. If the Program as you received it, or any part of it, contains a notice stating that it is governed by this License along with a term that is a further restriction, you may remove that term. If a license document contains a further restriction but permits relicensing or conveying under this License, you may add to a covered work material governed by the terms of that license document, provided that the further restriction does not survive such relicensing or conveying.

If you add terms to a covered work in accord with this section, you must place, in the relevant source files, a statement of the additional terms that apply to those files, or a notice indicating where to find the applicable terms.

Additional terms, permissive or non-permissive, may be stated in the form of a separately written license, or stated as exceptions; the above requirements apply either way.

8. Termination.

You may not propagate or modify a covered work except as expressly provided under this License. Any attempt otherwise to propagate or modify it is void, and will automatically terminate your rights under this License (including any patent licenses granted under the third paragraph of section 11).

However, if you cease all violation of this License, then your license from a particular copyright holder is reinstated (a) provisionally, unless and until the copyright holder explicitly and finally terminates your license, and (b) permanently, if the copyright holder fails to notify you of the violation by some reasonable means prior to 60 days after the cessation.

Moreover, your license from a particular copyright holder is reinstated permanently if the copyright holder notifies you of the violation by some reasonable means, this is the first time you have received notice of violation of this License (for any work) from that copyright holder, and you cure the violation prior to 30 days after your receipt of the notice.

Termination of your rights under this section does not terminate the licenses of parties who have received copies or rights from you under this License. If your rights have been terminated and not permanently reinstated, you do not qualify to receive new licenses for the same material under section 10.

9. Acceptance Not Required for Having Copies.

You are not required to accept this License in order to receive or run a copy of the Program. Ancillary propagation of a covered work occurring solely as a consequence of using peer-to-peer transmission to receive a copy likewise does not require acceptance.

However, nothing other than this License grants you permission to propagate or modify any covered work. These actions infringe copyright if you do not accept this License. Therefore, by modifying or propagating a covered work, you indicate your acceptance of this License to do so.

10. Automatic Licensing of Downstream Recipients.

    Each time you convey a covered work, the recipient automatically receives a license from the original licensors, to run, modify and propagate that work, subject to this License. You are not responsible for enforcing compliance by third parties with this License.

    An "entity transaction" is a transaction transferring control of an organization, or substantially all assets of one, or subdividing an organization, or merging organizations. If propagation of a covered work results from an entity transaction, each party to that transaction who receives a copy of the work also receives whatever licenses to the work the party's predecessor in interest had or could give under the previous paragraph, plus a right to possession of the Corresponding Source of the work from the predecessor in interest, if the predecessor has it or can get it with reasonable efforts.

    You may not impose any further restrictions on the exercise of the rights granted or affirmed under this License. For example, you may not impose a license fee, royalty, or other charge for exercise of rights granted under this License, and you may not initiate litigation (including a cross-claim or counterclaim in a lawsuit) alleging that any patent claim is infringed by making, using, selling, offering for sale, or importing the Program or any portion of it.

11. Patents.

    A "contributor" is a copyright holder who authorizes use under this License of the Program or a work on which the Program is based. The work thus licensed is called the contributor's "contributor version".

    A contributor's "essential patent claims" are all patent claims owned or controlled by the contributor, whether already acquired or hereafter acquired, that would be infringed by some manner, permitted by this License, of making, using, or selling its contributor version, but do not include claims that would be infringed only as a consequence of further modification of the contributor version. For purposes of this definition, "control" includes the right to grant patent sublicenses in a manner consistent with the requirements of this License.

    Each contributor grants you a non-exclusive, worldwide, royalty-free patent license under the contributor's essential patent claims, to make, use, sell, offer for sale, import and otherwise run, modify and propagate the contents of its contributor version.

    In the following three paragraphs, a "patent license" is any express agreement or commitment, however denominated, not to enforce a patent (such as an express permission to practice a patent or covenant not to sue for patent infringement). To "grant" such a patent license to a party means to make such an agreement or commitment not to enforce a patent against the party.

    If you convey a covered work, knowingly relying on a patent license, and the Corresponding Source of the work is not available for anyone to copy, free of charge and under the terms of this License, through a publicly available network server or other readily accessible means, then you must either (1) cause the Corresponding Source to be so available, or (2) arrange to deprive yourself of the benefit of the patent license for this particular work, or (3) arrange, in a manner consistent with the requirements of this License, to extend the patent license to downstream recipients. "Knowingly relying" means you have actual knowledge that, but for the patent license, your conveying the covered work in a country, or your recipient's use of the covered work in a country, would infringe one or more identifiable patents in that country that you have reason to believe are valid.

If, pursuant to or in connection with a single transaction or arrangement, you convey, or propagate by procuring conveyance of, a covered work, and grant a patent license to some of the parties receiving the covered work authorizing them to use, propagate, modify or convey a specific copy of the covered work, then the patent license you grant is automatically extended to all recipients of the covered work and works based on it.

A patent license is "discriminatory" if it does not include within the scope of its coverage, prohibits the exercise of, or is conditioned on the non-exercise of one or more of the rights that are specifically granted under this License. You may not convey a covered work if you are a party to an arrangement with a third party that is in the business of distributing software, under which you make payment to the third party based on the extent of your activity of conveying the work, and under which the third party grants, to any of the parties who would receive the covered work from you, a discriminatory patent license (a) in connection with copies of the covered work conveyed by you (or copies made from those copies), or (b) primarily for and in connection with specific products or compilations that contain the covered work, unless you entered into that arrangement, or that patent license was granted, prior to 28 March 2007.

Nothing in this License shall be construed as excluding or limiting any implied license or other defenses to infringement that may otherwise be available to you under applicable patent law.

12. No Surrender of Others' Freedom.

    If conditions are imposed on you (whether by court order, agreement or otherwise) that contradict the conditions of this License, they do not excuse you from the conditions of this License. If you cannot convey a covered work so as to satisfy simultaneously your obligations under this License and any other pertinent obligations, then as a consequence you may not convey it at all. For example, if you agree to terms that obligate you to collect a royalty for further conveying from those to whom you convey the Program, the only way you could satisfy both those terms and this License would be to refrain entirely from conveying the Program.

13. Use with the GNU Affero General Public License.

    Notwithstanding any other provision of this License, you have permission to link or combine any covered work with a work licensed under version 3 of the GNU Affero General Public License into a single combined work, and to convey the resulting work. The terms of this License will continue to apply to the part which is the covered work, but the special requirements of the GNU Affero General Public License, section 13, concerning interaction through a network will apply to the combination as such.

14. Revised Versions of this License.

The Free Software Foundation may publish revised and/or new versions of the GNU General Public License from time to time. Such new versions will be similar in spirit to the present version, but may differ in detail to address new problems or concerns.

Each version is given a distinguishing version number. If the Program specifies that a certain numbered version of the GNU General Public License "or any later version" applies to it, you have the option of following the terms and conditions either of that numbered version or of any later version published by the Free Software Foundation. If the Program does not specify a version number of the GNU General Public License, you may choose any version ever published by the Free Software Foundation.

If the Program specifies that a proxy can decide which future versions of the GNU General Public License can be used, that proxy's public statement of acceptance of a version permanently authorizes you to choose that version for the Program.

Later license versions may give you additional or different permissions. However, no additional obligations are imposed on any author or copyright holder as a result of your choosing to follow a later version.

15. Disclaimer of Warranty.

    THERE IS NO WARRANTY FOR THE PROGRAM, TO THE EXTENT PERMITTED BY APPLICABLE LAW. EXCEPT WHEN OTHERWISE STATED IN WRITING THE COPYRIGHT HOLDERS AND/OR OTHER PARTIES PROVIDE THE PROGRAM "AS IS" WITHOUT WARRANTY OF ANY KIND, EITHER EXPRESSED OR IMPLIED, INCLUDING, BUT NOT LIMITED TO, THE IMPLIED WARRANTIES OF MERCHANTABILITY AND FITNESS FOR A PARTICULAR PURPOSE. THE ENTIRE RISK AS TO THE QUALITY AND PERFORMANCE OF THE PROGRAM IS WITH YOU. SHOULD THE PROGRAM PROVE DEFECTIVE, YOU ASSUME THE COST OF ALL NECESSARY SERVICING, REPAIR OR CORRECTION.

16. Limitation of Liability.

    IN NO EVENT UNLESS REQUIRED BY APPLICABLE LAW OR AGREED TO IN WRITING WILL ANY COPYRIGHT HOLDER, OR ANY OTHER PARTY WHO MODIFIES AND/OR CONVEYS THE PROGRAM AS PERMITTED ABOVE, BE LIABLE TO YOU FOR DAMAGES, INCLUDING ANY GENERAL, SPECIAL, INCIDENTAL OR CONSEQUENTIAL DAMAGES ARISING OUT OF THE USE OR INABILITY TO USE THE PROGRAM (INCLUDING BUT NOT LIMITED TO LOSS OF DATA OR DATA BEING RENDERED INACCURATE OR LOSSES SUSTAINED BY YOU OR THIRD PARTIES OR A FAILURE OF THE PROGRAM TO OPERATE WITH ANY OTHER PROGRAMS), EVEN IF SUCH HOLDER OR OTHER PARTY HAS BEEN ADVISED OF THE POSSIBILITY OF SUCH DAMAGES.

17. Interpretation of Sections 15 and 16.

    If the disclaimer of warranty and limitation of liability provided above cannot be given local legal effect according to their terms, reviewing courts shall apply local law that most closely approximates an absolute waiver of all civil liability in connection with the Program, unless a warranty or assumption of liability accompanies a copy of the Program in return for a fee.

END OF TERMS AND CONDITIONS

How to Apply These Terms to Your New Programs

If you develop a new program, and you want it to be of the greatest possible use to the public, the best way to achieve this is to make it free software which everyone can redistribute and change under these terms.

To do so, attach the following notices to the program. It is safest to attach them to the start of each source file to most effectively state the exclusion of warranty; and each file should have at least the "copyright" line and a pointer to where the full notice is found.

```
one line to give the program's name and a brief idea of what it does.
Copyright (C) year name of author

This program is free software: you can redistribute it and/or modify
it under the terms of the GNU General Public License as published by
the Free Software Foundation, either version 3 of the License, or (at
your option) any later version.

This program is distributed in the hope that it will be useful, but
WITHOUT ANY WARRANTY; without even the implied warranty of
MERCHANTABILITY or FITNESS FOR A PARTICULAR PURPOSE.  See the GNU
General Public License for more details.

You should have received a copy of the GNU General Public License
along with this program.  If not, see http://www.gnu.org/licenses/.
```

Also add information on how to contact you by electronic and paper mail.

If the program does terminal interaction, make it output a short notice like this when it starts in an interactive mode:

```
program Copyright (C) year name of author
This program comes with ABSOLUTELY NO WARRANTY; for details type 'show w'.
This is free software, and you are welcome to redistribute it
under certain conditions; type 'show c' for details.
```

The hypothetical commands '`show w`' and '`show c`' should show the appropriate parts of the General Public License. Of course, your program's commands might be different; for a GUI interface, you would use an "about box".

You should also get your employer (if you work as a programmer) or school, if any, to sign a "copyright disclaimer" for the program, if necessary. For more information on this, and how to apply and follow the GNU GPL, see `http://www.gnu.org/licenses/`.

The GNU General Public License does not permit incorporating your program into proprietary programs. If your program is a subroutine library, you may consider it more useful to permit linking proprietary applications with the library. If this is what you want to do, use the GNU Lesser General Public License instead of this License. But first, please read `http://www.gnu.org/philosophy/why-not-lgpl.html`.

Concept Index

#

%

-

.

:

\

~

A

B

W

Function Index

A

B

C

D

E

F

G

H

I

J

K

L

M

N

O

P

Q

R

S

T

U

V

W

X

Y

Z

Operator Index

www.ingramcontent.com/pod-product-compliance
Lightning Source LLC
Chambersburg PA
CBHW082103220526
45472CB00009B/2020
* 9 7 8 9 8 8 1 3 2 7 7 4 1 *